Physical Foundations of the Millimeter and
Submillimeter Waves Technique
Volume 1. Open Structures

Physical Foundations of the Millimeter and Submillimeter Waves Technique

Volume 1
Open Structures

V.P. Shestopalov

Utrecht, The Netherlands, 1997

VSP BV
P.O. Box 346
3700 AH Zeist
The Netherlands

© VSP BV 1997

First published in 1997

ISBN 90-6764-215-0

Printed in The Netherlands by Ridderprint bv, Ridderkerk.

Contents

FOREWORD

In 1985, the Naukova Dumka Publishing House of the Ukrainian Academy of Sciences in Kiev published a two-volume monograph "Physical foundations of millimeter and submillimeter technology" (the first volume is entitled "Open structures" and the second – "Sources. Element Base. Radio-systems.") where my original, extensive theoretical and experimental studies in collaboration with my co-workers are presented.

It was proposed to publish this book for western readers by means of direct translation. The variety of new results obtained in the last decade made it necessary to create practically a new monograph, although having the same title. This idea was approved by Dr Jan Reijer F.Th. Groesbeek, and after his obliging proposal I gave my agreement to publish the renovated book with his company, VSP.

While writing the book, I had to consider the questions: with what purpose is it written and who will be its readers? These questions are very complicated and one can find answers to them only after attentively reading the whole book. However, I will risk trying to answer them. The essence of the present monograph is that it is an attempt to destroy stereotypes established during the long years of applications of large-scale modeling in the millimeter and submillimeter wavelength ranges and to develop new conceptions. I tried to make this book accessible for many specialists who require a detailed description of the mathematical groundwork and the physical interpretation of the results obtained. If these pages give readers at least some ideas about the presented methods and make them think over their possible roles in solving these new problems, the aim of the book will be achieved.

The development of physics, biology and astronomy has led to a sharp increase in the investigations of electromagnetic millimeter and submillimeter waves with the lengths 10–1 and 1–0.1 mm. Stimulation of the work in this area has been produced also by inner demands of radiophysics and electronics themselves, especially in the fields of radar and communication technology, remote sensing and spectroscopy.

The millimeter and submillimeter wavelength ranges had been almost completely mastered by the beginning of the sixties: the element base had been created and various radio-systems had been elaborated (Refs. [1–4]). These achievements became possible through the use of large-scale modeling. Regularities in the centimeter wavelength range are valid in the cases of linear variation (decreasing) of geometrical dimensions of such electrodynamic structures as closed resonators, waveguides and periodical systems. This allows one to solve the problems of generation, amplifying, canalization, transmission and receiving of waves in the millimeter and submillimeter wavelength ranges.

However, shortening the wavelength yields a decrease in the resonator q-factor. At the same time, the oscillation spectrum condenses, attenuation in empty waveguides grows and the transmission band diminishes. The manufacturing technology of units and elements then demands a high level of precision. Hence, large-scale modeling does not allow us to solve the problem of creating millimeter and submillimeter devices.

It is important that, together with the classical approach (large-scale modeling), new physical ideas were applied in the millimeter and submillimeter wavelength ranges to eliminate the faults of earlier methods. In this monograph, which naturally continues the works [5–11], the treatment is given from the single viewpoint of the analysis of physical processes occurring in various devices which are applied in the millimeter and submillimeter technology. The studies showed that one has to use open resonators as effective electrodynamic structures in this wavelength range. By means of open resonators the fields are concentrated in the given spatial domains. Open waveguides are used for creating slot (surface) and tearing off (volume) waves. Diffractional gratings transform surface waves into volume waves, and *vice versa*.

Knowledge of physical properties of open structures raises the possibility of constructing two- and multi-mirror, one- and multi-mode resonators, diffractional gratings and open waveguides, and to work out high-coherent sources of the millimeter and submillimeter waves, generators of diffractional radiation (GDR) [6, 12, 13], various slotted micro-miniature lines, nonlinear solid-state devices and specially oriented types of antennas. They are widely used in spectroscopic, interferometric, polarimetric, holographic, radar and communication technology.

This monograph contains the results of the studies carried out by the author and his co-workers S.D. Andrenko, I.M. Balaklitskij, A.V. Brovenko, V.V. Veremey, A.A. Vertij, V.N. Derkach, A.P. Evdokimov, G.P. Ermak, I.V. Ivanchenko, A.A. Kirilenko, A.G. Kovalenko, G.I. Komar', V.K. Korneenkov, V.M. Kurin, K.A. Lukin, S.A. Masalov, P.N. Melezhik, V.S. Miroshnichenko, A.I. Nosich, I.D. Revin, A.A. Petrushin, S.A. Provalov, A.E. Poyedinchuk, N.A. Popenko, I.V. Pochanina, A.S. Svezhentsev, Yu.A. Svishchev, Yu.K. Sirenko, B.K. Skrynnik, S.I. Tarapov, Yu.A. Tuchkin, G.I. Khlopov, A.I. Tsvik, V.V. Shcherbak, V.V. Yatsik, and N.P. Yashina and consists of two volumes.

The results of theoretical and experimental studies of open electrodynamic structures (open waveguides, open resonators, diffractional gratings) allowing determination of the characteristics of various devices used in the millimeter and submillimeter technology, are presented in the first volume.

Introduction

Development of radiophysics and electronics is first of all connected with mastering radio waves having a length from hundreds to tens of meters. The element base in this wavelength range consists of one- and two-wire transmission lines, resonant, scattering and radiating contours made of their sections and also electron-vacuum lamps which are used for excitation of these devices. It is essential that the given structures are open and that their quasi-statical behaviour is described by electrical circuits with concentrated parameters.

Such open structures turn out to be less effective in the process of shortening the wavelength and transferring to the centimeter wavelength range. Overcoming the difficulties in this wavelength range became possible only with the use of closed structures: resonators and waveguides and various types of devices made on their basis. Theoretical and experimental studies of closed structures with distributed electromagnetic parameters required new approaches based on the study of dynamical processes in spatial volumes bounded by closed surfaces.

The main idea of this book is that millimeter and submillimeter wavelength technology may be completely constructed by means of open electrodynamical structures: open resonators, waveguides and diffractional gratings. Of course, they differ significantly from those which were used at the first stages of radio-engineering, and this transition takes place on a qualitatively new physical basis (let us note that open structures may be used also in the centimeter and longer wavelength ranges).

Millimeter and submillimeter waves are close to coherent optical waves. Open resonators, waveguides and grating are widely used in the optical wavelength range but a number of essential differences arise. The latter requires a change of theoretical methods and the establishment of new approaches for experimental studies of physical phenomena and construction of radio systems.

When open resonators are investigated in the optical wavelength range, it is assumed that their geometrical dimensions are much greater than the working wavelength, the field is strictly transversal-electromagnetic and resonance wave beams are paraxial. The creation of open resonators for lasers requires the establishment of methods for selection of resonant waves and obtaining single-mode oscillations [14].

In the millimeter and submillimeter wavelength ranges, open resonators have dimensions comparable with the wavelength, and wave beams usually have greater diffractional divergence in resonators with small curvature radii. Here, the paraxial property is violated and the field is not strictly transversal-electromagnetic because non-zero longitudinal field components appear. One can also observe substantial influence of spherical and astigmatic mirrors which implies complication of polarization properties of open resonators. High q-factor multi-mode open resonators are of great importance in this wavelength range. Background results presented in this book allow us to determine not only the characteristics of open resonators in this wavelength range but also to construct different devices for applications in the physics of elementary particles and generation of high-coherent electromagnetic oscillations.

An acceptable method has not yet been established for canalization of millimeter and submillimeter waves which is physically adequate for the given wavelength range. A variety of transmission lines, namely, single- and multi-mode metallic and dielectric waveguides, strip and fin-lines, etc., is evidence of ambiguous solution to this problem.

In this monograph we consider a new type of open electrodynamical structure, a cylindrical slot waveguide (CSW) formed by a dielectric pivot whose external surface is covered by a metallic film with a longitudinal slot. The modification of CSW, the image slot waveguide (ISW), is of great interest for constructing effective transmission sections in the millimeter wavelength range. It consists of a metallic coating and part of cylindrical screen with a slot formed by the remaining screen edge and the coating.

Optimal transformation of electromagnetic waves may be carried out with the help of diffractional gratings. At first, gratings were used for transforming volume waves to surface and volume waves. Unlike volume waves, surface waves are localized in space, and propagate with a phase velocity v_ϕ less than the light velocity c. Volume waves exponentially decrease with respect to the distance from the grating. Transformation of surface waves to surface waves by a grating is also widely applied.

It is critical that it becomes necessary in the millimeter and submillimeter wavelength ranges to use multi-act wave transformation. Successive surface–volume and volume–surface wave transformation by a diffractional grating with intermediate volume–volume and surface–surface transformation turns out to be possible when a nonlinear element is placed close to the grating (for example, an electron flow). A complicated electrodynamical structure is thus created, which consists not only of a grating but of an open resonator and a waveguide.

Hence, one possible physical foundation for millimeter and submillimeter wave technology may be the necessity to carry out a continuous process of wave transformation by different open electrodynamical structures, which must be effectively excited. Each of these elements has to be coordinated with neighboring ones. Such a sequence of elements with corresponding passive and active nonlinear objects (electron flows, solid-state generators, super-gratings, etc.) become the initial units to make an assembly for a complicated multi-purpose radio system.

As a conclusion, one can state that open electrodynamical structures should be used as key elements in the millimeter and submillimeter technology. Knowledge of physical conformities of the wave interaction with these objects forms the basis for constructing more complicated radio-engineering devices. The wavelength in this range is comparable with the size of metallic screens and this produces a need to elaborate new fundamental methods for solving the boundary value problems of electrodynamics describing the wave propagation and scattering in waveguides, resonators and gratings. Rigorous theory may be developed for infinitely thin, perfectly conducting screens having definite geometrical shape. One can take into account the finite conductivity, presence of different obstacles and inclusions, excitation by nonlinear and other sources.

Experiment plays a crucial role for studying the properties of open structures. It is convenient to explore various characteristics of these structures in the millimeter and submillimeter wavelength ranges using open resonators, waveguides, and gratings themselves as measuring devices. One can then determine the most delicate parameters which simply cannot be registered by classical methods.

It has been established theoretically and experimentally that the millimeter and submillimeter wavelength ranges may be effectively mastered with the help of open structures. Modern methods of the theory of diffraction are discussed here together with new approaches for measuring the parameters of considered objects. The results give new possibilities for constructing original elements, units and systems involving millimeter and submillimeter technology.

Specific features of the considered wavelength ranges do not allow us to reject the creation of rigorous spectral theory of open resonators, including the problems of their excitation by concentrated and distributed sources. It is connected, in turn, with working out special methods of functional analysis, operator theory, linear algebra, etc., which are considered in this book.

Similar situations occur in experimental studies and the manufacturing of the element base and radio systems. In this case, new ways of organizing experiments are described which have no analogous approaches in the microwave and optical ranges.

The question arises: is this conception justified? One can assert that the fundamental physical phenomena which are the basis of the processes occurring in open structures in the considered wavelength ranges definitely determine the methods of study and, hence, their adequate description as presented in this book.

Chapter 1

Spectral theory and excitation of open resonators

The principles of open resonators are different from the closed ones because of radiation loss, edges, multi-connected cross-sections, and the behaviour of the electromagnetic field at infinity. That is why the spectrum of eigenoscillations is not real, additional demands to the energy relations in different space domains appear, and the characteristics of the spectral problem change. Such problems of spectral oscillations for open resonators form a new class of non-self-adjoint spectral problems with a spectral parameter entering in a nonlinear way. These nonlinear spectral problems in mathematical physics require working out new methods of solution, among which, together with the Fourier method, the integral equations method, etc., is the method of operator-valued functions of one or several complex variables, which has been applied for the first time in Refs. [15–20] for the construction of spectral theory of transmission lines (a variety of open waveguides). This approach has been further developed in Refs. [21, 22, 10, 11, 23, 24]. The method of operator-valued functions puts together the homogeneous (spectral) and inhomogeneous (excitation) problems in electrodynamics because it allows to determine simultaneously the domains of spectra localization and the resolvent set of operators, i.e., to prove the existence and the uniqueness of the solutions to boundary value problems.

Foundations of rigorous spectral theory for two-dimensional and three-dimensional open resonators with Dirichlet and Neumann boundary conditions are developed in Refs. [10, 11, 23–27]. Spectral problems for these resonators are reduced to the study of characteristic numbers of operator-valued functions of special types (like the Fredholm and kernel finite-meromorphic operator-valued functions), acting in a Hilbert space, with the application of partial inversion of operators (in particular, by means of the method of Riemann–Hilbert problem [5]).

Thus, the initial physical problem is directly connected with the spectral theory of the canonical Fredholm operator-valued function. Once given the possibility to prove the discreteness and finite multiplicity of the spectrum of the Maxwell operator in the functional space determined by the radiation condition [10], a mathematically justified method for calculating the spectra of open empty and partially filled resonators can be constructed and realized as a package of applied computer programs.

The study of the spectral characteristics of open resonators is reduced to the description of solutions to the equation of the second kind [10], which allows to construct the algorithm for spectral computation based on approximate solutions of the characteristic equation [28]. Basic qualitative features of the spectra of free oscillations of open resonators are studied. In particular, it is proved that they are discrete on the corresponding Riemann surfaces and their points may accumulate only at infinity; they continuously vary as functions of all non-spectral parameters (describing the shape of a resonator, its dielectric filling, etc.).

The study of the dispersion relations for open resonators in Refs. [10, 11, 23–27] shows that the smoothness of spectral characteristics as functions of non-spectral parameters may be violated in the "interaction" zone of free oscillations corresponding to different fundamental frequencies, i.e., when they approach each other in the complex space. When introducing the notion of interaction we assume that it is the change in qualitative and quantitative composition of partial components of the eigenoscillation under the conditions close to the existence of other eigenoscillations.

If the open resonator and the values of frequency parameters are such that the spectral problem is symmetrical, then the homogeneous operator equation splits into several independent equations for symmetrical and anti-symmetrical free oscillations with respect to the planes (axes or points) of symmetry. It allows to observe degenerated free oscillations in a resonator, in situations where two linearly independent free fields from different classes of symmetry correspond to one fundamental frequency. There are examples in [29] illustrating the effect of degeneration of free electromagnetic fields; their instability, i.e., intertype coupling (interaction) is shown, when the symmetry of the problem is violated.

Roots with multiplicities greater than one have not been found, while carrying out numerical simulations for the homogeneous boundary value problems of symmetry as considered above. Absence of degeneration points in the spectrum of fundamental frequencies corresponding to the free oscillations of the same symmetry class leads to the violation of regular behaviour of the spectral curves in the domain of interaction. The latter shows that spectral curves approach the critical points.

Description of interaction of free oscillations in open resonators by means of the theory of singularities of smooth mappings enables the complete analysis of intertype oscillations. One of the most important results here is that in the vicinity of an isolated and non-degenerated (Morse-type) critical point the dispersion relation, with an accuracy up to the cubic terms, is presented in the form of a quadratic equation with respect to the unknown frequency parameters. If the left-hand side of the dispersion relation does not vanish at the Morse critical point, then this quadratic equation gives the dispersion law for interacting oscillations. If the dispersion relation vanishes at the Morse critical point, then its corresponding root has double degeneration and the electromagnetic eigenfields do not interact. It means that the value which the dispersion relation takes at the Morse point, determines the degree of mutual "influence" of oscillations when the fundamental frequencies draw together.

Degenerated (linearly independent) eigenfields of open resonators may be found by the simultaneous solution of independent operator equations with a definite operator-valued function, producing Koch matrices [28] which are obtained after splitting the operator of the initial boundary value problem (due to its symmetry).

Using the properties of Koch matrices and the results of the analytical description of spectral characteristics in the vicinity of isolated non-degenerated critical points, one can show that the degeneration points, corresponding to free field oscillations of different symmetry classes, are Morse points of the corresponding mapping and that their spectral sets have the form of intersecting hyperplanes in the vicinity of these points. One can completely understand the qualitative features of interaction processes by considering various spectral surfaces and their intersections with the planes arbitrarily oriented in space.

The mathematically justified and effective solutions to the boundary value problems for free and forced oscillations in two-dimensional and three-dimensional open resonators and two-dimensional resonators with layered-radial inclusions are presented in the first chapter.

For the two-dimensional open resonators the spectral problem is reduced in an equivalent way by separation of variables in local coordinates (using the addition theorems for the cylindrical functions) and by the method of the Riemann-Hilbert problem [5] to a problem of characteristic numbers for the matrix operator-valued function, which is finite-meromorphic with respect to the spectral parameter. It is established by the methods of the spectral theory of operator-valued functions that the calculation of the spectrum of fundamental frequencies can be carried out by a mathematically justified algorithm for which the estimation of the convergence rate is obtained.

Spectral data of the considered two-dimensional open resonators allow us to solve the excitation (diffraction) problem for the open structures with different types of inclusions by monochromatic sources. The initial excitation problem is solved by separation of variables in local coordinates, and it is reduced to the system of coupled functional equations with a trigonometrical kernel. There exists its left equivalent regularizator, effectively determined by the method of the Riemann-Hilbert problem. The corresponding system of Hilbert–Schmidt equations of the second kind defines the excitation problem, and there exists a unique solution to this system. For numerical computations the method of reduction is applied and an asymptotic estimate for the accuracy of approximate solutions is obtained.

Spectral characteristics of a resonator with inclusions in the presence of the Morse critical point are studied. Electromagnetic field distributions near these points are constructed. A special algorithm for the search of the Morse critical points of the dispersion relation is worked out.

The conception of singularities of smooth mappings of complex-analytical functions with respect to several variables gives possibilities to develop the theory of Morse points in the dispersion relations of open resonators with inclusions. The analytical nature of Morse points is studied and the resonance restructuring of the spectral characteristics of an open resonator is given, in a way different from the so-called "Wien graphs". Basic results are obtained in the works of the author and his co-workers A.V.Brovenko, P.N.Melezhik, A.E.Poedinchuk and Yu.A.Tuchkin.

As one can see, the rigorous solution to the boundary value problem of determining the spectral and spatial characteristics of the electromagnetic field in open resonators is connected with serious mathematical difficulties, which gives the possibility to analyze the properties of resonators only in the simplest and idealized situations [10]. Therefore, we assume a special character of the experiments where specific demands are kept for the open resonators in the millimeter and submillimeter wavelength range. These experiments are designed to study open structures under the special conditions when mirror dimensions, the distance between mirrors and their curvature are commensurable with the wavelength. In this case the multi-mode character of the eigenoscillations of a resonator becomes apparent and, correspondingly, its resonance and intertype states for excitation problems. When mastering the millimeter and submillimeter wavelength range, the biggest difficulty is in working out an experimental technique to study the phase, amplitude and polarization properties of the field in open resonators. In this chapter the basic results of the experimental radiophysics of open resonators in the millimeter wavelength range are stated, and are connected with applications in various fields of science and technology. Delicate methods are necessary for the detailed analysis which allow one to obtain, in addition to the qualitative information (for example, by the visualization method [6]), the quantitative information about the phase, amplitude and polarization. Such methods have been developed at the Institute of Radiophysics and Electronics of the Ukrainian Academy of Sciences. Traditional ways to study the field's spatial structure are reduced to the measurement of the amplitude of electric and magnetic field components [30–39] based on the thermography and reactive sensing. The thermographic method of field image registration allows one to control field processes in quasi-real time [34, 35], which is its main advantage. However, these data have a qualitative character and impede the use of such information for the measurement of detailed field structure. Linking the thermo-indicators with modern computers gives the possibility of recording and analyzing field information during the process of experiment. The methods of field measurements are widely used by means of a reactive probe [31] placed in the open resonator, when either the displacement of the resonance frequency or the change of the resonator's transmission coefficient are measured. Probe characteristics are of great importance. If it is small enough and one can neglect its disturbance, then for the resonant field measurement one can use active probing with a connecting element placed on the mirror's surface or by controlling the field scattered by the probe out of the open resonator.

Studying the works in the area of thermography, active and reactive sensing in the millimeter and submillimeter wavelength range allows to conclude that there were practically no existing methods for investigating the phase and polarization structures of resonant and intertype fields. It naturally hampered the study of specific mechanisms of field forming in open resonators and lowered the possibilities of their application.

The new ways to study open resonators, worked out by the author and his co-workers, open the new possibilities in the experimental treatment of resonant and intertype fields in the millimeter and submillimeter wavelength range. It especially concerns phase and polarization measurements, because with their help one can find a way to tighten together rigorous theoretical models for open resonators with their real prototypes and also to understand how different types of inclusions influence the measured values. Information about the vector structure of an electromagnetic field in an open resonator is of extraordinary importance in designing new types of resonators. As an example of these studies one can take experimental data about the influence of a connecting slot on the characteristics of an open resonator. It turns out that connecting slot not only leads to phase displacements of field components, but to a strong deformation of the resonance curve which is not connected with the so-called splitting caused by destroying polarization degeneration. Another important problem is the effective open resonator's excitation in distributive way by the quasi-optical transmission line (when the radii of the spots of the excitation and resonance beams and the curvature radii of their phase surfaces must be coordinated).

Experimental studies of open resonators are systematically carried out at the Institute of Radiophysics and Electronics by the author and his co-workers A.A. Vertij, V.N. Derkach, I.V. Ivanchenko, V.K. Korneenkov, I.K. Kuzmichev, V.M. Kurin, I.D. Revin, A.A. Petrushin, N.A. Popenko, B.K. Skrynnik, G.I. Khlopov and A.I. Tsvik.

§ 1. Spectral theory of open resonators with material inclusions

Let us consider the open resonator formed by a pair of infinitely thin, perfectly conducting, unclosed circular cylindrical surfaces containing inclusions in the form of an inhomogeneous isotropic circular pivot with the axis parallel to that of the resonator's mirrors (the permittivity and permeability of the medium, $\varepsilon(x,y)$ and $\mu(x,y)$, depend on spatial coordinates). The resonator itself is placed in a homogeneous medium with complex permittivity and permeability, ε_0 and μ_0 (Im $\varepsilon_0 \geq 0$, Im $\mu_0 \geq 0$). Here x,y are the Cartesian coordinates where the structure is considered. It is assumed that the open resonator with inclusions is infinite and uniform along the z-axis.

Let $S = \cup_{j=1}^{3} S_j$ be the structure's cross-section in the xy-plane, where S_1 and S_2 are the mirror cross-sections and S_3 is the cross-section of the inclusion. Let us introduce the polar coordinate systems (r_j, ϕ_j) where $j = 1,2,3$ in such a way that S_j $(j = 1,2)$ coincide with $r_j = $ const, and ∂S_3 with $r_3 = $ const.

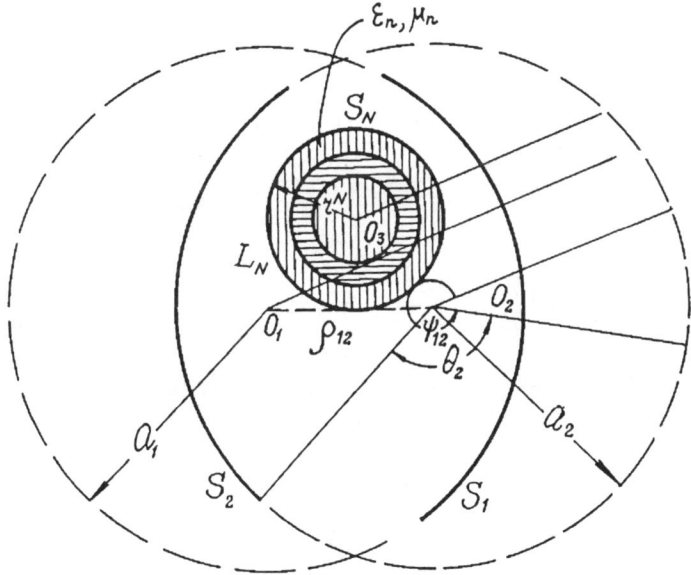

Figure 1.

Then the structure's geometry $S = \cup_{j=1}^{3} S_j$ (see Fig.1) may be described by the following parameters: a_1 and a_2 are the curvature radii of S_1 and S_2, respectively, ρ_{jn}, ψ_{jn} are the coordinates of the origin of the nth coordinate system with respect to the jth coordinate system $(n, j = 1,2,3)$, and $2\theta_j$ $(j = 1,2)$ is the angle dimension of $\hat{S}_j$ which is the supplement of S_j to a complete circle and ψ_j $(j = 1,2)$ is the orientation angle of the center of S_j in the jth coordinate system. Axes of all three coordinate systems are parallel to each other. We also assume that $[S_j \cup \hat{S}_j] \cap S_3 = \emptyset$ when $j = 1,2$ and $[S_j \cup \hat{S}_j] \cap S_n = \emptyset$ when $j \neq n$, $j,n = 1,2$.

Let $\Omega = [R^2 \setminus \cup_{j=1}^2 S_j] \setminus \partial S_3$ and Λ is the Riemann surface of the analytical continuation of the fundamental solution to the two-dimensional Helmholtz equation with respect to the spectral parameter $\kappa = ka$ ($k = \omega/c$, ω is the fundamental frequency of S, c is the velocity of light in vacuum, $a = \max\{a_1, a_2\}$).

We will study the spectrum of eigenoscillations of this electrodynamic structure S for two-dimensional oscillations $\left(\frac{\partial}{\partial z} = 0\right)$. Taking this into account, together with the presence of inclusions inside the resonator, one can reduce the problem in terms of a homogeneous system of the Maxwell equations (the time dependence here is chosen in the form of $e^{-i\omega t}$) to the two independent problems for E- and H-type oscillations.

In other words, one has to determine the values of the spectral parameter $\kappa \in \Lambda$ for which there exist non-trivial solutions of the homogeneous differential equation

$$Bu(x,y) + k^2\lambda(x,y)u(x,u) = 0, \quad (x,y) \in \Omega, \tag{1.1}$$

satisfying homogeneous boundary conditions

$$u\,|_{S_j} = 0, \quad (E\text{-oscillations}), \qquad \frac{\partial u}{\partial N}\bigg|_{S_j} = 0, \quad (H\text{-oscillations}), \tag{1.2}$$

the conjugation conditions

$$(u^+ - u^-)\,|_{\partial S_j} = 0, \qquad \left[\frac{1}{\gamma(x,y)}\frac{\partial u^+}{\partial N} - \frac{1}{\gamma_0}\frac{\partial u^-}{\partial N}\right]\bigg|_{S_3} = 0, \tag{1.3}$$

the Meixner conditions

$$\iint_M \left(|u|^2 + \left|\frac{\partial u}{\partial x}\right|^2 + \left|\frac{\partial u}{\partial y}\right|^2\right)dx\,dy < \infty \tag{1.4}$$

for every compactum $M \subset R^2$ and the Reichardt radiation condition [40]: there exists an $R_0 > 0$ such that for all x, y: $(x^2 + y^2)^{1/2} \geq R_0$ the following representation holds for the function $u(x,y)$:

$$u(x,y) = \sum_{m=-\infty}^{+\infty} \eta_m H_m^{(1)}(kr\sqrt{\varepsilon_0\mu_0})e^{im\phi}. \tag{1.5}$$

B is the differential operator in $C^2\{[0,a] \times [0,2\pi]\}$, defined as follows:

$$Bu(x,y) = \begin{cases} \frac{\partial^2 u}{\partial r^2} + \left[\frac{1}{r} - \frac{\gamma'(r)}{\gamma(r)}\right]\frac{\partial u}{\partial r} + \frac{1}{r}\frac{\partial^2 u}{\partial \phi^2}, & (x,y) \in S_3, \\ \Delta u, & (x,y) \bar{\in} S_3, \end{cases}$$

$k = \kappa/a$, $r = x/\cos\phi = y/\sin\phi$, N is the unit normal to the boundary of the domain $R^2 \setminus S$; $u^\pm$ and $\frac{\partial u^\pm}{\partial N}$ are the limiting values of u and $\frac{\partial u}{\partial N}$ on the boundary, where "+" denotes the values taken from inside the domain and "−" from outside the domain.

The Hankel function $H_m^{(1)}(x)$ of the first kind and mth order is considered here on the Riemann surface Λ.

$$\lambda(x,y) = \begin{cases} \varepsilon(x,y)\mu(x,y), & (x,y) \in S_3, \\ \varepsilon_0\mu_0, & (x,y) \overline{\in} S_3, \end{cases}$$

$$\gamma(x,y) = \begin{cases} \mu(x,y), & (E\text{-oscillations}), \\ \varepsilon(x,y), & (H\text{-oscillations}) \end{cases}$$

$$\gamma_0 = \begin{cases} \mu_0, & (E\text{-oscillations}), \\ \varepsilon_0, & (H\text{-oscillations}) \end{cases}$$

The radiation condition in the form (1.5) is connected with the fact that the value of $\kappa = \kappa' + i\kappa''$ is complex. In case of real κ one can take the Sommerfeld radiation condition [41] and for complex κ with positive imaginary part the Kupradze condition [42]. The condition (1.5) is the generalized Sommerfeld (and Kupradze) condition for arbitrary complex κ, or, in other words, the continuation of the Sommerfeld condition from the real axis and in the case of the Kupradze condition it is the continuation from the upper half plane of the complex plane to the Riemann surface Λ.

Let us note that in Refs. [22, 43, 44] the Hankel functions are considered on the complex plane with the branch cut along the negative real axis. This branch cut is taken for convenience, although one can make it arbitrarily, "cutting out" from the origin to an infinitely far point.

The Riemann surface Λ has the following structure: $\Lambda = \cup_{j=-\infty}^{+\infty}\Lambda_j$ where $\Lambda_j = \{\kappa : -\pi + 2\pi j < \arg\kappa < \pi + 2\pi j\}$. The sheet Λ_0 is called "physical". There will be fundamental frequencies both with $\operatorname{Im}\kappa < 0$ and with $\operatorname{Im}\kappa > 0$ on all sheets of Λ and, consequently, attenuating and growing eigenoscillations corresponding to these frequencies. Therefore, this condition picks out "outgoing" and "incoming" solutions for $\kappa \in \Lambda_0$. For $\kappa \in \Lambda_j$

$$(-1)^j H_m^{(1)}(\kappa) + (-1)^{1+j}(1-j)H_m^{(1)}(\kappa) + (-1)^{m+j}jH_m^{(2)}(\kappa) = 0,$$

which is transformed into the identity $H_m^{(1)}(\kappa) - H_m^{(1)}(\kappa) = 0$ for $\kappa \in \Lambda_0$. Here $H_m^{(2)}(\kappa)$ is the Hankel function of the second kind and mth order.

Since Λ and the Riemann surface for the function $\operatorname{Ln}(\kappa)$ are homeomorphic, the set of fundamental frequencies of the structure S may be considered on the Riemann surface for $\operatorname{Ln}(\kappa)$ which we will denote by Λ.

The radiation condition in the form (1.5) has been introduced by Reichardt [40] and it was used in the spectral theory of the open electrodynamic structures in Refs. [8–11, 24–27].

The function $u(x,y)$ is the longitudinal component of the electromagnetic eigenfield, H_z in the case of H-polarization and E_z in the case of E-polarization. Other components of E- and H-eigenfields can be determined from the system of the Maxwell equations from $u(x,y)$.

Let us consider the two-mirror resonator with an inclusion in the form of layered-radial circular pivot parallel to one of the mirrors. The layers are filled with homogeneous isotropic media with complex permeabilities ε_n and μ_n, $n = 1, 2, \ldots, N$ (N is the number of layers).

Let us denote by K_n the sets of inner points of pivot layers, where $L_1 = \partial K_1$, $L_1 \cup L_2 = \partial K_2$, $L_2 \cup L_3 = \partial K_3$, $\ldots$, $L_{N-1} \cup L_N = \partial K_N$. Here ∂K_n is the layer's boundary and L_n is the circle with radius r_n. In this case $\varepsilon(x,y)$ and $\mu(x,y)$ are piece-wise constant functions:

$$\varepsilon(x,y) = \begin{cases} \varepsilon_n, & (x,y) \in K_n, \\ \varepsilon_0, & (x,y) \overline{\in} S_3, \end{cases} \qquad \mu(x,y) = \begin{cases} \mu_n, & (x,y) \in K_n, \\ \mu_0, & (x,y) \overline{\in} S_3. \end{cases}$$

Then conditions (1.3) will have the form

$$(u^+ - u^-)\,|_{L_n} = 0, \qquad \left.\left(\frac{1}{\mu_n}\frac{\partial u^+}{\partial N} - \frac{1}{\mu_{n+1}}\frac{\partial u^-}{\partial N}\right)\right|_{L_n} = 0,$$

for E-oscillations, and

$$(u^+ - u^-)\,|_{L_n} = 0, \qquad \left.\left(\frac{1}{\varepsilon_n}\frac{\partial u^+}{\partial N} - \frac{1}{\varepsilon_{n+1}}\frac{\partial u^-}{\partial N}\right)\right|_{L_n} = 0,$$

for H-oscillations.

These conditions assure that tangential components of eigenfields are continuous across the boundaries between different media. Since E- and H-oscillations are analogous in terms of boundary value problems, we will consider only the case of E-oscillations.

We look for the solution of the boundary value problem (1.1)–(1.5) in the case of E-oscillations (under the assumption of its existence) in the form

$$u(x,y) = \sum_{1 \leq n \leq 3} u_n(r_n, \phi_n), \tag{1.6}$$

where $u_1(r_1, \phi_1)$, $u_2(r_2, \phi_2)$, $u_3(r_3, \phi_3)$, are represented in the form of series of Bessel and Neumann functions in different domains of the space, i.e., when $r_1 \gtrless a_1$, $r_2 \gtrless a_2$, $r_3 > r^N$ and $0 < r_3 < r^1$, where r^N, r^1 are the radii of dielecric inclusions (see Fig. 1), $\tilde{k} = k\sqrt{\varepsilon_0\mu_0}$, $k_n = k\sqrt{\varepsilon_n\mu_n}$ and the principal values of the roots are chosen in such a way that their imaginary part is greater than or equal to zero (in the latter case we assume that the real part is positive).

In accord with condition (1.4), the solution of the boundary value problem (1.1)–(1.5) must belong to the Sobolev space $W_2^1(M)$ for every compactum $M \subset R^2$ and

$$\|u(x,y)\|_{W_2^1(M)} = \iint_M \left[|u(x,y)|^2 + \left|\frac{\partial u}{\partial x}\right|^2 + \left|\frac{\partial u}{\partial y}\right|^2\right]\,dx\,dy.$$

To satisfy this condition we demand that the series

$$\sum_m z_m^3 J_m(\tilde{k}r^N)H_m^{(1)\prime}(\tilde{k}r^N)e^{im\phi_3}, \qquad \sum_m B_m^1 J_m^{1\prime}(k_1 r^1)H_m^{(1)}(k_1 r^1)e^{im\phi_3},$$

$$\sum_m [A_m^n J_m(k_n r^{n-1})H_m^{(1)}(k_n r^n) + B_m^n H_m^{(1)}(k_n r^n)J_m'(k_n r^{n-1})]e^{im\phi_3},$$

$$(n = 2, 3, \ldots, N)$$

converge uniformly with respect to ϕ_3 on ∂K_n; and

$$\sum_m a_m J_m'(\tilde{k}a_1)e^{im\phi_1}, \qquad \sum_m b_m H_m^{(1)\prime}(\tilde{k}a_1)e^{im\phi_1},$$

$$\sum_m c_m J_m'(\tilde{k}a_2)e^{im\phi_2}, \qquad \sum_m d_m H_m^{(1)\prime}(\tilde{k}a_2)e^{im\phi_2},$$

converge uniformly with respect to ϕ_n on $S_n \cup \hat{S}_n$ except in the vicinities of endpoints S_n ($n = 1, 2$).

Besides, the last series must define absolutely integrable functions with respect to ϕ_n on $S_n \cup \hat{S}_n$. z_m^3, B_m^1, A_m^n, B_m^n, a_m, b_m, c_m are constants to be determined.

The properties of cylindrical functions yield infinite differentiability of the function $u(x, y)$ with respect to x and y. The function $u(x, y)$ in $R^2 \setminus \cup_{n=1}^2 S_n \cup \hat{S}_n \setminus \cup_{n=1}^N L_n$ will satisfy differential equation (1.1) and radiation condition (1.5) for k on Λ.

Thus, we will demand, using representation (1.6) and the properties of convergence of the series, that $u(x, y)$ satisfies the boundary conditions (1.2) and (1.3). We restrict ourselves to the case when the mirrors of the open resonator (Fig.1) have the same direction of concavity and together with the inclusion are contained in a circle with a sufficiently large radius.

Fulfillment of the boundary conditions for $u(x, y)$ allows us to obtain a system of functional (summation-type) equations of the first kind on surface S_1 (Fig.1) with the kernel formed by trigonometrical functions:

$$\begin{cases} \sum_m y_m^1 e^{im(\phi_1-\psi_1)} = 0, & |\phi_1 - \psi_1| > \theta_1, \\ \sum_m |m| y_m^1 e^{im(\phi_1-\psi_1)} = \frac{i}{\pi} x_0^1 + \sum_m f_m e^{im(\phi_1-\psi_1)}, & |\phi_1 - \psi_1| < \theta_1, \end{cases} \quad (1.7)$$

where

$$f_m = x_m^1 e^{im\psi_1} J_m(\tilde{k}a_1) H_m^{(1)}(\tilde{k}a_1) + |m|\beta_m.$$

Satisfying the boundary conditions (1.2) similarly yields the summation-type equations

$$\begin{cases} \sum_m y_m^2 e^{im(\phi_2-\psi_2)} = 0, & |\phi_2 - \psi_2| > \theta_2, \\ \sum_m |m| y_m^2 e^{im(\phi_2-\psi_2)} = \frac{i}{\pi} x_0^2 + \sum_m g_m e^{im(\phi_2-\psi_2)}, & |\phi_2 - \psi_2| < \theta_2, \end{cases} \quad (1.8)$$

where

$$g_m = x_m^2 e^{im\psi_2} J_m(\tilde{k}a_2) H_m^{(1)}(\tilde{k}a_2)\varepsilon_m + |m|\hat{\beta}_m,$$

$$\hat{\beta}_m = J_m(\tilde{k}a_2) e^{im\psi_2} \sum_n [x_n^1 J_{n-m}(\tilde{k}\rho_{12}) H_n^{(1)}(\tilde{k}a_1) e^{i(n-m)\psi_{12}} + R_n H_{n-m}^{(1)}(\tilde{k}\rho_{32}) e^{i(n-m)\psi_{32}}],$$

$$\varepsilon_m = |m| + i[\pi J_m(\tilde{k}a_2) H_m^{(1)}(\tilde{k}a_2)]^{-1}.$$

The following notations are used in (1.7) and (1.8):

$$\beta_m = \gamma_m^{21} e^{im\psi_1} J_m(\tilde{k}a_1) + \gamma_m^{31} e^{im\psi_1} H_m^{(1)}(\tilde{k}a_1),$$

$$\gamma_m^{21} = \sum_n d_n H_{n-m}^{(1)}(\tilde{k}\rho_{12}) e^{i(n-m)\psi_{21}},$$

$$\gamma_m^{31} = \sum_n R_n J_{n-m}(\tilde{k}\rho_{31}) e^{i(n-m)\psi_3},$$

$$y_m^j = x_m^1 e^{im\psi_1} J_m(\tilde{k}a_1) H_m^{(1)}(\tilde{k}a_1) + \beta_m, \quad j = 1, 2,$$

$$R_m = z_m^3 J_m(\tilde{k}r^N).$$

One can see that systems of functional equations (1.7) and (1.8) may be regularized by the method of the Riemann-Hilbert problem [5]. Applying this procedure to these systems and taking into account the relations between y_m^n and x_m^n, where $n = 1, 2$, relations $z_m^n = x_m^n J_m(\tilde{k}a_n) H_m^{(1)}(\tilde{k}a_n)$ $(n = 1, 2)$, we obtain for $k \in \Lambda$ the homogeneous system of linear algebraic equations

$$z_m^n = \sum_{s=1}^{3} \sum_q a_{mq}^{ns}(k) z_q^s, \quad (n = 1, 2), \tag{1.9}$$

with matrix elements depending on the spectral parameter $k \in \Lambda$;

$$a_{mq}^{ss}(k) = F_q^m(u_s) \varepsilon_q(\tilde{k}a_s) e^{i(q-m)\psi_s}, \quad (s = 1, 2),$$

$$a_{mq}^{12}(k) = \{H_q^{(1)}(\tilde{k}a_2)\}^{-1} \sum_n Q_n^m(u_1) R_{q-n}(\tilde{k}\rho_{12}) T_n(\tilde{k}a_1) \chi_n^{mq}(\psi_1, \psi_{21}),$$

$$a_{mq}^{21}(k) = \{J_q(\tilde{k}a_2)\}^{-1} \sum_n Q_n^m(u_2) J_{q-n}(\tilde{k}\rho_{21}) J_n(\tilde{k}a_2) \chi_n^{mq}(\psi_2, \psi_{12}),$$

$$a_{mq}^{s3}(k) = J_q(\tilde{k}r^N) \sum_n Q_n^m(u_s) R_{q-n}(\tilde{k}\rho_{s3}) T_n(\tilde{k}a_s) \chi_n^{mq}(\psi_s, \psi_{3s}),$$

where

$$u_s = \cos\theta_s, \quad \chi_n^{mq}(\alpha, \beta) = (-1)^m e^{i[(q-n)\beta + (n-m)\alpha]},$$

$$F_q^m(x) = \begin{cases} -\ln\frac{1+x}{2}, & q = 0, \ m = 0, \\ \frac{1}{q} V_{q-1}^{-1}(x), & q \neq 0, \ m = 0, \\ \frac{1}{m} V_{m-1}^{q-1}(x), & q \neq 0, \ m \neq 0, \end{cases}$$

$$Q_n^m(x) = \begin{cases} (-1)^{n+1} V_{n-1}^{m-1}(-x), & m \neq 0, \\ \frac{|n|}{n} V_{n-1}^{-1}(-x), & n \neq 0, \ m \neq 0, \\ -1, & n = 0, \ m = 0, \end{cases}$$

$$R_{q-n}(x) T_n(y) = \begin{cases} H_{q-n}^{(1)}(x) J_n(y), & |x| > |y|, \\ J_{q-n}(x) H_n^{(1)}(y), & |x| < |y|. \end{cases}$$

Functions $V_p^m(x)$, which are defined by means of Legendre polynomials, are computed in Ref. [5].

Let us demand now that the function $u(x, y)$ satisfies the conjugation conditions (1.3), which may be rewritten in the case of E-oscillations as

$$u^n = u^{n+1}, \quad \frac{1}{\mu_n} \frac{\partial u^n}{\partial r} = \frac{1}{\mu_{n+1}} \frac{\partial u^{n+1}}{\partial r}, \tag{1.10}$$

where $u^n = u_3(r^n, \phi)$, $n = 1, 2, \ldots, N$.

As a result of bulky transformations we obtain the system of linear algebraic equations

$$z_n^3 = \sum_{s=1}^{3} \sum_m a_{nm}^{3s}(k) z_m^s, \tag{1.11}$$

where

$$a_{nm}^{31}(k) = \frac{\mu_n - \mu_0}{\mu_n + \mu_0} \frac{J_{n-m}(\tilde{k}\rho_{31})}{J_m(\tilde{k}a_1)H_n^{(1)}(\tilde{k}r^N)} e^{i(m-n)\psi_{13}},$$

$$a_{nm}^{32}(k) = \frac{\mu_N - \mu_0}{\mu_N + \mu_0} \frac{H_{n-m}(\tilde{k}\rho_{32})}{H_m(\tilde{k}a_2)H_n^{(1)}(\tilde{k}r^N)} e^{i(m-n)\psi_{23}},$$

$$a_{nm}^{33}(k) = \left\{ 1 - \frac{\mu_N - \mu_0}{\mu_N + \mu_0} [\Pi_n(\tilde{k}r^N; \Phi)]^{-1} \right\} \delta_{mn},$$

$$\Pi_n(x; \Phi) = \frac{H_n^{(1)}(x)\{w_n J_n(x) - w_0 J_n'(x)\Phi_n(k; N)\}}{J_n(x)\{w_0 H_n^{(1)'}(x)\Phi_n(k; N) - w_n H_n^{(1)}(x)\}}.$$

Here δ_{mn} is the Kronecker delta function and $\Phi_n(k; N)$ is defined in the following way:
if $N = 1$ then $\Phi_n(k; N) = [J_n(k_1 r^1)]/[J_n'(k_1 r^1)]$, otherwise

$$\Phi_n(k; N) = \frac{T_n(k; N-1)J_n(k_N r^{N-1})H_n^{(1)}(k_N r^N) + H_n^{(1)}(k_N r^N)J_n(k_N r^N)}{T_n(k; N-1)J_n(k_N r^{N-1})H_n^{(1)'}(k_N r^N) + H_n^{(1)}(k_N r^N)J_n'(k_N r^N)},$$

and for $T_n(k; N-1)$ one has recurrence relations

$$T_n(k; j+1) = \frac{T_n(k; j)A_{nj+1} + B_{nj+1}}{1 + T_n(k; j)D_{nj+1}},$$

where

$$A_{nj+1} = K_n^{j+1} \times (N_n^{j+1})^{-1},$$

$$B_{nj+1} = L_n^{j+1} \times (N_n^{j+1})^{-1},$$

$$D_{nj+1} = M_n^{j+1} \times (N_n^{j+1})^{-1}.$$

Expressions for K_m^n, L_m^n, M_m^n and N_m^n have the form:

$$K_m^n = F_{mn}^1\{w_{n+1}J_m'(k_{n+1}r^n)H_m^{(1)}(k_n r^n) - w_n J_m(k_{n+1}r^n)H_m^{(1)'}(k_n r^n)\},$$

$$L_m^n = F_{mn}^2\{w_{n+1}J_m(k_n r^n)J_m'(k_{n+1}r^n) - w_n J_m'(k_n r^n)J_m(k_{n+1}r^n)\},$$

$$L_m^n = F_{mn}^3\{w_n H_m^{(1)}(k_{n+1}r^n)H_m^{(1)'}(k_n r^n) - w_{n+1}H_m^{(1)'}(k_{n+1}r^n)H_m^{(1)}(k_n r^n)\},$$

$$N_m^n = F_{mn}^4\{w_n H_m^{(1)}(k_{n+1}r^n)J_m'(k_n r^n) - w_{n+1}H_m^{(1)'}(k_{n+1}r^n)J_m(k_n r^n)\},$$

$$F_{mn}^1 = \xi_n \frac{J_m(k_n r^{n-1})}{J_m(k_{n+1}r^n)}, \qquad F_{mn}^2 = \xi_n \frac{H_m^{(1)}(k_n r^{n-1})}{J_m(k_{n+1}r^n)},$$

$$F_{mn}^3 = \xi_n \frac{J_m(k_n r^{n-1})}{H_m^{(1)}(k_{n+1}r^n)}, \qquad F_{mn}^4 = \xi_n \frac{H_m^{(1)}(k_n r^{n-1})}{H_m^{(1)}(k_{n+1}r^n)},$$

$$w_n = \begin{cases} \sqrt{\frac{\mu_n}{\varepsilon_n}}, & H\text{-oscillations,} \\ \sqrt{\frac{\varepsilon_n}{\mu_n}}, & E\text{-oscillations,} \end{cases}$$

$$\xi_n = \frac{i\pi k \gamma_{n+1}r^n}{2},$$

$$\gamma_n = \begin{cases} \varepsilon_n, & H\text{-oscillations,} \\ \mu_n, & E\text{-oscillations.} \end{cases}$$

Finally, combining the systems of linear algebraic equations (1.9) and (1.11) we obtain an infinite system of linear algebraic equations of the second kind

$$z_n^j = \sum_{s=1}^{3} \sum_m a_{nm}^{js}(k) z_m^s, \quad (j = 1, 2, 3). \tag{1.12}$$

Similar considerations for the case of H-oscillations give an infinite homogeneous system of linear algebraic equations in the form

$$x_n^j = \sum_{s=1}^{3} \sum_m b_{nm}^{js}(k) x_m^s, \quad (j = 1, 2, 3). \tag{1.13}$$

Here the matrices $\|b_{nm}^{js}(k)\|$ are the same as in Ref. [10] and elements of the matrices $\|b_{nm}^{3s}(k)\|$ and $\|b_{nm}^{s3}(k)\|$ have the form

$$b_{nm}^{s3}(k) = \pi i(\tilde{k} a_s) J_m(\tilde{k} r^N) \sum_p T_{m-p}(\tilde{k} \rho_{3s}) R_p(\tilde{k} a_s) F_p^n(u_s) \tilde{\chi}_p^{nm}(\psi_s, \psi_{3s}),$$

$$b_{nm}^{3s}(k) = \frac{\varepsilon_N - \varepsilon_0}{\varepsilon_N + \varepsilon_0} \frac{T_{m-p}(\tilde{k} \rho_{s3}) R_m(\tilde{k} a_s)}{H_m^{(1)}(\tilde{k} r^N)} e^{i(m-n)\psi_{s3}}, \quad (s = 1, 2),$$

$$b_{nm}^{33}(k) = \left\{ 1 - \frac{\varepsilon_N - \varepsilon_0}{\varepsilon_N + \varepsilon_0} \hat{\Pi}_n(\tilde{k} r^N; \Phi) \right\} \delta_{mn},$$

where $u_s = -\cos\theta_s$ in $F_p^n(u_s)$, and δ_{mn} is the Kronecker delta function.

$$T_{m-n}(\tau) R_n(\xi) = \begin{cases} J_n'(\xi) H_{m-n}^{(1)}(\tau), & |\xi| < |\tau|, \\ H_n^{(1)}(\xi) J_{m-n}(\tau), & |\xi| > |\tau|, \end{cases}$$

$$\tilde{\chi}_p^{nm}(\alpha) = (-1)^p \chi_p^{mn}(\alpha, \beta),$$

$$\hat{\Pi}_n(x, \Phi) = \frac{J_n(x)\{\hat{w}_0 H_n^{(1)\prime}(x) \hat{\Phi}(k; N) - \hat{w}_N H_n^{(1)}(x)\}}{H_n^{(1)}(x)\{\hat{w}_N J_n(x) - \hat{w}_0 J_n'(x) \hat{\Phi}(k; N)\}},$$

$$\hat{w}_n = \sqrt{\frac{\mu_n}{\varepsilon_n}}, \quad (n = 0, 1, 2, \dots, N).$$

$\hat{\Phi}(k; N)$ is obtained in this case by the following transformation of the function $\Phi_n(k; N)$, for E-oscillations: $\hat{\Phi}(k; N) = \sigma_{\varepsilon,\mu} \Phi_n(k; N)$. The inverse transform holds: $\Phi_n(k; N) = \sigma_{\varepsilon,\mu}^{-1} \hat{\Phi}(k; N)$. Here $\sigma_{\varepsilon,\mu}$ is the operator determining the transposition of the values ε_n and μ_n when $n = 0, 1, \dots$. Finally, let us note that the matrices $\|a_{mn}^{js}(k)\|_{j,s=1}^3$ for E-oscillations and $\|b_{mn}^{js}(k)\|_{j,s=1}^3$ for H-oscillations depend not only on spectral but also on several non-spectral parameters describing the structure's geometry and the properties of the media. Analytical properties of these matrices will be studied later.

Let us investigate the behaviour of the matrix elements of the system (1.12), (1.13) for large indices ($|n| \gg |\kappa|, |m| \gg |\kappa|$). We will determine their analytical properties with respect to both spectral and non-spectral parameters, in particular, the permittivities and permeabilities of the media of the inclusion and the opening angle of the mirrors.

The presence of an inhomogeneous inclusion in the form of a layered pivot will cause the fact that only $\|a_{nm}^{33}(k)\|$ (and $\|b_{nm}^{33}(k)\|$ in the case of H-oscillations) will differ from the matrices obtained in Ref. [10]. The following assertions hold for H-oscillations: for $|n| \to \infty$ and $|m| \to \infty$, when an open resonator with homogeneous dielectric cylinder is considered, $\sum_{n,m} |b_{nm}^{js}(k)| < \infty$ uniformly with respect to $k \in \Lambda_0$. Hence, it is sufficient here to study only $\|b_{nm}^{33}(k)\|$ for $|n| \gg |\kappa|$ and $|m| \gg |\kappa|$ and, more precisely, the expression $\hat{\Phi}_n(k; N)$ (in the case of E-oscillations, for $\Phi_n(k; N)$).

In the case of E-oscillations other than $\|a_{nm}^{33}(k)\|$ for $|n| \to \infty$ and $|m| \to \infty$ one has to consider also the matrices $\|a_{nm}^{j3}(k)$ and $\|a_{nm}^{3s}(k)\|$, $(s, j = 1, 2)$ because in Ref. [10] estimations have been obtained for $\|a_{nm}^{js}(k)\|_{j,s=1}^2$ which are sufficient for the convergence of $\sum_{n,m} |a_{nm}^{js}(k)| < \infty$, $(j, s = 1, 2)$ uniformly with respect to $k \in \Lambda$.

For $\Phi_n(k; N)$ (in the case of H-oscillations, for $\hat{\Phi}_n(k; N)$) when $|n| \to \infty$ the asymptotic representation

$$\Phi_n(k; N) \sim \frac{1}{n} kr^N (\varepsilon_N \mu_N)^{1/2}, \quad \operatorname{Im} \sqrt{\varepsilon_N \mu_N} \geq 0, \tag{1.14}$$

holds. In other words,

$$\lim_{n \to \infty} \left| \Phi_n(k; N) \frac{|n|}{kr^N (\varepsilon_N \mu_N)^{1/2}} \right| = 1.$$

In accordance with (1.14) and asymptotics of cylindrical functions for large index [46], we have

$$|a_{nn}^{33}(k)| < R(F) \frac{1}{n^2}, \qquad |b_{nn}^{33}(k)| < \tilde{R}(F) \frac{1}{n^2},$$

where $R(F)$ and $\tilde{R}(F)$ are constants depending on the structure parameters and on the compactum F in Λ which does not contain singularities of corresponding matrices.

Let us consider now matrices $\|a_{nm}^{j3}(k)\|_{j=1,2}$, assuming that $n \neq 0$ and $\rho_{s3} < a_s$ (the case for $\rho_{s3} > a_s$ is treated similarly). Then $a_{nm}^{s3}(k)$ may be expressed as

$$a_{nm}^{s3}(k) = J_m(\tilde{k}r^N) \sum_{q=-\infty}^{+\infty} (-1)^{n+q+1} V_{q-1}^{n-1}(-u_s) J_{m-q}(\tilde{k}\rho_{s3})$$

$$\times H_q^{(1)}(\tilde{k}a_s) e^{i[(q-n)(\pi s + \psi_s) + (m-q)\psi_{3s}]}.$$

Assume that

$$a_{nm}^{s3} = \sum_q A_{nq} D_{qn}(k),$$

where

$$A_{nq} = \frac{1}{q} V_{q-1}^{n-1}(-u_s) e^{i[\pi(1-s) - \psi_s]n}, \quad u_s = \cos \theta_s,$$

and

$$D_{qm}(k) = q J_{m-q}(\tilde{k}\rho_{s3}) \tilde{H}_q^{(1)}(\tilde{k}a_s) \gamma_{qm},$$

$$\gamma_{qm} = \exp\{i[\pi(q+1) + (m-q)\psi_{3s} + q(\pi s + \psi_s)]\}.$$

Consistent with Ref. [5] the series $\sum_{q,n} \left| (1/q) V_{q-1}^{n-1}(-u_s) \right|^2$ converges for all values of $u_s \neq \pm 1$. Estimating $D_{qm}(k)$ for $|q| \to \infty$ and $|m| \to \infty$ we have $\sum_{q,n} |D_{qm}(k)| < \infty$ uniformly with respect to k on every compactum $F \subset \Lambda$.

Since $\sum_{q,n}\left|(1/q)V_{q-1}^{n-1}(-u_s)\right|^2 < \infty$ and $\sum_{q,n}|D_{qm}(k)| < \infty$, matrices $\|A_{nq}(k)\|$ and $\|D_{qm}(k)\|$ determine Hilbert-Schmidt operators in l^2.

Then the representation $a_{nm}^{s3}(k) = \sum_q A_{nq}D_{qm}(k)$ holds in an operator sense due to [10] and the matrices $\|a_{nm}^{s3}(k)\|$ for $s = 1,2$ determine a kernel operator-valued function in l^2. Using the asymptotics of cylindrical functions for large index, the Cauchy inequality and the inequality [46]

$$|H_m^{(1)}(\kappa)| \prec C(\kappa)(|m|-1)!\left(\frac{2}{|\kappa|}\right)^{|m|},$$

where

$$C(\kappa) = \frac{|\kappa|}{2}\{I_1(|\kappa|)|H_0^{(1)}(|\kappa|)| + I_0(|\kappa|)H_1^{(1)}(|\kappa|)\},$$

and $I_{0,1}(|\kappa|)$ are the modified Bessel functions, one can show that for $|m| \gg |\kappa|$ and $|n| \gg |\kappa|$ the estimations for matrices $\|a_{nm}^{3s}(k)\|$ ($s = 1,2$), with an accuracy up to a constant proportional to $|\mu_N - \mu_0| \cdot |\mu_N + \mu_0|^{-1}$ (when the inclusion is a layered-radial pivot), are similar to the estimations obtained for $\|b_{nm}^{3s}(k)\|$ ($s = 1,2$) (the case of H-oscillations) in the references indicated above. The only difference is that in the case of H-oscillations the constant is proportional to $|\varepsilon_N - \varepsilon_0| \cdot |\varepsilon_N + \varepsilon_0|^{-1}$.

Thus, for $s = 1,2$ we have that $\sum_{n,m}|a_{nm}^{3s}(k)| < \infty$ holds uniformly with respect to k on every compactum $F \subset \Lambda$ which does not contain poles $f(k) = a_{nm}^{3s}(k)$.

Let us introduce the Hilbert space

$$L = \left\{\hat{z} = \left\|\begin{matrix} z^1 \\ z^2 \\ z^3 \end{matrix}\right\|, \ z^n = \{z_m^n\}_{m=-\infty}^{\infty} \in l^2, \ n = 1,2,3\right\},$$

related to the inner product $\langle z_1, z_2\rangle = \sum_{n=1}^3 \sum_m z_{1m}^n \bar{z}_{2m}^n$ ("$-$" denotes complex conjugation). Let us relate to the system of operators $\{A^{sj}(k)\}$ the operator $A(k) = \|a_{mm}^{sj}(k)\|_{s,j=1}^3$, acting on L and defined as follows:

$$A(k)\hat{z} = \left\|\begin{matrix} \sum_{j=1}^3 \sum_m a_{nm}^{1j}(k) \\ \sum_{j=1}^3 \sum_m a_{nm}^{2j}(k) \\ \sum_{j=1}^3 \sum_m a_{nm}^{3j}(k) \end{matrix}\right\|.$$

In accordance with this, $\sum_{n,m}|a_{nm}^{js}(k)| < \infty$ for almost all values of $k \in \Lambda$ ($j,s = 1,2,3$). Therefore and in line with the results proved in Ref. [10], the following statement is true:

$A(k)$ is a kernel operator-valued function; it is finite-meromorphic with respect to $k \in \Lambda$, and analytical with respect to ε and μ in the case when an inclusion has the form of a layered-radial pivot.

Here ε and $\mu \in \Lambda_0$, besides, $\mathrm{Re}\,\varepsilon \geq 1$, $\mathrm{Re}\,\mu \geq 1$ and $\mathrm{Im}\,\varepsilon \geq 0$, $\mathrm{Im}\,\mu \geq 0$.

Then the obtained infinite system of linear homogeneous algebraic equations (1.12) may be considered as an operator equation in L,

$$\{I - A(k)\}\hat{z} = \emptyset, \tag{1.15}$$

where $A(k) = \|a_{nm}^{js}(k)\}_{j,s=1}^3$, $\hat{z} = \{z_m^s\}_{m=-\infty}^\infty$, I is the unit operator in L and $\emptyset$ is the zero element of L. Let $f(k)$ be an analytical function defined on Λ and with a set of values in L. Then we will call $k_0 \in \Lambda$ the characteristic number of the operator-valued function $I - A(k)$ if $f(k_0) \neq 0$ and $\{I - A(k_0)\}f(k_0) = \emptyset$.

Let us denote by $\sigma(A)$ the set of all characteristic numbers of the operator-valued function $I - A(k)$. The following statement holds: the spectral problem (1.1)–(1.5) and the problem on characteristic numbers for the operator-valued function $I - A(k)$ are equivalent. In other words, if $k_0 \in \sigma(A)$ and $\hat{z}$ is a non-zero solution to the operator equation (1.15), then the function $u(x, y)$ (1.6) is not identically zero and it is the solution to the problem (1.1)–(1.5) for $k = k_0$, i.e., k_0 is the fundamental frequency of the open resonator ($k_0 = \omega_0/c$). The opposite is true: if $u(x, y)$ (1.6) is the solution of the problem (1.1)–(1.5) for $k = k_0$, then $k_0 \in \sigma(A)$.

The kernel and finite-meromorphic properties of the operator-valued function $A(k)$ allow to conclude, in accordance with the Fredholm theorem for the finite-meromorphic operator-valued functions [47], that $\sigma(A)$ is a discrete and finite-multiple set on Λ (i.e., consisting of isolated points with the only possible accumulation point at infinity).

Here the finite multiplicity is understood in such a way that no more than a finite number of solutions of the problem (1.1)-(1.6) are allowed for each value of $k \in \sigma(A)$. The kernel property of the operator-valued function allows to introduce for $I - A(k)$ the characteristic determinant

$$F(k) = \exp\{\sum_{n=1}^\infty \ln[1 - \lambda_n(k)]\}. \tag{1.16}$$

Here $\lambda_n(k)$ are the eigenvalues of the operator $A(k)$. The determinant $F(k)$ together with the operator-valued function $A(k)$ depend on non-spectral parameters of the electrodynamic structure, in addition to the spectral parameter $k \in \Lambda$.

In line with the properties of $A(k)$ formulated above and Ref. [28], the function $F(k)$ is meromorphic with respect to $k \in \Lambda$, and analytical with respect to ε and μ in $\tilde{\Omega} \subset \Lambda_0$, where $\tilde{\Omega} = \{z \in \Lambda_0 : \text{Re}\, z \geq 1, \text{Im}\, z \geq 0\}$. Using the estimations [5] for $V_m^n(u)$, one can show that $F(k)$ is continuous as a function of θ_s, except for the points $\theta_s = \pi n$, where n is an integer ($s = 1, 2$). The Liouville theorem [48] yields that $F(k)$ cannot be analytically continued on Λ_0 with respect to θ_s. And finally we note that the set $\sigma(A)$ has the following property due to (1.16): if $k_0 \in \sigma(A)$, then there exists an integer n_0 such that $\lambda_{n_0}(k) = 1$, where $\lambda_{n_0}(k_0)$ is an eigenvalue of the operator $A(k_0)$.

Thus, everything is prepared to construct the algorithm for the computation of the set $\sigma(A)$. One has now to take the "truncated" systems in place of the corresponding infinite systems (1.12), (1.13) or to replace the series in (1.16) by a finite sum with a sufficient number of terms to provide the calculations of $\sigma(A)$ with prescribed accuracy.

Let us construct the algorithm to calculate the values of $\sigma(A)$. We will limit ourselves to those $k \in \Lambda$ which are in Λ_0.

Let us introduce a family of finite-dimensional operator-valued functions $A_N(k) = I - V_N A(k) V_N$, where V_N is an ortho-projector onto the finite-dimensional subspaces $L_1 \subseteq L_2 \subseteq \cdots \subseteq L_p \subseteq \ldots$ of the space L, defined as follows:

$$V_N(L) = \left\{ \hat{z} = \left\| \begin{matrix} z^1 \\ z^2 \\ z^3 \end{matrix} \right\|, \ z^j = \{z_q^j\} = 0, \ |q| > N, \ j = 1,2,3 \right\}, \quad N = 1,2,\ldots.$$

The results from [10, 21, 22, 26] were used in Ref. [49] for the justification of the computational algorithms to calculate spectral characteristics of such electrodynamic structures. Here we will use the operator generalization of Rouchet's theorem which may be formulated in the following way: let γ be a smooth or piece-wise smooth, simple closed curve on Λ_0, and the operator-valued function $P(k)$ is normal to the curve γ, and for all values of $k \in \gamma$ the inequality holds: $\|P^{-1}(k)D(k)\|_L < 1$. Then the operator-valued function $B(k) = P(k) + D(k)$ is also normal to γ and the total multiplicities of the characteristic numbers of $P(k)$ and $B(k)$ coincide in the domain bounded by γ.

In addition to Rouchet's theorem the following statements hold: the operator-valued function $\hat{A}(k) = A(k) - V_N A(k) V_N$ satisfy the condition $\lim_{N \to +\infty} \|A(k) - V_N A(k) V_N\| = 0$, and:

(1) if $k_N \in \sigma(A_N)$, $\lim_{N \to +\infty} |k_N - \hat{k}| = 0$, and $\hat{k} \in \Lambda_0$ is not a pole of $A(k)$, then $\hat{k} \in \sigma(A)$ ($\sigma(A_N)$ is the set of characteristic numbers of operator-valued functions $A_N(k) = I - V_N A(k) V_N$);

(2) if $\hat{k} \in \sigma(A)$, then there exist $\{k_N\}_{N=1}^{\infty} \in \sigma(A_N)$ such that $\lim_{N \to \infty} |k_N - \hat{k}| = 0$.

Hence, the characteristic numbers of the operator-valued function $I - A(k)$ are approximated by the characteristic numbers of finite-dimensional operator-valued functions $I - V_N A(k) V_N$. Because the operator-valued function $I - V_N A(k) V_N$ is finite-dimensional, the fundamental frequencies of the initial structure may be determined from the equation

$$\det\{I - V_N A(k) V_N\} = 0 \qquad (1.17)$$

by means of numerical methods (like Newton's method, the secant method or their modifications). As has been shown by computer simulations, one can take $N = [|\kappa|] + \tau$ as the truncation order N where $\tau > 5$ is the entire part.

Equation (1.17) may be written in the following way:

$$\kappa = f(\varepsilon, \mu, \theta_s, \dots), \qquad (1.18)$$

where f generally is a function of several complex variables. When applying Newton's method $f(\varepsilon, \mu, \theta_s, \dots) = \kappa - F_N(\kappa)[F_N'(\kappa)]^{-1}$ where $F_N'(\kappa) = \frac{\partial F_N(\kappa)}{\partial \kappa}$ and we assume that $F_N'(\kappa) \neq 0$. Thus, varying one of the non-spectral parameters in (1.18), with all others fixed, we get, with the help of a computer, the dependence of κ ($\mathrm{Re}\,\kappa, \mathrm{Im}\,\kappa$) on $\mathrm{Re}\,\varepsilon$ or other non-spectral parameters.

Now one can assert that the study of fundamental frequencies is reduced by means of the boundary problem for open resonators with inhomogeneous inclusions to a problem of characteristic numbers of the matrix Fredholm operator-valued function having the form $I - A(\varepsilon, \mu, \theta_s, \dots)$ and its further analysis is connected with the study of operator-valued function $A(\varepsilon, \mu, \theta_s, \dots)$. In particular, it is connected with the proof that this operator-valued function is kernel for almost all values of $k \subset \Lambda$, finite-meromorphic and analytical with respect to ε and μ in $\tilde{\Omega} \subset \Lambda_0$, and continuous as a function of θ_s, $s = 1, 2$.

Furthermore, the determination of the fundamental frequencies of an open resonator with inclusions depends on the properties of the characteristic determinant (1.16). In particular, one has to establish that it is meromorphic with respect to k, analytical with respect to ε and μ, and continuous with respect to θ_s.

As a result we show that the set of characteristic numbers of the operator-valued function $I - A(\varepsilon, \mu, \theta_s, \dots)$ is discrete and it has finite multiplicity.

In other words, the spectrum of the fundamental frequencies of an open resonator with inclusions of the types considered above is discrete and has finite multiplicity. For numerical determination of the fundamental frequencies the mathematically correct algorithm is constructed.

§ 2. Excitation of open resonators with inhomogeneous inclusions

In this section the solution to the problem of the excitation of two-dimensional open resonators with different inclusions by a monochromatic source is presented. Green's functions are constructed for E- and H-oscillations; the existence and uniqueness of corresponding solutions are proved when the parameter k is such that $\operatorname{Im} k > 0$ ($k = (\omega/c)\sqrt{\varepsilon_0\mu_0}$, ω is the frequency of the monochromatic source, ε and μ are the permittivity and permeability of the medium, where $\operatorname{Im}\varepsilon_0 > 0$ and $\operatorname{Im}\mu_0 > 0$). Green's functions are analyzed with respect to the parameters k, ε, μ, θ_s ($s = 1, 2$) and, as a result, the rigorous mathematical justification for the possibility of their analytical continuation with respect to k from the upper half of the complex plane to a meromorphic function on the Riemann surface Λ is given and in such a way that poles coincide with the spectrum of the fundamental frequencies for the structure S (see corresponding definitions in section 1).

Assume that the monochromatic source is concentrated in a point $M \in R^2$ with coordinates $(x_0, y_0) \bar{\in} S_3$, and assume also that

$$G(k;r) = \begin{cases} -\frac{i}{4}H_0^{(1)}(kr) + g^-(k;r), & (x,y) \bar{\in} S_3 \\ g^+(k;r), & (x,y) \in S_3 \end{cases}$$

where $\operatorname{Im} k > 0$, $r = \sqrt{(x-x_0)^2 + (y-y_0)^2}$, $\lim_{r\to\infty} |g^-(k;r)| = 0$. We will call $G(k;r)$ the Green function if the following conditions hold:

(1) $g^\pm(k;r)$ satisfies the Meixner condition [41] on every compactum in R^2 which does not contain the points (x_0, y_0),

(2) $g^\pm(k;r)$ satisfies the homogeneous operator equation of the type given by (1.1),

(3) $G(k;r)$ satisfies the boundary condition on the mirrors in case of E-oscillations

$$\iint_F G\Delta f\, dx dy + \iint_F (\nabla G, \nabla f) dx dy = \oint_{\partial F} G\frac{\partial f}{\partial n}\, dS_F$$

for every twice continuously differentiable function f in $\bar{\Omega}$ satisfying the condition $\iint_F (|\nabla f|^2 + |f|^2) dx dy < \infty$. In case of H-oscillations

$$\iint_F G\Delta f\, dx dy + \iint_F (\nabla G, \nabla f) dx dy = \oint_{\partial F} \frac{\partial G}{\partial n} f\, dS_F$$

for every twice continuously differentiable function f in $\bar{\Omega}$ satisfying the condition $\iint_F (|\nabla f|^2 + |f|^2) dx dy < \infty$ and the conjugation conditions on ∂S_F in the form

$$(G^+ - G^-)\,|_{\partial S_3} = 0,$$

(4)

$$\left\{ \frac{1}{\gamma(x,y)}\frac{\partial^+ G}{\partial \hat{n}} - \frac{1}{\gamma_0}\frac{\partial^- G}{\partial \hat{n}} \right\}\bigg|_{\partial S_3} = 0.$$

Here we assume that $(x_0, y_0) \overline{\in} F$ where $F = \cup_{n=1}^{2} F_n$, F_n is an arbitrary compactum containing the mirror S_n, n and $\hat{n}$ are the external normals to ∂F and ∂S_3, respectively, ∂S_F are elements of the curve ∂F. $G^{\pm}$ and $\frac{\partial^{\pm} G}{\partial n}$ are the limiting values of the function G and its normal derivative on ∂S_3 where the "+" sign denotes approaching ∂S_3 from within and the "−"sign from the outside.

Conditions (3) are the usual specification of the boundary conditions $G \mid_{S_m} = 0$ and $\frac{\partial G}{\partial n} \mid_{S_m} = 0$ for the cases of E- and H-oscillations [51] ($m = 1, 2$).

Let us construct the Green function using the results of section 1.

We rewrite the representation of $G(k; r)$ in the form

$$G(k; r) = \begin{cases} \frac{1}{4i} H_0^{(1)}(kr) + g^{\pm}(x, y; x_0, y_0; k), & (x, y) \overline{\in} S_3, \\ \sum_{n=-\infty}^{+\infty} B_n^1 H_n^{(1)}(k_1 r^1) J_n(k_1 r_3) e^{in\phi_3}, & 0 < r_3 < r^1, \\ \sum_{n=-\infty}^{+\infty} [\hat{A}_n^j H_n^{(1)}(k_j r_3) + \hat{B}_n^j J_n(k_j r_3)] e^{in\phi_3}, & r^{j-1} < r_3 < r^j, \end{cases}$$

when inclusion S_3 is a circular layered cylinder. Here $g^{-}(\dots)$ is the function to be determined, $\hat{A}_n^j = A_n^j J_n(k_j r^{j-1})$, $\hat{B}_n^j = B_n^j H_n^{(1)}(k_j r^j)$, where A_n^j and B_n^j are unknown coefficients to be determined.

Due to the conditions stated above, the function $g^{-}(\dots)$ satisfies the homogeneous Helmholtz equation and the boundary conditions

$$g^{-}(\dots) \mid_{S_m} = (i/4) H_0^{(1)}(kr), \qquad (E\text{-oscillations})$$

$$\left\{ \frac{\partial g^{-}}{\partial n} - \frac{i}{4} \frac{\partial}{\partial n} H_0^{(1)}(kr) \right\} \Bigg|_{S_m} = 0, \quad (H\text{-oscillations}),$$

where $m = 1, 2$. The function $H_0^{(1)}(kr)$ is considered here in the mth coordinate system, where n is the normal vector to S_m.

Let us consider at first E-oscillations. We write the representation in the form of a series of eigenfunctions of the Helmholtz operator:

$$g^{-}(\dots) = \sum_{j=1}^{3} \sum_{n=-\infty}^{+\infty} z_n^j \hat{G}_n^j(\dots) e^{in\phi_j}, \qquad (1.19)$$

where for $j = 1, 2$

$$\hat{G}_n^j(\dots) = \begin{cases} \frac{J_n(kr_j)}{J_n(ka_j)}, & r_j < a_j, \\ \frac{H_n^{(1)}(kr_j)}{H_n^{(1)}(ka_j)}, & r_j > a_j, \end{cases}$$

and for $j = 3$ (when the inclusion is a layered-radial pivot) the representations are similar.

Because $g^{-}(\dots)$ must satisfy the Meixner condition for every compactum $M \subset R$, (1.19) holds and one only has to determine z_n^j ($j = 1, 2$) from the conditions for $g^{-}(\dots)$, and to find z_n^3, A_n^j and B_n^j from the conjugation conditions. After determining these coefficients we will construct the Green function.

The conditions for the resonator mirrors stated in the previous section and the results of [10] allow us to apply the technique based on the method of the Riemann-Hilbert problem [5], using the separation of variables in local coordinates (r_n, ϕ_n)

$(n = 1, 2, 3)$ and the addition theorems for cylindrical functions. Finally, one can obtain for z_n^j $(j = 1, 2)$ an infinite homogeneous system of linear algebraic equations,

$$z_n^j = \sum_{s=1}^{3} \sum_{m=-\infty}^{+\infty} a_{nm}^{js}(k) z_m^s + F_n^j(k), \quad (j = 1, 2), \tag{1.20}$$

with the same matrix elements $a_{nm}^{js}(k)$ as in the previous section, and on the right-hand side $F_n^j(k)$ have the form

$$F_n^j(k) = \frac{1}{4i} \sum_{m=-\infty}^{+\infty} Q_m^n(u_j) \begin{Bmatrix} J_m(ka_j) H_m^{(1)}(k\rho_j), & a_j < \rho_j \\ J_m(k\rho_j) H_m^{(1)}(ka_j), & a_j > \rho_j \end{Bmatrix} e^{i[(m-n)(\pi+\psi_j)-m\psi_{4j}]},$$

$$\rho_j = \text{dist}\{(x_0, y_0), Oj\}, \quad j = 1, 2.$$

Using the conjugation conditions for $G(k; r)$ we obtain the following inhomogeneous infinite system of linear algebraic equations

$$z_n^3 = \sum_{s=1}^{3} \sum_{m=-\infty}^{+\infty} a_{nm}^{3s}(k) z_m^s + F_n^3(k), \tag{1.21}$$

where

$$F_n^3(k) = \frac{1}{4i} \Pi_n(kr^N, \Phi) \frac{H_n^{(1)}(k\rho_{43})}{H_n^{(1)}(kr^N)} e^{in(\pi-\psi_{43})},$$

when the inclusion is a layered circular cylinder ("multi-layer"). Here (ρ_{43}, ϕ_{43}) are the coordinates of the origin S_3 in the coordinate system of the "source" (Fig.1). Combining the infinite inhomogeneous systems (1.20) and (1.21) we have

$$z_n^j = \sum_{s=1}^{3} \sum_{m=-\infty}^{\infty} a_{nm}^{js}(k) z_m^s + F_n^j(k), \quad (j = 1, 2, 3). \tag{1.22}$$

After similar considerations for the case of H-oscillations we get

$$x_n^j = \sum_{s=1}^{3} \sum_{m=-\infty}^{\infty} b_{nm}^{js}(k) x_m^s + \hat{F}_n^j(k), \quad (j = 1, 2, 3) \tag{1.23}$$

with the elements on the right-hand side

$$\hat{F}_n^j(k) = \frac{\pi}{4}(ka_j)^2 \sum_{m=-\infty}^{\infty} Q_m^n(-u_j) \begin{Bmatrix} J_m'(ka_j) H_m^{(1)}(k\rho_j), & \rho_j > a_j \\ J_m(k\rho_j) H_m^{(1)'}(ka_j), & \rho_j < a_j \end{Bmatrix} e^{i[m(\psi_j-\psi_{4j})-n(\pi+\psi_j)]},$$

$$\hat{F}_n^3(k) = \frac{1}{4i} \Pi_n(kr^N, \hat{\Phi}) \frac{H_n^{(1)}(k\rho_{43})}{H_n^{(1)}(kr^N)} e^{in(\pi-\psi_{43})},$$

where $j = 1, 2$. All matrix elements are the same as in (1.22) and (1.23).

Let us consider the behaviour of $F_n^j(k)$ for $n \to \infty$ ($j = 1, 2, 3$). Applying the Cauchy inequality for the Bessel functions, an approximation for the Hankel function of large index, and using estimates for $Q_m^n(u_j)$, in line with Ref. [5], we get

$$|F_n^j(k)| \leq \bar{k} \left\{ \frac{\tau^{|n|}}{|n|} - \frac{\ln(1 + |n|)}{1 + |n|} \right\}. \tag{1.24}$$

$\bar{k}$ are coefficients depending on the geometrical and material parameters of the structure, and

$$\tau_j = \begin{cases} a_j \rho_j^{-1}, & a_j < \rho_j, \\ \rho_j a_j^{-1}, & a_j > \rho_j. \end{cases}$$

It follows from the asymptotics of the Bessel functions of large indices that

$$\lim_{|n| \to \infty} \Pi_n(kr^N; \Phi) = \frac{\mu_N - \mu_0}{\mu_N + \mu_0}.$$

Using the values of these limits, we have for $|n| \to \infty$,

$$H_n^{(1)}(\kappa) \approx (n-1)! \left(\frac{2}{\kappa} \right)^n, \quad |F_n^3(k)| \leq \tilde{L} \left| \frac{\mu_N - \mu_0}{\mu_N + \mu_0} \right| \left(\frac{r^N}{\rho_{43}} \right)^{|n|},$$

when S_3 is the radially layered pivot ($\tilde{L} = \text{const}$).

Hence, one can consider the system (1.22) as an operator equation in L:

$$\{I - A(k)\}\bar{z} = \bar{F}, \tag{1.25}$$

where $\bar{F} = \{F_n^j\}_{n=-\infty}^{+\infty}$, $j = 1, 2, 3$.

In the case of H-oscillations the system (1.23) may be represented similarly:

$$\{I - B(k)\}\hat{z} = \hat{F}, \tag{1.26}$$

where $\hat{F} = \{\hat{F}_n^j\}_{n=-\infty}^{+\infty}$, $j = 1, 2, 3$.

After solving (1.25) and (1.26) we get

$$\hat{z} = R_A(k)\bar{F}, \quad \hat{x} = R_B(k)\hat{F}, \tag{1.27}$$

where $R_A(k)$ and $R_B(k)$ are the inverses of the corresponding operator-valued functions, having the form

$$R_A(k) = \{I - A(k)\}^{-1}, \quad R_B(k) = \{I - B(k)\}^{-1}.$$

Justification of these transformations will be proved below. One can see that they hold, if there exist bounded operators $R_A(k)$ and $R_B(k)$, for k: $\text{Im}\, k > 0$. Then, since $\hat{z} = (z_n^1, z_n^2, z_n^3)$ and $\hat{x} = (x_n^1, x_n^2, x_n^3)$, there will be the following representation for the Green function:

in the case of E-oscillations:

$$G(k;r) = \begin{cases} \frac{1}{4i}H_1^0(kr) + \sum_{j=1}^3 \sum_n z_n^j \hat{G}_n^j(k;r)e^{in\phi}, & (x,y) \overline{\in} S_3, \\ \sum_n B_n^1 H_n^{(1)}(k_1 r^1) J_n(k_1 r_3)e^{in\phi_3}, & 0 < r_3 < r^1, \\ \sum_n [\hat{A}_n^j H_n^{(1)}(k_j r_3) + \hat{B}_n^j J_n(k_j r_3)]e^{in\phi_3}, & r^{j-1} < r_3 < r^j, \end{cases} \quad (1.28)$$

$$B_n^1 = z_n^3 \frac{2i(\pi k_0 r^N \mu_0)^{-1} J_n(kr^N)}{w_n \lambda_{2n}^N J_N(kr^N) - w_0 \lambda_{1n}^N J_n'(kr^N)},$$

$$\hat{A}_n^j = B_n^1 \hat{L}_n^{j-1} J_n(k_j r^{j-1}), \quad \hat{B}_n^j = B_n^1 \hat{N}_n^{j-1} H_n^{(1)}(k_j r^j),$$

and in the case of H-oscillations

$$G(k;r) = \begin{cases} \frac{1}{4i}H_0^{(1)}(kr) + \sum_{j=1}^2 \sum_n x_n^j \left\{ \begin{matrix} J_n(kr_j)H_m^{(1)'}(ka_j), & r_j < a_j \\ J_n'(ka_j)H_m^{(1)}(kr_j), & r_j > a_j \end{matrix} \right\} e^{in\phi_j} + \\ \quad + \sum_n x_n^3 J_n(kr^N) H_n^{(1)}(kr_3)e^{in\phi_3}, & (x,y) \overline{\in} S_3, \\ \sum_n \tilde{B}_n^1 H_n^{(1)}(k_1 r^1) J_n(k_1 r_3)e^{in\phi_3}, & 0 < r_3 < r^1, \\ \sum_n [\tilde{A}_n^j H_n^{(1)}(k_j r_3) + \tilde{B}_n^j J_n(k_j r_3)]e^{in\phi_3}, & r^{j-1} < r_3 < r^j, \end{cases} \quad (1.29)$$

$$\tilde{B}_n^1 = x_n^3 \frac{2i(\pi k_0 r^N \varepsilon_0)^{-1} J_n(kr^N)}{\hat{w}_n \bar{\lambda}_{2n}^N J_N(kr^N) - \hat{w}_0 \bar{\lambda}_{1n}^N J_n'(kr^N)},$$

$$\tilde{A}_n^j = \tilde{B}_n^1 \bar{L}_n^{j-1} J_n(k_j r^{j-1}), \quad \tilde{B}_n^j = \tilde{B}_n^1 \bar{N}_n^{j-1} H_n^{(1)}(k_j r^j).$$

Here $k_0 = \omega/c$, $\bar{\lambda}_{1n}^N$ and $\bar{\lambda}_{2n}^N$ differ from λ_{1n}^N and λ_{2n}^N by $\bar{\lambda}_{1n}^N = \sigma_{\varepsilon,\mu}\lambda_{1n}^N$, and $\bar{\lambda}_{2n}^N = \sigma_{\varepsilon,\mu}\lambda_{2n}^N$. Similar relations exist for $\hat{L}_n^{j-1}$, $\hat{N}_n^{j-1}$, $\bar{L}_n^{j-1}$, $\bar{N}_n^{j-1}$; namely, $\bar{L}_n^{j-1} = \sigma_{\varepsilon,\mu}\hat{L}_n^{j-1}$, $\bar{N}_n^{j-1} = \sigma_{\varepsilon,\mu}\hat{N}_n^{j-1}$ $(j = 2,3,\ldots,N)$.

The questions arise: is it possible in general to construct the Green function satisfying the conditions stated above, and, if so, will the representations (1.28) and (1.29) be unique for these functions?

To get correct answers to these questions one has to establish first of all the dependence between the existence and uniqueness of the solution of the operator equation (1.25) and the Green function $G(k;r)$ (1.28) in the upper half of the complex plane (in general, in the upper half-plane Λ_0). Therefore, let us consider first the operator equation (1.25). Properties of equation (1.26) may be obtained in the same way. The following statement holds: the operator-valued function $A(k) = A(k,\varepsilon,\mu,\theta_s,\ldots)$ is analytical with respect to parameters k, ε, μ, when $\text{Im}\,k > 0$, $\text{Im}\,\varepsilon > 0$, $\text{Im}\,\mu > 0$, except, perhaps, at an infinitely far point and when $k = 0$, $\varepsilon = 0$ and $\mu = 0$. To prove this, one has to point out that the singularities of the function $A(k)$ may appear for values of κ_0 at which $J_n(\kappa_0) = 0$, or $H_n^{(1)}(\kappa_0) = 0$, and also at the zeros of $\Pi_n(\tilde{k}r^N;\Phi)$. All zeros of the function $J_n(\kappa)$ lie on the real axis and $H_n^{(1)}(\kappa) = 0$ for $\text{Im}\,\kappa < 0$. Since $A(k)$ is a kernel operator-valued function, then, in line with [28] and the Riesz theorem about functional representation in Hilbert space, we may conclude that $A(k,\varepsilon,\mu,\theta_s,\ldots)$ is an analytical function with respect to k when $\text{Im}\,\kappa > 0$, except, perhaps, at the points k_0 where for a fixed value of n: $\Pi_n(\tilde{k}_0 r^N;\Phi) = 0$. The study of the properties of $A(k,\varepsilon,\mu,\theta_s,\ldots)$ as a function of ε and μ is similar, because $\tau = \sqrt{\varepsilon\mu}$ is a simple analytical function for $\text{Im}\,\varepsilon > 0$, $\text{Im}\,\mu > 0$ and $A(k,\varepsilon,\mu,\theta_s,\ldots)$ is a kernel operator-valued function (see the structure of $A(k,\varepsilon,\mu,\theta_s,\ldots)$).

Now one can assert that for every k: $\text{Im}\,k > 0$ there exists an unique solution $\hat{z}$ of the operator equation (1.25) in L.

Really, due to the Fredholm alternative [52] and the results of [7], equation (1.25) has a unique solution if and only if two conditions hold simultaneously: $A(k)$ for every k: $\operatorname{Im} k > 0$ completely determines a continuous operator in L, and the homogeneous equation corresponding to (1.25) has only trivial solution. Let us check the validity of these conditions. Since $A(k)$ is a kernel operator-valued function, the first condition holds. One can easily solve the operator equation $\hat{z} = A(k)\hat{z}$ for k: $\operatorname{Im} k > 0$ because the uniqueness theorem for the excitation problem holds [10]. It implies that for every k: $\operatorname{Im} k > 0$ there exists a bounded operator $R_A(k)$. Besides, $\hat{z} = R_A(k)\bar{F}$ and the inequality $\|\hat{z}\|_L \leq T\|\bar{F}\|_L$ holds, where T is a constant. In other words, $\hat{z} \in L$. Since $\hat{z} \in L$, every vector-function's component $\hat{z} = (\hat{z}_n^1, \hat{z}_n^2, \hat{z}_n^3)$ is in the space l^2. The Cauchy-Bunyakovski-Schwarz inequality for the series [59] and asymptotic behaviour of the functions $J_n(\kappa)$ and $H_n^{(1)}(\kappa)$ for $|n| \gg |\kappa|$ result in that the representation for the Green function $G(k; r)$ (1.28) holds for all k: $\operatorname{Im} k > 0$. The inverse statement holds: if the representation (1.28) is valid for the function $G(k; r)$ for all k: $\operatorname{Im} k > 0$, i.e., the corresponding series converge uniformly with respect to k, then the vector-function $\hat{z} = (\hat{z}_n^1, \hat{z}_n^2, \hat{z}_n^3) \in L$ satisfies (1.25). This statement follows from the correct way of the transition from the problem of constructing $G(k; r)$ to the operator equation (1.25). Hence, we have shown that the Green function $G(k; r)$ exists and it is unique for all k: $\operatorname{Im} k > 0$ for E-oscillations.

The proof of the existence and uniqueness of the Green function $G(k; r)$ for H-oscillations is similar.

Let us consider the properties of $G(k; r)$ (only for E-oscillations) with respect to k: $\operatorname{Im} k > 0$; $\varepsilon, \mu \in \bar{\Omega}$, and $\theta_s \in R^1 \setminus \cup_n \tau_n$ where $\tau_n = \pi n$, $(s = 1, 2; n \in Z)$.

The following statement holds: Green's function $G(k; r)$ is analytical with respect to k when $\operatorname{Im} k > 0$ and it admits analytical continuation in Λ with the exception of the poles which coincide with the spectrum $\sigma(A)$ of the problem (1.1)-(1.5).

This statement is based on the analysis of the operator equation $\hat{z} = R_A(k)\bar{F}(k)$. As it has been pointed out, the operator $R_A(k)$ is a kernel in the pair of spaces $L \to L$. Then the Fredholm theorem for analytical operator-valued functions [47] yields that $R_A(k)$ is analytical with respect to k: $\operatorname{Im} k > 0$. Since $\bar{F}(k) \in L$ and the vector-function $\bar{F}(k)$ is analytical with respect to k, $G(k; r)$ is analytical with respect to k in the upper half-plane Λ_0 due to (1.28). The operator-valued function $R_A(k)$ is meromorphic in Λ with respect to k, in line with the Fredholm theorem for finite-meromorphic operator-valued functions [47]. Hence, $G(k; r)$ is finite-meromorphic as a function of k on Λ.

Let $\lambda(k)$ be an eigenvalue of the operator $A(k)$ corresponding to the eigenvector $\hat{z}$:

$$A(k)\hat{z} = \lambda(k)\hat{z}. \tag{1.30}$$

One gets from (1.25) and (1.30)

$$\hat{z} = \{I - \lambda(k)\}^{-1}\bar{F}(k). \tag{1.31}$$

It follows from (1.31) that singularities of the vector-function $\hat{z}(k)$ may only take the values of $k_0 \in \Lambda$ which satisfy the equation $\lambda_n(k_0) - 1 = 0$ for a fixed n, i.e., $k_0 \in \sigma(A)$, in accordance with results from section 1.

In other words, the poles of the function $G(k; r)$ on the Riemann surface Λ coincide with the set $\sigma(A)$.

In addition to this, the properties of the operator-valued function $A(k, \varepsilon, \mu, \theta_s, \dots)$, the representation (1.25) for $G(k; r)$, and the conclusions in [10] give us the following statements.

$G(k; r)$ is analytical with respect to ε and μ when $\varepsilon \in \bar{\Omega}$ and $\mu \in \bar{\Omega}$,

$G(k; r)$ is continuous with respect to k, ε and μ when $\operatorname{Im} k = \operatorname{Im} \varepsilon = \operatorname{Im} \mu = 0$,

$G(k; r)$ is continuous with respect to θ_s when $\theta_s \in R^1 \setminus \cup_n \tau_n$.

Let us consider one more property of the function $G(k; r)$ for its analytical continuation on Λ. Let us introduce two sets

$$\Gamma^1 = \{k \in \Lambda : J_n(ka_s) = 0, \ s = 1, 2, \ n = 0, 1, 2, \dots\},$$
$$\Gamma^2 = \{k \in \Lambda : H_n^{(1)}(ka_s) = 0, \ s = 1, 2, \ n = 0, 1, 2, \dots\},$$

and let us form the set $\Gamma = \Gamma^1 \cup \Gamma^2$.

The following statement holds: the function $G(k; r)$ has removable singularities at points $k \in \Gamma \subset L$ and

$$G(k; r) = -\frac{i}{4} H_0^{(1)}(kr) + H_{n_0}^{(1)}(kr) \frac{Z_{n_0}(ka_1)}{X_{n_0}(ka_1)} e^{in_0 \phi_1} + \sum_{n \neq n_0} z_n^1 \frac{H_n^{(1)}(kr_1)}{H_n^{(1)}(ka_1)} e^{in\phi_1}, \quad (1.32)$$

where n_0 is an integer, $X_{n_0}(ka_1) \neq 0$, $Z_{n_0}(ka_1) \neq 0$.

Equation (1.32) implies that $\lim_{k \to k_0} G(k; r)$ exists and it is finite, i.e., k_0 is a removable singularity of $G(k; r)$ on Λ.

Since $\lim_{r \to \infty} |g^-(k; r)| = 0$, one can make sure that for all k with $\operatorname{Im} k > 0$: $\lim_{r \to \infty} G(k; r) = 0$ using the asymptotical representation for $H_0^{(1)}(\kappa)$ when $|\kappa| > 0$.

In other words, the Green function $G(k; r)$ satisfies the Maluzhinec liquidation condition [41].

The possibility of analytical continuation of the Green function from the upper half-plane Λ_0 onto the whole Riemann surface Λ allows to replace the Maluzhinec condition for $k \in \Lambda$ by the generalized condition in infinity, the Reichardt condition: there exists an $R_0 > 0$ such that for all $r > R_0$ and for all $k \in \Lambda$ the following equation holds:

$$G(k\sqrt{(x - x_0)^2 + (y - y_0)^2}) = \sum_{m=-\infty}^{+\infty} \eta_m H_m^{(1)}(k\sqrt{(x - x_0)^2 + (y - y_0)^2}) e^{im\phi}.$$

Here the series $\sum_{m=-\infty}^{+\infty} \eta_m H_m^{(1)}(k\sqrt{(x - x_0)^2 + (y - y_0)^2})$ converges absolutely outside a circle containing the cross-section of the electrodynamic structure S and the source point.

In conclusion let us make some remarks:

1. Construction of the Green function $G(k; r)$ (for E- and H-oscillations) for the problems of diffraction of the cylindrical wave $-(i/4) H_0^{(1)}(kr)$, where $r = \sqrt{(x - x_0)^2 + (y - y_0)^2}$, from the electrodynamic structure S is reduced to the solution of an inhomogeneous operator equation with δ-type right-hand side

$$BG(x, y) + k^2 \lambda(x, y) G(x, y) = -\delta(x - x_0, y - y_0),$$

with homogeneous boundary conditions: $G|_{S_m} = 0$ for E-oscillations and $\frac{\partial G}{\partial \bar{n}}|_{S_m} = 0$ for H-oscillations ($m = 1, 2$).

2. For a finite function $f(x,y)$ in $\bar{\Omega}$ and differential equations

$$Bu(x,y) + k^2\lambda(x,y)u(x,y) = f(x,y),$$

the following relation holds

$$u(x,y) = \iint G(k;x,y;x_0,y_0)f(x_0,y_0)dx_0 dy_0.$$

3. The Green function exists and it is unique for the considered types of resonators.

4. The Green function $G(k;r)$ may be continued from the upper half-plane of Λ_0 onto the Riemann surface Λ, and the continuation will be a meromorphic function with the set of poles on Λ coinciding with $\sigma(A)$.

§ 3. Calculation of the characteristics of open resonators

The study of the physical properties of open resonators is first of all connected with the determination of their spectral characteristics and qualitative appropriateness based on the theory of analytical operator-valued functions in Hilbert (Banach) spaces, and the theory of critical points of analytical functions of several complex variables.

If the distance d between the mirrors in an open resonator varies with all the other parameters fixed, a certain Fredholm operator-valued function which describes the resonator spectral properties will uniquely (to the accuracy of isomorphism) corresponds to it for every value of d. The construction of this operator-valued function presents a rather complicated problem, the solution of which is, as a rule, based on the regularization procedure (e.g., the Riemann-Hilbert method, the method of integral equations, etc.). The operator-function to be found depends on the spectral parameter (frequency) and isolated non-spectral parameters describing physical and geometrical properties of the open resonator.

In a number of cases (or nearly always for the spectral parameter) the operator-valued function is an analytical (finite-meromorphic) parameter function in a certain complex domain (in the general case, this domain is determined on a certain Riemann surface). A characteristic (infinite) determinant (dispersion relation), regularized in the general case and analytical with respect to the spectral parameter, corresponds to this operator-valued function. Consequently, a characteristic determinant which depends on several parameters (control parameters), describes this resonator and corresponds to it.

Qualitative changes in the spectral parameters (resonator frequencies) of the open resonator can be easily observed in the vicinity of those values of non-spectral parameters at which the characteristic determinant has critical points. Therefore, the family of characteristic determinants has to be analyzed to reveal the critical points. The simplest of them are the Morse critical points. The problem of analyzing the behaviour of fundamental frequencies and eigenoscillations with changing parameters in the vicinity of the Morse points is fundamental in linear electrodynamics, since it determines a qualitatively new state of the open resonator, i.e., the intertype (intermode) oscillation.

Direct dispersion relations obtained from the rigorous spectral theory (see section 1) can be used to determine practically all the required characteristics of an open resonator for arbitrary non-spectral parameters, or in the vicinity of the Morse point.

Naturally, the search for this point involves much effort and, in fact, requires computer-aided selection of dispersion curves. Odd experimental data and symmetry properties of empty open resonators present extra references to a "feel for" those regions of non-spectral parameters where the Morse points are expected to be observed. This is the case when the geometry is used as the resonator non-spectral parameter. The study of direct dispersion relations together with the analysis of the Morse points becomes much more difficult under variations of other non-spectral, e.g., material, parameters. Here the existence of the critical points for an empty resonator appears to be the only "prompt".

According to section 1, a matrix Fredholm operator-valued functions for the simplest open resonator (free of any inclusions) formed by two cylindrical strip mirrors may be represented in the following form for H-oscillations:

$$\left\| \begin{array}{cc} B^{11}(ka_1) - I & B^{12}(ka_1) \\ B^{21}(ka_2) & B^{22}(ka_2) - I \end{array} \right\| : l_2^{(2)} \to l_2^{(2)}. \tag{1.33}$$

$B^{ij}(ka_i) : l_2 \to l_2$ are kernel operator-valued functions with matrix realization defined by the corresponding formulas in section 1 in the absence of any inclusions in the resonator. The operator-valued functions $B^{ij}(ka_i)$ ($i \neq j$; $i,j = 1,2$) describe the interaction between the mirrors.

In section 1, proceeding from the discreteness of the resonant frequencies for an open resonator, we have constructed a computational algorithm which is implemented as the computer codes to be subsequently used in calculations of the spectral characteristics of empty resonators. Resonant frequencies are calculated to an accuracy of up to four or five significant digits. Systematic calculations have shown that it is sufficient to take the reduction order $P_i = [|ka_i|] + \tau$, $\tau \geq 5$ for the operator-valued function $B^{ij}(ka_j)$ (where $[\cdot]$ means taking the integer).

A detailed analysis of the spectral characteristics (resonant frequency spectrum, types of resonant oscillations) for a two-dimensional empty open resonator formed by two parallel circular cylindrical-shaped mirrors having ideal conductance and infinite length, is given in Ref. [10].

A confocal symmetrical open resonator is known to exhibit, for $b/a \ll 1$ ($a = l$ is the distance between the mirrors, $2b$ is the mirrors aperture) and in the frequency range $|\kappa| \gg 1$ ($\kappa = ka$, $\mathrm{Re}\,\kappa = 2\pi a/\lambda$), a resonant frequency spectrum which corresponds to the resonance oscillations of "bouncing ball" type with high q-factor. Open resonators are typically used in the millimeter and submillimeter wavelength ranges where $1 < |\kappa| < 20$ and $b/a \sim 1$ [6]. Systematic calculations using the results of sections 1 and 2 have shown that, in the frequency range $1 < |\kappa| < 20$, the confocal resonator for $b/a \sim 1$ has definite resonance properties. Several fundamental frequencies of this resonator (with the highest q-factor) are given in Table 1 with respect to increasing modulus of κ for $b/a = 0.63$.

For classification of the inner resonant oscillations corresponding to these frequencies, the problem of the excitation of an open resonator by the unit-amplitude H-polarized plane wave has been solved at a frequency equal to $\mathrm{Re}\,\kappa$. Distribution of the lines of equal values of $|H_z|$ and $\arg H_z$ inside the resonator has been studied. The oscillations may be classified by the number of zeros of the magnetic field on the two perpendicular axes of symmetry of the resonator. An oscillation is by definition of H_{mn}-type if the magnetic field vanishes n times on resonator's axis and m times in the perpendicular direction. The same classification is used in heuristic models of open resonators [14]. However, such models allow to describe only a part of the resonator spectrum corresponding to the "bouncing ball" oscillations [53]. In our notation these oscillations is of H_{mn}-type with $n \gg m$. One can see from Table 1 that there exist H_{41} and H_{51} oscillations which cannot be described within the frame of the "bouncing ball" model. It turns out that in some situations H_{41} and H_{51} oscillations effectively interact with H_{03} and H_{13} oscillations. It leads to a sharp increase of diffractional losses and to other phenomena, which will be discussed below.

Table 1.

	H_{00}	H_{01}	H_{11}	H_{02}	H_{12}	H_{03}	H_{22}
Re κ	1,4947	3,7172	5,4903	6,9948	8,6491	10,142	10,201
$-$ Im κ	0,2571	$2,028 \cdot 10^{-2}$	0,1053	$3,738 \cdot 10^{-2}$	0,2339	$6,08 \cdot 10^{-4}$	0,5666

	H_{41}	H_{13}	H_{51}	H_{04}	H_{14}	H_{05}	H_{24}
Re κ	11,231	$4,119 \cdot 10^{-3}$	12,837	13,298	14,894	16,450	16,515
$-$ Im κ	0,8115	$4,119 \cdot 10^{-3}$	0,8514	$7,929 \cdot 10^{-4}$	$3,572 \cdot 10^{-2}$	$5,369 \cdot 10^{-4}$	$1,944 \cdot 10^{-2}$

Calculations show how the change of the mirror aperture influences the spectrum of fundamental frequencies. It turns out that the fundamental frequencies of H_{0n} oscillations (Re κ) practically do not depend on b/a for $n \geq 4$ (there are more than two resonance wavelengths in the resonator's curvature radius). This is connected with the fact that while index n is growing, the fields of these oscillations concentrate close to the resonator axis and hence the vicinities of the mirror edges do not take part in forming the inner resonance field. However, the sharp dependence of the fundamental frequency when the mirror aperture varies is still characteristic of oscillations with $n < 4$. Here the edge diffraction phenomenon plays a crucial role.

The parameter b/a influences the fundamental frequencies of higher order H_{mn} oscillations with $m \neq 0$ in the following way. For $n > m > 0$ one has the same situation as in the case of H_{0n} oscillations, namely, the fundamental frequencies practically do not depend on b/a for $n \geq 4$. If $m \geq n$, then, while index n is growing, the fundamental frequencies more and more depend on b/a. They also increase when the aperture of the mirrors increases. The fundamental frequencies for higher order oscillations (H_{mn}, $m \geq n$) behave in such a way because the inner resonant field are concentrated close to the mirrors and the field amplitude has local maxima in the vicinity of the mirror edges. Hence, diffraction phenomena play a basic role in creating the resonant fields of these oscillations. That is why even a small variation in b/a leads to sharp changes of the fundamental frequency. It is especially characteristic for H_{m1} oscillations ($m \geq 4$).

For the types of oscillations considered the dependence of the q-factor $Q = -(\text{Re}\,\kappa)/(2\,\text{Im}\,\kappa)$ on the parameter b/a has a monotonical character (H_{03} and H_{13} oscillations are exceptions). When b/a grows the q-factors of all oscillations increase (note that for $b/a = \sin(\pi/6) \approx 0.87$ the open resonator becomes a closed one).

Let us compare the data obtained here for H_{mn} oscillations ($n > m$) of a symmetrical confocal open resonator in the frequency range $1 < \kappa < 20$ with the results of heuristic theories [14]. It will allow us to establish the limits of applying the approximate methods for calculating the fundamental frequencies of open resonators. The real parts of the fundamental frequencies calculated by means of an asymptotical formula

[14] have the form

$$\mathrm{Re}\,\kappa = \pi\left(n + \frac{m}{2} + \frac{1}{4}\right) + O(n^{1-\varepsilon}) + O(n^{-1}). \tag{1.34}$$

It is shown in Ref. [54] that this formula is uniform with respect to m for $0 \le m \le Cn^{1-\varepsilon}$ where $\varepsilon \in (0,1)$, and $C > 0$ is an arbitrary constant. As one can see, (1.34) does not take into account that $\mathrm{Re}\,\kappa$ depends on the parameter b/a. However, this formula describes the real part of the fundamental frequency sufficiently good for H_{mn} oscillations ($m < n$) beginning from $n \ge 4$. Here the relative error rapidly decreases when index n grows (for H_{05} oscillations it is 0.3%). The asymptotical formula (1.36) gives the upper estimate for the real part of the fundamental frequency. Hence, one can use this formula for calculations of real parts of the fundamental frequencies of H_{0n} oscillations with a relative error of 0.3% beginning from $n = 5$.

Together with the study of fundamental frequencies and the structure of the inner resonant fields for an open resonator in the $1 < \kappa < 20$ frequency range, it is of great interest to consider the diffractional (outside) fields of open resonators [6]. The most important characteristic here is the far-field pattern in the problem of an H-polarized plane wave diffraction by the resonator mirrors when the frequency is equal to the real part of one of the fundamental frequencies. As is shown above, this problem is reduced to the solution of infinite systems of linear algebraic equations. These systems may be effectively solved numerically by the method of reduction with prescribed accuracy. The normalized far-field patterns are calculated based on the solutions of these systems. Far-field patterns have a rather complicated character. The H_{00} oscillation is an exception, and its far-field pattern is practically isotropic (this can be explained by the long-wave property of this oscillation with $\lambda_{\mathrm{res}}/a \approx 10$). Other types of oscillations have several (two or more) distinct slabs together with shadow slabs. For H_{0n} oscillations the shadow slab becomes narrower with increasing index n. For example, for H_{05} oscillations it is almost three times smaller than for H_{01} oscillations. In fact, oscillations with larger n have greater fundamental frequencies and, hence, the geometrical-optical properties of the scattered field start to play more a more noticeable role. It is confirmed by the appearance of a strong diffraction of the far-field pattern in the region of a shadow slab.

Let us study the symmetrical non-confocal open resonator with mirror aberration. In practical use mirrors with different curvature radii can move along the resonator axis [6]. Therefore, it is important to study how these factors influence the fundamental frequencies as well as the inner resonant fields. A number of works [14, 55, 56] is devoted to the corresponding analysis of the behaviour of open resonators in the frequency domain.

The fundamental frequencies and the q-factor of the symmetrical open resonator decrease monotonically with respect to the distance l between the mirrors $(b/l \sim 1$ and $1 < \kappa < 20)$ on the interval $1 \le l/a \le 2$, for all types of oscillations, while going from the confocal resonator $(l/a = 1)$ to the concentric one $(l/a = 2)$. The q-factor decreases less intensively for H_{0n} oscillations with even index than for odd index on the interval $1.5 \le l/a \le 1.8$. Clearly different picture is seen in the case $l/a \le 1$. The real parts of the fundamental frequencies increase monotonically with decreasing l/a. For some values of l/a, $\mathrm{Re}\,\kappa$ of H_{03} oscillations coincide with the real parts of the fundamental frequencies of H_{22} and H_{41} oscillations. However, the q-factors of these oscillations, corresponding to the cross-points, considerably differ from the others. Hence, there is only a partial spectrum's degeneration. The q-factors of H_{mn} oscillations $(m < n)$ have local maxima on the interval $0.8 \leq l/a \leq 1$ and increase monotonically. A further decrease of l/a leads to a monotonical increase of the q-factors for all types of oscillations (for $l/a \approx 0.44$, it becomes a closed resonator).

The dependence of the real parts of the H_{mn} fundamental frequencies $(n > m)$ on the parameter l/a, calculated by asymptotical formulas, give a maximal relative error of about 1.5% even for the H_{22} oscillation [14]. With increasing index n the relative error rapidly decreases. For H_{04} oscillations it is about 0.4% on the whole interval, $1 \leq l/a \leq 2$.

Hence, one can assert that H_{mn} oscillations $(m < n, n \geq 3)$ are similar to the "bouncing ball" oscillations $(H_{mn}, n \gg m, |\kappa| \gg 1)$ if to consider them as functions of the relative distance between the mirrors in the frequency range $1 < \kappa < 20$.

Let us consider how the fundamental frequencies and inner resonant fields of an open resonator are influenced by the varying aperture and the mirror position taking H_{03} and H_{13} oscillations as examples. Calculations show that for before-confocal $(l/a < 1)$, confocal $(l/a = 1)$, and over-confocal $(l/a > 1)$ resonators the dependence is weak for $\operatorname{Re}\kappa$ of the H_{03} oscillation with respect to the mirror apertures. For H_{13} oscillations, this interval becomes narrow and the q-factors of both types of oscillations sharply decrease. Namely, for the before-confocal resonator the q-factor is 100, and the real part of the fundamental frequency practically has not changed. The study of the equal contour values for $|H_z|$ of the inner resonant field of H_{03} oscillations in the confocal resonator shows that the fields move in the direction of the mirror with smaller aperture. Decreasing the mirrors' aperture leads to the destruction of the inner resonant fields and to the forming of low q-factor oscillations by the cylindrical strip. It follows from the fact that the operator-valued function $B^{21}(ka)$ which describes the interaction of resonator mirrors and the operator-valued function $(B^{22}(ka) - I)$ characterizing the resonance properties of the mirror weakly converge to zero and unit operators, respectively, with decreasing aperture. The dispersion relation degenerates to the equation for fundamental frequencies of the mirror with the aperture b.

The dependence of $\operatorname{Re}\kappa$ and the logarithms of the q-factors of H_{03} and H_{13} oscillations on the position angle of one of the mirrors is less critical for the confocal resonator. The real part of the fundamental frequency of such an open resonator practically does not depend on the position angle δ on the interval $0° \div 20°$, the q-factor decreases almost by a factor of two (for $\delta = 20°$, $Q = 185$), the fields move to the ends of the mirrors, and the distance is greater between them (this explains the decreasing q-factor because greater diffractional losses correspond to greater field amplitudes at the end of the mirror).

For H_{03} oscillations in a non-confocal open resonator, the real part of the fundamental frequency considerably depends on the position angle, while the inner resonant fields are characterized by the same features as for a confocal resonator.

Let us clear up how the spectrum of fundamental frequencies and the inner resonant fields of an open resonator deform with decreasing curvature radius of the mirror for fixed aperture dimension and fixed distance between the mirrors. Systematical calculations show that for before-confocal geometry, the real part of H_{mn} fundamental frequencies $(m < n)$ monotonically grows with decreasing curvature radius. For H_{0n} and H_{mn} oscillations $(m \neq 0)$ the growth of the corresponding curves is clearly different.

The behaviour of the q-factors differs considerably. In case of a before-confocal resonator the q-factor of H_{03} oscillations decreases much more rapidly than that for a resonator with greater curvature radius. One can explain this phenomena by considering the fields of H_{03} oscillations as well as the phase distribution of the inner resonant field. There is some concentration of the $|H_z|$ =constant contours in the direction of the resonator's axis close to the mirror with smaller curvature radius and their expansion in the vicinity of the other mirror. It leads to a field increase near the edge of the mirror with greater curvature radius and, hence, to a decrease of the diffractional losses.

It is interesting to point out that within the frames of heuristic models of open resonators [30], the size of the field spot on one of the mirrors is equal to infinity and it vanishes on the other mirror when the distance between the mirrors is equal to one of the curvature radii. As follows from the results presented above, this approximation does not take into account diffraction by the mirror edges and gives a rather inadequate picture of this phenomenon.

Let us consider intertype oscillations in two-dimensional open resonators. This effect cannot be discovered in principle by the "bouncing ball" approach. The first attempt to explain the intertype oscillations in volume (closed) resonators with the help of the perturbation theory of self-adjoint operators has been carried out in Refs. [57, 58].

As has been shown in section 1, boundary value problems describing the spectral characteristics of open two-dimensional resonators are non-self-adjoint, because for unbounded domains one has to formulate conditions in infinity [59, 60]. Hence, the fundamental frequencies of an open resonator are complex and the corresponding eigenoscillations cannot form a basis [14, 61]. The heuristic models of closed resonators [57, 58], where intertype oscillations may exist, does not allow to construct a theory for this phenomenon. A similar situation is found in case of the heuristic theory [14]: it allows to obtain information only about a part of the spectrum of the fundamental frequencies which corresponds to the short-wave oscillations having the greatest q-factor. It turns out that this information is not complete enough to construct a mathematical model for the phenomenon of intertype interaction. This problem has been considered for the first time in Refs. [7, 62] for the resonator in the form of a perfectly conducting circular cylinder with a longitudinal slot (and has been solved by the mathematically rigorous method of the Riemann-Hilbert problem [5]). It has been shown that there exist such values of the diameter of the slot (points of coincidence) where some fundamental frequencies approach each other (with respect to the topology of complex plane) and the corresponding eigenoscillations interact in a sense of intertype interaction, i.e., they exchange energy. After the point of coincidence these oscillations transform one to another.

We limit ourselves here to the study of the H-polarized intertype oscillations in a symmetrical confocal open resonator whose curvature radii a, distance between the mirrors l and the mirror apertures are equal to each other. Since this resonator is symmetrical with respect to two perpendicular axes, the equation for fundamental frequencies splits into four independent equations

$$\det \| \delta_p^m - [(b_{pm}^{11}(ka) \pm b_{pm}^{12}(ka)) \pm (b_{p-m}^{11}(ka) \pm b_{p-m}^{12}(ka))] \|_{p,m=0}^{+\infty} = 0. \qquad (1.35)$$

Here δ_m^p is the Kronecker delta function, $b_{pm}^{ij}(ka)$ are the matrix elements of the operator-valued function $B^{ij}(ka)$; signs $+$ and $-$ vary independently inside and between the parentheses. The matrix elements have the form

$$b_{pm}^{ii}(ka_i) = \delta(ka_i)F_p^m(u_i)e^{i(\psi_i+\pi)m},$$
$$b_{pm}^{21}(ka_2) = i\pi(ka_2)^2 H_m^{(1)\prime}(ka_1)J_m'(ka_2)F_p^m(u_2)e^{i(\psi+\pi)m},$$
$$b_{pm}^{12}(ka_1) = i\pi(ka_1)^2 H_m^{(1)}(ka_1)J_m'(ka_2)F_p^m(u_2)e^{i(\psi+\pi)m}, \qquad (1.36)$$
$$\delta_p(x) = |m| + i\pi \dot{x}^2 J_m'(x)H_m^{(1)\prime}(x),$$

and expressions for $F_p^m(u_i)$ are given in Ref. [5]. Equations with the signs "$+$" or "$-$" between parentheses correspond to the eigenoscillations of type $H_{2m\,p}$ $(H_{2m+1\,p})$ and the equation inside the brackets corresponds to the eigenoscillations of type $H_{m\,2p}$ $(H_{m\,2p+1})$. It is clear that the intertype connection is possible only for oscillations of the same class of symmetry, when the resonator mirrors are deformed in such a way the symmetry with respect to the two perpendicular axes is kept.

Computer simulations show that the H_{03} and H_{13} oscillations interact with the H_{41} and H_{51} oscillations, respectively, when the mirror aperture b/a varies. One can note that oscillations H_{03} and H_{41} (H_{13} and H_{51}) belong to the same class of symmetry $H_{2m\,2p+1}$ $(H_{2m+1\,2p+1})$.

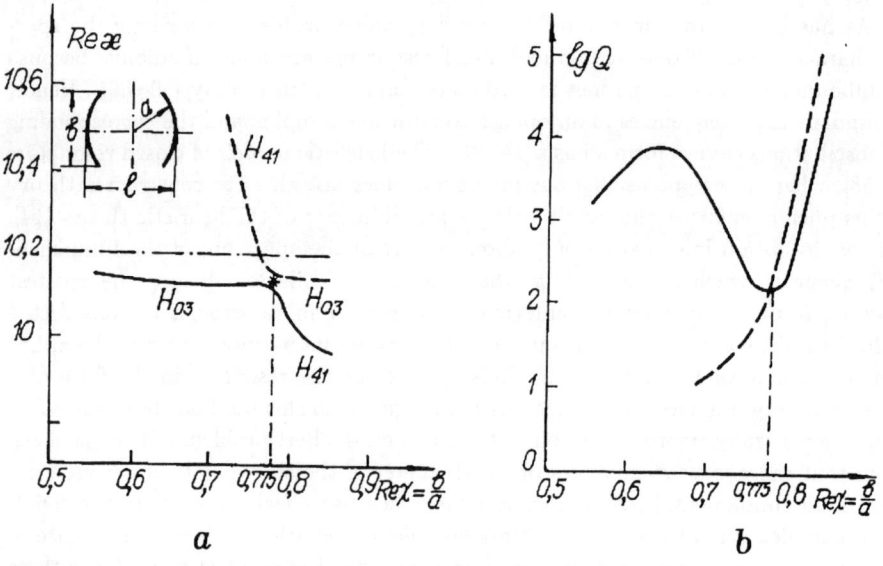

Figure 2.

The real parts of the fundamental frequencies and the logarithms of the q-factors of H_{03} (bold line) and H_{41} (bar line) oscillations as functions of $\mathrm{Re}\,\chi = b/a$ (the ratio of the mirror aperture and its curvature radius) are presented in Fig.2,a,b. One can see that the q-factors strongly depend on b/a. Increasing b/a the q-factor of the H_{03} oscillation (with three field variations along the resonator axis) monotonically increases up to $Q \approx 10^4$ ($b/a = 0.65$), and then falls (about two orders of magnitude) and becomes comparable with the q-factor of the H_{41} oscillation in the vicinity of $b/a = 0.775$, which rapidly increases with increasing b/a. The real parts of the fundamental frequencies (Fig.2a) for $0.73 \leq b/a \leq 0.83$ form the dispersion curves similar to the Wien graph in a coupled system [57, 58], where b/a changes the frequency of the resonant system. There exists such a value of b/a (= 0.775, point of coincidence) where the modulus of the difference between two fundamental frequencies has minimal value, and the q-factors of H_{03} and H_{41} oscillations coincide (see Fig.2b). The dash-dotted line in Fig.2a shows the fundamental frequency of the H_{03} oscillation determined in [14]. One can see that an asymptotical theory does not take into account the intertype oscillations, although it describes the fundamental frequency of H_{03} oscillations sufficiently good in the vicinity of $b/a = 0.775$.

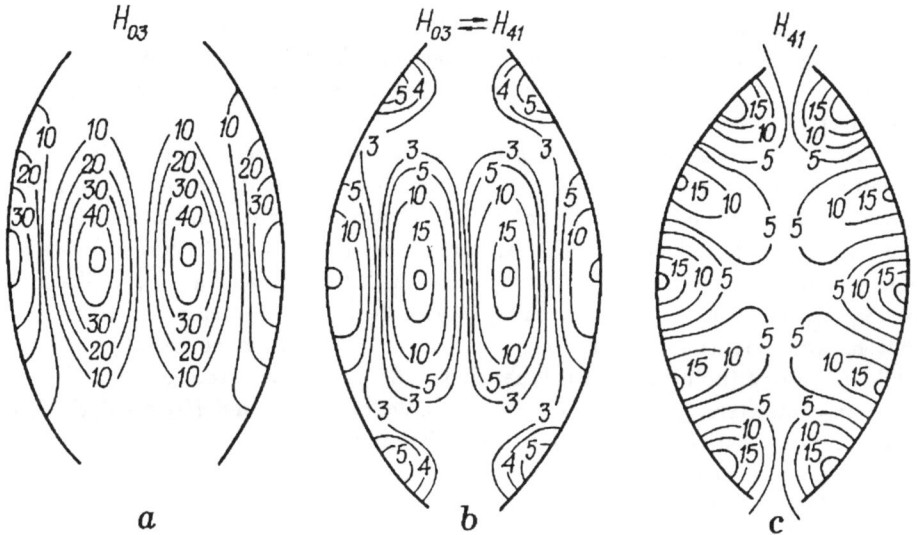

Figure 3.

There is a considerable interest in determining the field structures for H_{03} and H_{41} oscillations in the vicinity of the point of coincidence. To do this one can use the results of section 2 where the problem for a resonator excited by H-polarized sources has been solved. Contours of equal amplitudes of the H_z-component of the field excited by the H-polarized wave of unit amplitude for the frequency coinciding with $\mathrm{Re}(ka)$ of the H_{03} oscillation for different values of b/a are presented in Fig.3,a–c. For $b/a = 0.69$ and $ka = 10.139$ the field, excited in the resonator, corresponds to the H_{03} oscillation, for $b/a = 0.775$ and $ka = 10.130$ the excited field is intertype (hybrid): H_{03} and H_{41} oscillations "mix" in a special way. For $b/a = 0.82$ and $ka = 10.001$ the excited field corresponds to the H_{41} oscillation. When the frequencies of the exciting plane wave coincide with the real part of the fundamental frequency of the H_{41} oscillation, a similar situation occurs: for $b/a > 0.775$ the excited field is connected with the H_{41} oscillation, for $b/a = 0.775$ the field has intertype character, and for $b/a < 0.775$ it corresponds to the H_{03} oscillation.

Now one can make basic conclusions: a rigorous spectral theory of the symmetrical confocal empty open resonator developed in sections 1 and 2 allows us to directly establish the definite spectral points from the dispersion relations. In the vicinity of these points the oscillations intensively interact (in particular, H_{03} and H_{41} oscillations) when non-spectral parameters of an open resonator vary. Diffractional losses of the high q-factor H_{03} oscillation sharply increase close to these points, up to the level of the low q-factor H_{41} oscillation. After reaching the point of coincidence the transformation of the types of oscillations H_{03} and H_{41} takes place.

One can follow the dynamics of this transformation considering the current distribution in a resonator as a function of the distance between its mirrors. It is shown that while approaching to $b/a = 0.775$ the amplitude of the current on the mirrors' ends sharply increases. The latter causes considerable diffractional losses of H_{03} oscillations. For $b/a > 0.775$ the current corresponding to the H_{03} oscillation slowly transforms to the current of the H_{41} oscillation.

Similar processes are characteristic for H_{13} and H_{51} oscillations when b/a varies. An intertype rate of these oscillations is determined by the value of non-spectral parameter $b/a = 0.79$. In the vicinity of this value H_{13} and H_{51} oscillations interact as the intertype ones.

Let us determine how the enlargement of the distance between the mirrors and dielectric bodies inside the resonator influences the intertype interaction. It has been shown above that the increase of the distance $(l/a > 1)$ between the mirrors leads to the growth of diffractional losses for all types of oscillations. Diffractional losses for higher order H_{mn} oscillations $(m > n)$ grow much faster than for H_{0n} oscillations. Hence, the variation of this distance may control the intertype interaction. It is shown that $\mathrm{Re}(ka)$ practically does not depend on b/a, and $\log Q$ monotonically increases while b/a grows (note that when $b/a = 0.88$, this resonator becomes a closed one). One can completely destroy the intertype interaction of H_{03} and H_{41} oscillations by increasing the relative distance b/a between the mirrors.

Presence of dielectric bodies inside a resonator may also destroy intertype interaction. In this case $\mathrm{Re}(ka)$ practically does not depend on b/a. However, in the vicinity of the "point of coincidence" $b/a = 0.775$ of H_{03} and H_{41} oscillations of an empty resonator, a decrease of the H_{03} oscillation is observed. Hence, one can weaken the intertype interaction of oscillations in a confocal open resonator by changing the permittivlity ε.

The existing heuristic models [14] cannot describe the intertype interaction in open resonators. In particular, none of H_{m1} oscillations for $m > 1$ in a confocal resonator (and, of course, H_{41} and H_{51} oscillations) can be described within the framework of such models. Here we have shown how one can discover interaction of H_{03} and H_{41} (as well as H_{13} and H_{51}) oscillations by means of rigorous mathematical theory. Formula (1.34) gives a sufficiently good approximation for the real parts of the fundamental frequencies of H_{03} and H_{13} oscillations (the relative error is $\approx 0.6\%$ for H_{03} oscillations and $\approx 0.12\%$ for H_{13} oscillations). At the same time, the diffractional losses and q-factors of these oscillations calculated by the formulas [14] are different from the results obtained by the methods of rigorous theory.

Let us analyze the spectral characteristics of H-polarized eigenoscillations for a cylindrical open resonator with dielectric inclusions in the form of a circular pivot (having the diameter d). The resonator aperture is $2b$, the curvature radius a and the distance between the mirrors is l. These values are chosen to be close to the wavelength. The permittivity of the pivot is complex: $\varepsilon = \mathrm{Re}\,\varepsilon + i\,\mathrm{Im}\,\varepsilon)$. We will consider the confocal symmetrical resonator assuming that the pivot center lies on its axis.

This open structure is characterized by the parameters ε, μ, d/a, $2b/a$, s/a (s is the distance from the pivot center to the origin of the "left" mirror). If the resonator eigenoscillations are continuously transformed to H_{mn} oscillations of an empty resonator when $\varepsilon \to 1$ and $\mu \to 1$, we will call them H_{mn} oscillations ($m, n \geq 0$).

The equation

$$\left\| \begin{matrix} B^{11}(k) - I & B^{12}(k) & B^{13}(k) \\ B^{21}(k) & B^{22}(k) - I & B^{23}(k) \\ B^{31}(k) & B^{32}(k) & B^{33}(k) - I \end{matrix} \right\| \left\| \begin{matrix} x^1 \\ x^2 \\ x^3 \end{matrix} \right\| = \emptyset \tag{1.37}$$

(where $\emptyset$ is a zero element of the space $l_2^{(3)}$, I is unit operator in l_2 and $B^{ns}(k) : l_2 \to l_2$ are the kernel finite-meromorphic operator-valued functions) implicitly determines the countable set of functions

$$(ka)_{mn} = f_{mn}\left(\varepsilon, \mu, \frac{d}{a}, \frac{s}{a}, \frac{b}{a}\right), \tag{1.38}$$

corresponding to the sought fundamental frequencies. Functions f_{mn} are calculated by the numerical algorithm presented in the previous section. Let us study the properties of these functions assuming that

a) $1 \leq \varepsilon \leq 10$, $\mu = 1$, $d/a = \text{const}$, $s/a = 1/2$, $b/a = \text{const}$;

b) $\varepsilon = \text{const}$, $\mu = 1$, $0 \leq d/a < 1$, $s/a = 1/2$, $b/a = \text{const}$;

c) $\varepsilon = \text{const}$, $\mu = 1$, $d/a = \text{const}$, $d/(2a) \leq s/a \leq 1 - d/(2a)$, $b/a = \text{const}$.

Let us clear up how the spectrum of the complex fundamental frequencies (ka) of the resonator with a pivot depends on $(\mathrm{Re}\,\varepsilon)$ for $s/a = 1$ and $\mu = 1$. One has to distinguish the cases $(d/\lambda)\sqrt{\mathrm{Re}\,\varepsilon} \ll 1$ and $(d/\lambda)\sqrt{\mathrm{Re}\,\varepsilon} \gg 1$ (here $\lambda = 2\pi a\,\mathrm{Re}(ka)^{-1}$).

Assume first that $(d/\lambda)\sqrt{\mathrm{Re}\,\varepsilon} \ll 1$ and $\mathrm{Im}\,\varepsilon = 0$. The results of computations show that the behaviour of $(ka)_{mn} = f_{mn}(\mathrm{Re}\,\varepsilon, 1, \text{const}, 1/2, \text{const})$ is determined by the pivot position inside the resonator: a pivot may be in the "maximum" or in the "knot" of the eigenfield. For H_{m2n} oscillations (the electric field has a minimum in the resonator center) the fundamental frequencies and the q-factors practically do not depend on $\mathrm{Re}\,\varepsilon$ in the sufficiently big interval $[1, \varepsilon_0]$ (the value of ε_0 grows when the parameter $E_0 = (d/a)\{[\mathrm{Re}(ka)]/(2\pi)\}\sqrt{\mathrm{Re}\,\varepsilon}$ decreases). Here the inequalities $|ka - k_0 a| < C \cdot 10^{-3}$, $\mathrm{Re}(ka) < \mathrm{Re}(k_0 a)$ hold uniformly with respect to $\mathrm{Re}\,\varepsilon \in [1, \varepsilon_0]$, where $k_0 a$ are the fundamental frequencies of H_{m2n} oscillations for an open resonator, $C \sim 1$ and it depends on ka. For $\mathrm{Re}\,\varepsilon > \varepsilon_0$ the real part of the fundamental frequencies monotonically decreases when $\mathrm{Re}\,\varepsilon$ grows. Calculations of the real part and the logarithm of the q-factor of H_{02} oscillations as functions of $\mathrm{Re}(ka)$ confirm the qualitative characteristics stated above. For the H_{02} oscillation $\varepsilon_0 = 4$, $C = 5$. However, even for $\mathrm{Re}\,\varepsilon = 10$ the magnitude $[\mathrm{Re}(k_0 a) - \mathrm{Re}(ka)]/[\mathrm{Re}(k_0 a)]$ is less than 0.2%.

For H_{m2n+1} oscillations (the electric field has a maximum in the resonator center) the real part of the fundamental frequencies monotonically decreases as $\mathrm{Re}\,\varepsilon \to \infty$ (here $\mathrm{Re}(ka) < \mathrm{Re}(k_0 a)$) and tends to the real part of the fundamental frequency of an open resonator with a perfectly conducting inclusion having the diameter of the dielectric pivot. Here $(ka)_{m2n+1} = f_{m2n+1}(\mathrm{Re}\,\varepsilon, 1, \text{const}, 1/2, \text{const})$. It turns out that for $\mathrm{Re}\,\varepsilon = 10$ the spectral characteristics of the resonators with a dielectric pivot and with a metallic pivot practically coincide.

If the pivot diameter is close to the resonance wavelength $((d/a)\{[\text{Re}(ka)]/(2\pi)\}\sqrt{\text{Re}\,\varepsilon} \approx 1)$, $\text{Re}(ka)$ rapidly decreases when $\text{Re}\,\varepsilon$ grows for H_{03} oscillations, and the sharp local maximum of $\log Q$ appears (for $\text{Re}\,\varepsilon = 1.5$). The increase of $\text{Re}\,\varepsilon$ (for example, in the interval $[3, 10]$) leads to the strong increase of the diffractional losses for H_{03} oscillations ($Q \approx 100$) caused by the scattering properties of the dielectric inclusion. Calculations show that such a behaviour of $\text{Re}(ka)$ and Q is characteristic for other H_{0n} oscillations.

Let us establish now how the spectral characteristics of an open resonator with a dielectric pivot depend on its diameter, i.e., let us consider the functions $(ka)_{mn} = f_{mn}(\text{const}, 1, d/a, 1/2, \text{const})$ for $0 \leq d/a \leq 1$. The $\text{Re}(ka)$ of H_{0n} oscillations monotonically decreases and the q-factors have distinct resonance character for $n \geq 2$ when d/a grows. If the losses in the dielectric are small enough ($\text{Im}\,\varepsilon \sim 10^{-4}$), the q-factor of the resonator with a pivot is much greater than the q-factor in an empty resonator. It becomes especially clear for H_{03} oscillations where one can observe the first ($d/a = 0.18$, $\text{Im}\,\varepsilon = 2 \cdot 10^{-4}$) and the second local maximuma for $\log Q$. The q-factor for the resonant values of d/a strongly depends on $\text{Im}\,\varepsilon$. It is maximal for $\text{Im}\,\varepsilon = 0$ and decreases when $\text{Im}\,\varepsilon$ grows. For $\text{Im}\,\varepsilon \geq 10^{-2}$ the q-factor of the H_{03} oscillation monotonically depends on d/a, while it is practically constant as a function of $\text{Im}\,\varepsilon$ (for example, if $\text{Im}\,\varepsilon = 0$ and $\text{Im}\,\varepsilon = 10^{-2}$, $\text{Re}(ka)$ differs in the third significant digit).

The reasons for such a behaviour of the q-factors become clear after the study of the fields excited by the H-polarized wave with the frequency $\text{Re}(ka)$ for $\text{Im}\,\varepsilon = 0$, when d/a takes the resonant values. The amplitude of the E_y field component sharply increases inside the dielectric inclusion, as well as the amplitude of the magnetic field H_z. It turns out that the distribution of $|H_z|$ inside the dielectric inclusion coincides with the distribution of the pivot eigenfield.

Hence, the resonance character of the q-factor of an open resonator with a pivot considered as a function of d/a is caused by the excitation of eigenoscillations of the inclusion. For the non-resonant value of d/a the field excited inside the pivot does not correspond to any eigenoscillation of the dielectric inclusion, and the q-factor varies due to the scattering of the resonator eigenfield by the pivot. If we consider the q-factor as a function of d/a, the disappearance of local maxima in the case dielectric losses is caused by the fact that the dielectric pivot has the low q-factor two-dimensional eigenoscillations even for $\text{Im}\,\varepsilon = 0$ (except for the "whispering gallery" oscillations, when $|kd| \gg 1$, and the "locked" oscillations , when $\text{Re}\,\varepsilon \gg 1$, $\text{Im}\,\varepsilon \ll 1$ and $|kd| \sim 1$; the presence of such types of oscillations are not characteristic in the considered frequency range). Therefore, for the big values of $\text{Im}\,\varepsilon$, the resonance properties of the dielectric pivot cannot be observed in the frequency range $(wd/a)\sqrt{\text{Re}\,\varepsilon} = O(1)$. As has been mentioned above, $\text{Re}(ka)$ practically does not depend on $\text{Im}\,\varepsilon$ for the resonator with an inclusion and monotonically decreases when d/a grows (while $\text{Im}(ka)$ sharply increases) due to the hit and diffractional losses in the pivot Hence, the q-factor of such a resonator monotonically decreases when d/a grows.

The sharp growth of the q-factor of the resonator containing a low-loss dielectric body is characteristic not only for H_{0n} oscillations in the symmetrical confocal open resonator. Resonant values of d/a exist also for H_{22} and H_{13} oscillations. One can observe this phenomenon in the symmetrical non-confocal resonators. The resonant value of d/a decreases for the given type of oscillation when $\text{Re}\,\varepsilon$ grows. A similar situation occurs for H_{0n} oscillations when the index n increases: the first resonant value for the H_{05} oscillation is $d/a = 0.06$, and, for the H_{02} oscillation, $d/a = 0.3$ at $\operatorname{Re}\varepsilon = 2.08$.

Let us establish how the fundamental frequencies and the q-factors of an open resonator depend on the pivot position, i.e., let us consider the functions $(ka)_{mn} = f_{mn}(\text{const}, 1, \text{const}, x/a, \text{const})$ for $d/(2a) \leq x/a \leq 1 - d/a$ (x/a determines the pivot position on the resonator axis). Here we will limit ourselves only to the cases of H_{0n} oscillations in the symmetrical confocal resonator containing

(1) a dielectric inclusion with big losses ($\operatorname{Im}\varepsilon \sim 10^{-1}$);
(2) a perfect dielectric pivot ($\operatorname{Im}\varepsilon = 0$);
(3) a perfectly conducting pivot $\left(\mu = 1/\varepsilon, \varepsilon \to \infty\right)$.

The behaviour of $\operatorname{Re}(ka)$ and $\log Q$ of H_{03} oscillations with respect to parameter x/a in the resonator with a lossy dielectric inclusion are described by oscillating functions. Maximal difference between $\operatorname{Re}(ka)$ and $\operatorname{Re}(k_0a)$, and Q and Q_0 (the zero index corresponds to the case of the empty resonator) occurs when the pivot is in the maximum (or in the knot) of the electric (magnetic) field. The oscillating amplitude of $\operatorname{Re}(ka)$ increases when d/a grows and the oscillations of the q-factor weaken (when $d/a \to 0$, $\operatorname{Re}(ka)$ and $\log Q$ practically do not oscillate, and the pivot weakly disturbs the eigenfield of an empty resonator when $(d/a)\{[\operatorname{Re}(ka)]/(2\pi)\} \ll 1$). The data obtained for $\operatorname{Re}(ka)$ and $\log Q$ may be used for the proper choice of the size of a sensor for the reactive sensing of the fields in an open resonator.

Let us consider the case of a perfect dielectric pivot ($\operatorname{Im}\varepsilon = 0$). The dependences $\operatorname{Re}(ka)$ and $\log Q$ for H_{03} and H_{04} oscillations are also oscillating functions (and the former coincides with the function calculated for a lossy pivot). When $x/a \gtrsim 0.5$, the q-factor of the H_{06} oscillation of the resonator with a lossy pivot is much less than the q-factor Q_0 of the empty resonator, but it is greater than Q_0 for the resonator with a perfect dielectric inclusion. Such an anomalous behaviour of the q-factor is connected with the excitation of the pivot eigenoscillations. If the pivot moves out of the resonator center ($x/a = 0.5$), the q-factor sharply falls and reaches its minimal value when the pivot is in the maximum of the electric field (for the H_{03} oscillation $x/a = 0.82$). The equal contour values of $|H_z|$ in the resonator with a pivot excited by the H-polarized wave at the frequency $\operatorname{Re}(ka)$ and $x/a = 0.82$ (when the q-factor is minimal and the pivot is in the maximum of the electric field) demonstrate substantial perturbations of the field structure of the resonant H_{03} oscillation of the empty resonator.

Let us consider the case of a perfectly conducting inclusion. The results of calculations of the $\operatorname{Re}(ka)$ and $\log Q$ of H_{03} oscillations as functions of x/a for the two chosen values of d/a are considerably different from the data discussed above: the value of $\operatorname{Re}(ka)$ may be greater (or less) or even equal to the value of $\operatorname{Re}(k_0a)$ for the empty resonator (in the case of the resonator with a dielectric inclusions, the inequality $\operatorname{Re}(ka) \leq \operatorname{Re}(k_0a)$ always holds). It is established that $\operatorname{Re}(ka) > \operatorname{Re}(k_0a)$, when the pivot is in the knot (maximum) of the electric (magnetic) field, and $\operatorname{Re}(ka) < \operatorname{Re}(k_0a)$, when it is in the maximum (knot) of the electric (magnetic) field. Calculations show that $\operatorname{Re}(ka) = \operatorname{Re}(k_0a)$, when the amplitudes of the electric and magnetic fields coincide at the point of the pivot position. The q-factor weakly oscillates. As the pivot diameter increases, the picture sharply changes: the q-factor substantially decreases when the pivot moves out of the resonator center, and the periodicity of $\operatorname{Re}(ka)$ is broken. The latter property is typical for the small values of d/a.

Let us consider the symmetrical confocal resonator containing the double-layered dielectric cylinder with the normalized radii $\tilde{r}_1 = r_1/a$, $\tilde{r}_2 = r_2/a$ and permittivities $\varepsilon_1 = \varepsilon_1' + i\varepsilon_1''$ and $\varepsilon_2 = \varepsilon_2' + i\varepsilon_2''$ ($\mu_1 = \mu_2 = 1$).

The spectral properties of this open structure are studied on the main physical sheet of the Riemann surface Λ and are determined by a Fredholm finite-meromorphic operator-valued functions of the spectral parameter. This operator-valued function has a more complicated structure than (1.33). In the simplest case when there is one dielectric cylinder inside the resonator, the operator-valued function and dispersion relation have the form (1.37).

Let us consider the real parts of the fundamental frequencies, the diffractional losses and the q-factors of the resonator eigenoscillations as functions of the layer thickness $(\tilde{r}_1)$ assuming that $\varepsilon_1' = 1$, $\varepsilon_1'' = 0$, $\varepsilon_2' = 2,08$, $\varepsilon_2'' = 10^{-3}$, $\theta_n = 125°$ and $\tilde{r}_2 = 0,4$.

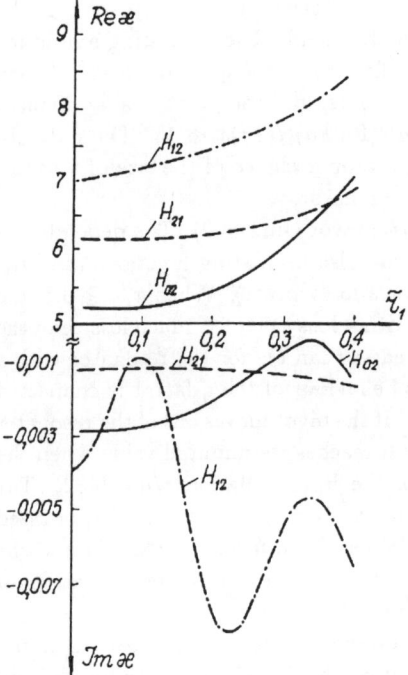

Figure 4.

The corresponding curves for Re κ and Im κ for H_{02}, H_{21} and H_{12} oscillations are shown in Fig.4. One can see that the presence of a dielectric tube inside the resonator leads to a decrease of its fundamental frequencies.

When $\tilde{r}_1$ grows, i.e., when the layer thickness decreases, the fundamental frequencies (and the curves Re κ) smoothly grow and approach the corresponding fundamental frequencies of the open resonator (for $\tilde{r}_1 = 0.4$ the layer thickness is equal to zero). A different situation occurs for the same oscillations, if to consider the behaviour of Im κ. For H_{02} and H_{12} oscillations the corresponding curves have a distinct oscillating character as functions of $\tilde{r}_1$. When $0.19 < \tilde{r}_1 < 0.38$, the $|\,\mathrm{Im}\,\kappa\,|$ of the H_{02} oscillation is less than that for the empty resonator.

Such a behaviour of the imaginary parts of the fundamental frequencies of the resonator with an inclusion is caused by the fact that a double-layered cylinder is an open resonance structure by itself and has its own sequence of fundamental frequencies. One can adjust this resonance structure with respect to its own resonances by varying the inner radius (in the given frequency range). Its inner field excites the resonant oscillations of the inclusion causing the increase of the q-factor of the system as a whole. For H_{21} oscillations the eigenoscillations of the inclusion are practically not excited, and the imaginary part of the fundamental frequency varies monotonically with respect to the layer thickness.

For the practical use of the open resonator with a double-layered dielectric inclusion, it is important to properly choose the working type of an oscillation and the dimensions of the inclusion (in order to increase the q-factor). For example, the inclusions with the inner radii $\tilde{r}_1 = 0.09$ and $\tilde{r}_1 = 0.345$ are optimal for H_{12} oscillations.

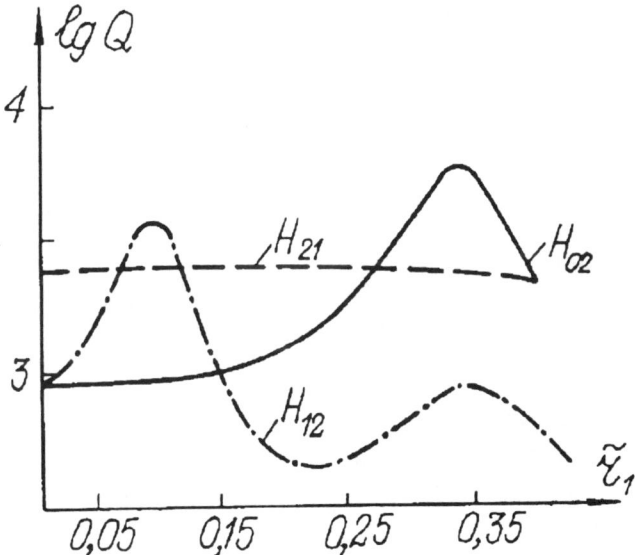

Figure 5.

The q-factors of H_{02}, H_{21} and H_{12} oscillations as function of the parameter $\tilde{r}_1$ are presented in Fig.5.

Consider now the resonator with a ring-type inclusion assuming that the permittivity ε_1 is not equal to 1. Let us study how the spectral characteristics of H_{12} oscillations depend on ε'_1 for the fixed values of $\tilde{r}_1$ situated close to the extreme values of $\mathrm{Im}\,\kappa$. Analysis of the spectral curves of the real parts of the fundamental frequencies $\mathrm{Re}\,\kappa$ and the diffractional losses $\mathrm{Im}\,\kappa$ as functions of ε'_1 (when $\varepsilon''_1 = 0$) for H_{12}, H_{02} and H_{21} oscillations shows that an increase of ε_1 leads to a decrease of the fundamental frequency. For $\varepsilon'_1 \neq 1$ the qualitative picture of the diffractional losses depends on ε_1 and on the oscillation type. For H_{21} oscillations the growth of ε'_1 yields an increase of the q-factor and greater the inner radius of the inclusion $\tilde{r}_1$, greater is the q-factor (the value $\tilde{r}_2 = 0.4$ is fixed).

For H_{02} and H_{12} oscillations the increase of ε_1' is accompanied, as a rule, by the decrease of the q-factor, except for several values of ε_1', when the q-factor increases (for the H_{12} oscillation $1 < \varepsilon_1' < 1.55$ when $\tilde{r}_1 = 0.23$ and $-1 < \varepsilon_1' < 1.6$ when $\tilde{r}_1 = 0.35$). This increase of the q-factor is connected with the resonance excitation of eigenoscillations in the inclusion.

Intersection of the spectral curves $\operatorname{Re}\kappa$ and $\operatorname{Im}\kappa$ occurring when $\varepsilon_1' = 2.08$ for the different values of $\tilde{r}_1$ (and for all types of the oscillations considered here) is caused by the coincidence of the layer permittivities (here one has almost a homogeneous cylinder because $\varepsilon_2'' \neq \varepsilon_1''$).

Solution to the problem of the plane wave diffraction by the open resonator with a ring-type inclusion at the frequencies equal to the real parts of the fundamental frequencies of the empty resonator (and of the resonator with a ring-type inclusion) allow us to construct the equal contour values of $|H_z|$ for the H_{12} oscillation. It turns out that for $\tilde{r}_1 = 0.09$ (when $\varepsilon_1 = 1$ and $\varepsilon_2 = 2,08 + 0,001i$) the field maximum has a displacement with respect to the resonator horizontal axis if to compare with the empty resonator. For $\tilde{r}_1 = 0.23$ and $\tilde{r}_2 = 0.345$ the field maximum is in the same position as in the empty resonator.

Assume now that the width of the resonator slot is $\theta_n = 130°$ ($n = 1, 2$) and take the values of the parameters $\tilde{r}_1 = 0.15$, $\tilde{r}_2 = 0.35$ and $\varepsilon_2 = 2.2 + 0.00055i$. Then the behaviour of the the the real parts of the fundamental frequencies and $\log Q$ as functions of ε_1' (for $\varepsilon_1'' = 0$) shows that for the H_{00} oscillation, the fundamental frequencies and the q-factor practically do not depend on ε_1'. For other types of oscillations (like H_{03}, H_{41}, H_{21} and H_{43} oscillations that belong to the same symmetry class H_{2n2p+1}) the fundamental frequency magnitudes increase as ε_1' grows.

The q-factors of these oscillations behave in a different way: the increase of ε_1' leads to the increase of the q-factors for H_{01} and H_{21} oscillations, while for H_{03} oscillations, such an increase occurs only when $1 < \varepsilon_1' < 1.15$. The q-factor of H_{03} oscillations decreases when $\varepsilon_1' > 1.15$. For all other oscillations the q-factors decrease as ε_1' grows.

The properties of $\operatorname{Re}\kappa$, $\operatorname{Im}\kappa$ and $\log Q$ as functions of ε_1' for the H_{01} oscillation determined at $\theta_n = 130°$, $r_1' = 0,45$, $\tilde{r}_2 = 0,49$ and $\varepsilon_2 = 2.08 + 0.001i$ (in the case of the low-loss medium of the first layer with $\varepsilon_1'' = 10^{-3}$) allow us to conclude that, for the H_{01} oscillation, a decrease or an increase of the q-factor (when ε_1' grows) are independent of the inclusion dimensions and the medium losses characterized by the parameter ε_1.

Let us consider another case of the variation of $\operatorname{Re}\kappa$, $\operatorname{Im}\kappa$ and $\log Q$ as functions of ε_2' ($\varepsilon_2'' = 0$) for the same geometrical parameters and $\varepsilon_1 = 2.08$. Computations show that the real parts of the fundamental frequencies decrease, when the permittivity of the first or of the second layer grows. The q-factors weakly increase because of the fall of the dielectric losses. Therefore, for H_{02} oscillations, one has to take into account not only the shape of a dielectric inclusion but also the medium losses.

Analysis of the equal contour values of $|H_z|$ for the dominant types of oscillations (obtained as a result of the solution to the problem of the plane unit-amplitude H-polarized wave diffraction at the resonance frequencies, that is, when the frequency of the incident wave is equal to $\operatorname{Re}\kappa$ corresponding to the values of $\varepsilon_1 = 1$, 1.1 and 3.2) shows that for $\varepsilon_1 = 1$ and $\varepsilon_1 = 1.1$ the fields are practically the same. A different field distribution is observed for $\varepsilon_1 = 3.2$, when the so-called "absorbing effect" takes place: a part of the equal contour lines come off the mirror and the lines become closed. The latter effect is explained by the fact that the permittivity of the first layer is greater than the permittivity of the second layer. Hence, a dielectric inclusion absorbs the resonator energy in a stronger way than in the cases considered above.

Let us investigate the behaviour of the real parts of the fundamental frequencies and the diffractional losses as functions of $\tilde{r}_1$ for the fixed parameters $\tilde{r}_1$, ε_1 and ε_2. We set $\theta_n = 125°$ and $\tilde{r}_1 = 0.04$ and consider three cases:

1) $\varepsilon_1 = 1$, $\varepsilon_2 = 2.08 + 0.001i$;
2) $\varepsilon_1 = 5.5 + 0.001i$, $\varepsilon_2 = 2.08 + 0.001i$;
3) $\varepsilon_1 = 2.08$, $\varepsilon_2 = 5.5 + 0.001i$.

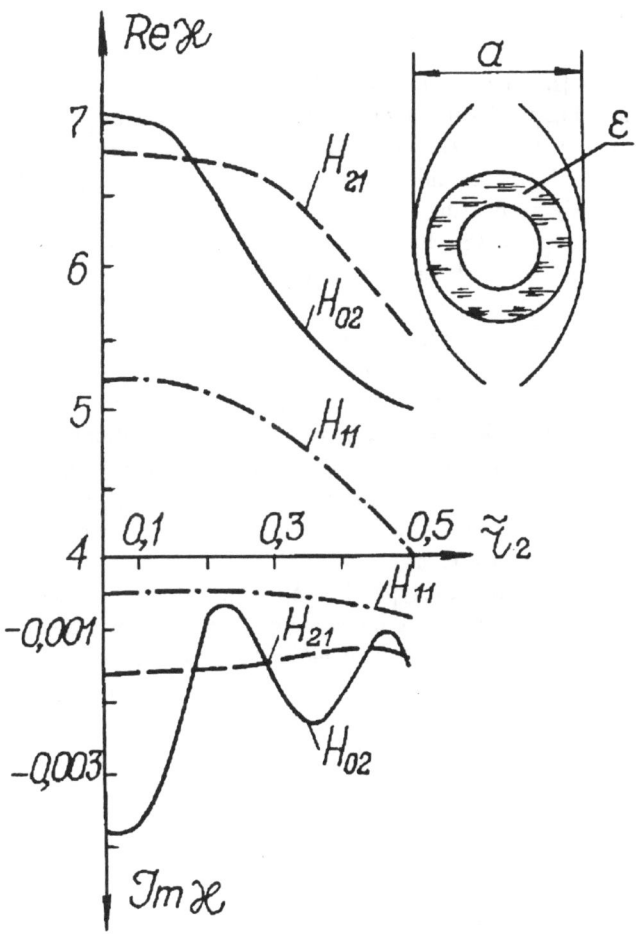

Figure 6.

The curves of $\operatorname{Re}\kappa$ and $\operatorname{Im}\kappa$ with respect to $\tilde{r}_2$ for H_{11}, H_{02} and H_{21} oscillations are presented in Fig.6. One can see that the real parts of the fundamental frequencies decrease while $\tilde{r}_2$ grows. The real parts of the fundamental frequencies of H_{02} and H_{11} oscillations approach each other when $0.23 < \tilde{r}_2 < 0.32$. Here the decrease of $\operatorname{Re}\kappa$ with increasing $\tilde{r}_2$ is faster than in the two previous cases.

The situation is different for $\mathrm{Im}\,\kappa$ considered as a function of $\tilde{r}_2$. One can see in Fig.6 that $\mathrm{Im}\,\kappa$ oscillates in the first two cases for H_{02} oscillations and practically does not depend on ε_1 (here the parameter $\tilde{r}_1$ is small compared with the resonance wavelength of the considered resonator). In the third case $\mathrm{Im}\,\kappa$ monotonically grows as $\tilde{r}_2$ increases. The interval $0.05 < \tilde{r}_2 < 0.11$ is an exception, and one can observe in this interval a considerable fall of the diffractional losses accompanied by the growth of the q-factor. Such a behaviour of $\mathrm{Im}\,\kappa$ for H_{02} oscillations is due to the excitation of the resonance oscillations of the dielectric inclusion by the resonator eigenfield. The presence of a perfect medium in the domain of the first layer, as well as of a medium with absorption, causes sharp changes of $\mathrm{Im}\,\kappa$ as a function of $\tilde{r}_2$. The optimal values of the q-factor are $\tilde{r}_2 = 0.195$ in the first case, $\tilde{R}_2 = 0.190$ in the second case and $\tilde{r}_2 = 0.11$ in the third case.

The equal contour values of $|H_z|$ have been studied for those values of $\tilde{r}_2$, at which the q-factor of H_{02} oscillations has the local maximum or the local minimum.

We have also considered the spectral characteristics of the open two-mirror resonator containing a double-layered radial cylinder whose external boundary is dielectric and internal boundary is perfectly conducting.

On the whole, the perfectly conducting boundary of the double-layered cylinder yields a decrease of the q-factor, although there exist the frequency ranges, where the q-factor is a little greater than the one of the resonator considered above. The imaginary parts $\mathrm{Im}\,\kappa$ of H_{11} and H_{21} oscillations are practically the same as for the resonator containing the cylinder with the internal dielectric boundary.

Assume that the value of ε is constant and consider the real parts of the fundamental frequencies and the diffractional losses of H_{02} oscillations as functions of $\tilde{r}_1$ for $\tilde{r}_2 = 0.465$ (at the point of the local maximum of $\mathrm{Im}\,\kappa$ and Q). When $\tilde{r}_1$ grows, the real part of the fundamental frequencies monotonically increases together with the diffractional losses and the q-factor falls due to the increase of the perfectly conducting internal boundary, which radiates the field energy.

Let us indicate some features of the intertype interaction in the open resonator with the layered circular dielectric cylinder. The intertype oscillations are characterized by the following conditions: the q-factors of the eigenoscillations sharply grow or fall with respect to a small perturbation of the aperture or the distance between the mirrors, and the hybrid intertype fields appear.

Consider the symmetrical confocal resonator with a ring-type inclusion having the parameters $\theta_n = 126°$, $\varepsilon_1 = 1$, $\varepsilon_2 = 1.05$, $\tilde{r}_1 = 0.05$ and $\tilde{r}_2 = 0.1$ and the behaviour of $\mathrm{Re}\,\kappa_{1,2}$ and $\mathrm{Im}\,\kappa_{1,2}$ for H_{03} and H_{41} oscillations when $0.8244 < \hat{l} < 2.0$. The values $\hat{l} = 0.8244$, $0.8244 < \hat{l} = l/a < 1$, $\hat{l} = 1$ and $\hat{l} > 1$ correspond, respectively, to the closed, before-confocal, confocal, and over-confocal resonators. The curves for the $\mathrm{Re}\,\kappa_{1,2}$ and $\mathrm{Im}\,\kappa_{1,2}$ for H_{03} and H_{41} oscillations as functions of $\hat{l}$ have the form of the Wien graphs [63]. The only deviation from the behaviour that is typical for the empty resonator is the splash of $\mathrm{Im}\,\kappa$ for H_{03} oscillations in the interval $0.9 < \hat{l} < 0.95$, when the modulus of the difference between the empty resonator fundamental frequencies and the fundamental frequencies of the inclusion reaches its maximal value.

Hence, the presence of an inclusion does not destroy the intertype communication between different types of oscillations. Similar phenomena can be observed, when the permittivity of one of the layers varies.

Let us consider the real parts of the fundamental frequencies, for H_{03} and H_{41}

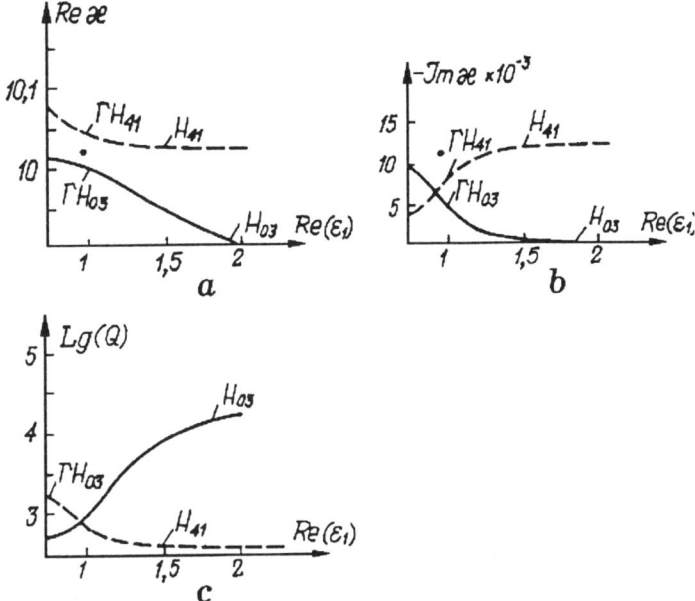

Figure 7.

oscillations as functions of ε_2', when $\varepsilon_1 = 1$, $\varepsilon_2'' = 0$ and the parameters of the mirror aperture and the inclusion are fixed. We will also present the results of calculations with respect to varying ε_1' when $\varepsilon_2 = 1.4 + 0.001i$ and $\varepsilon_1'' = 0$ (see Fig.7). One can see that H_{03} and H_{41} oscillations interact when $1 < \varepsilon_1' < 1.8$, and their behaviour is described by the Wien graphs. These oscillations exchange the energy, when there is a shortest distance between the curves, i.e., when $H_{03} \to H_{41}$ and $H_{41} \to H_{03}$. In both cases, when the curves $\mathrm{Re}\,\kappa_{1,2}$ approach each other, the corresponding graphs of $\mathrm{Im}\,\kappa_{1,2}$ and $\log Q_{1,2}$ intersect. When high q-factor H_{03} oscillations is transformed to H_{41} oscillation, one can observe the growth of the diffractional losses accompanied by the fall of the q-factor. The decrease of the diffractional losses is observed in the case of $H_{41} - H_{03}$ transformation, when the q-factor increases (see Fig.7).

Excitation of this structure by the unite-amplitude plane wave at the resonance frequencies corresponding to ε_2' for $1 < \varepsilon_2' < 1.8$ and $1 < \varepsilon_2' < 1.75$, and for $0.25 < \varepsilon_1' < 1.45$ and $0.4 < \varepsilon_1' < 1.8$ at $\varepsilon_2 = 1.4 + 0.001i$ allows us to construct the equal contour values of $|H_z|$ and to determine the transformation dynamics $H_{03} \rightleftarrows H_{41}$ with respect to varying ε_2' and ε_1'. For example, for $\kappa' = 10.0176$ and $\varepsilon_2' = 1$ one has "pure" H_{41} oscillations; and for $\kappa' = 10.0721$ and $\varepsilon_2' = 1.25$ one observes the oscillations having the structure of both the H_{41} and H_{03} oscillations. The H_{41} oscillation is transformed to the H_{03} oscillation for $\kappa' = 9.929$ and $\varepsilon_2' = 1.80$; and for $\varepsilon_2' = 1$ and $\kappa' = 10.1503$ one observes the oscillations containing the elements of both the H_{03} and H_{41} oscillations, which are transformed to the H_{41} oscillation at $\varepsilon_2' = 1.75$ and $\kappa' = 10.0261$.

We will call the intertype or hybrid the oscillations, whose fields contain the elements of two or more oscillations. The hybrid intertype H_{03} oscillations are observed, when the parameter ε_1' varies in the intervals $0.25 \leq \varepsilon_1' \leq 0.40$ and $1 \leq \varepsilon_1' \leq 1.5$, and H_{41} oscillations are observed in the interval $0.85 \leq \varepsilon_1' \leq 1$.

Hence, H_{03} and H_{41} oscillations of the open resonator with a dielectric inclusion have the intertype character caused by the small variations of the layer parameters.

The curves $\text{Re}\,\kappa_{1,2}$ may intersect in the frequency range where the intertype oscillations exist but do not exchange the types and do not form the Wien graphs in the vicinities of the intersection points.

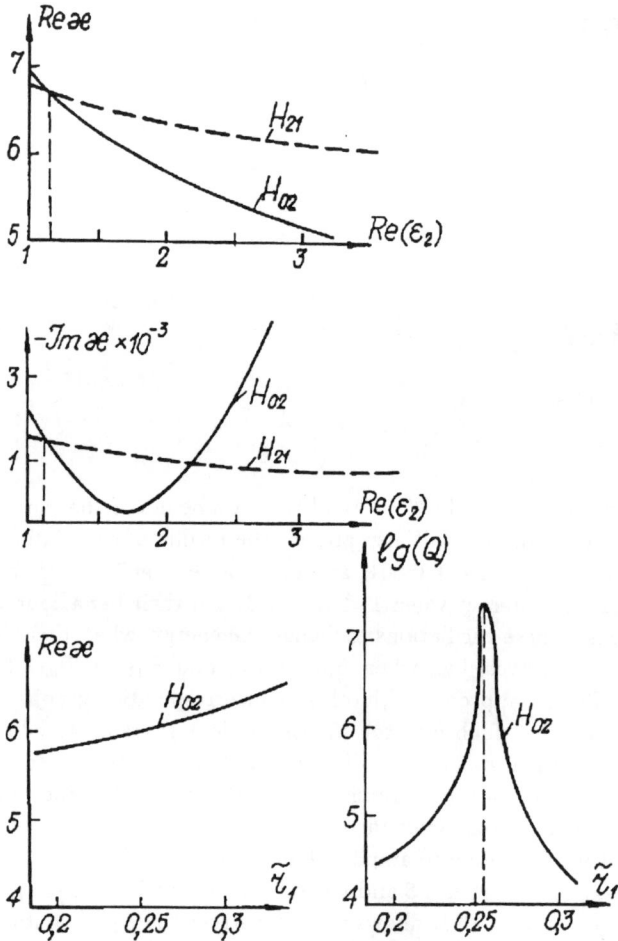

Figure 8.

To confirm this fact let us consider how $\text{Re}\,\kappa$ and $\text{Im}\,\kappa$ depend on the permittivity ε_2' for non-symmetrical H_{02} and H_{21} oscillations (belonging to the classes H_{2m2p} and H_{2m2p+1}, respectively) taking the values $\theta_n = 125°$, $\tilde{r}_1 = 0.25$, $\tilde{r}_2 = 0.4$ and $\varepsilon = 1$ (Fig.8). One can see that they monotonically decrease and intersect at the point $\varepsilon_2 = 1.1599$. Hence, such a type of inclusion leads to the decrease of the fundamental frequencies (i.e., to the decrease of $\text{Re}\,\kappa$) and the greater the permittivity in the second layer, the smaller are the real parts of the fundamental frequencies.

The curves $\text{Im}\,\kappa_{1,2}$ also intersect at the point $\varepsilon_2 \neq 1.1599$ shifted to the left, if to the compare with the intersection points of the curves $\text{Re}\,\kappa_{1,2}$. This effect is similar to the degeneration of oscillations, when the curves $\operatorname{Re}\kappa_{1,2}$ and $\operatorname{Im}\kappa_{1,2}$ of two different orthogonal oscillations intersect at a certain value of a non-spectral parameter. The graph for $\operatorname{Im}\kappa$ of H_{21} oscillations monotonically grows as ε_2 increases in the interval $1 < \varepsilon_2 < 1.72$ (i.e., the diffractional losses decrease in this range). The different situation occurs for the H_{02} oscillations: the diffractional losses fall in the range $1 < \varepsilon_2 < 1.72$, and, hence, the q-factors grow. For $\varepsilon_2 = 1.72$ the q-factors of the eigenoscillations reach the maximum value of about 10^6. In this the case the fundamental frequencies of the empty resonator are very close to the fundamental frequencies of a resonant dielectric tube.

Note that when $\varepsilon_2 = 1.72$, $\tilde{r}_1$ is increased and the other geometrical parameters are fixed, the q-factors of H_{02} oscillations sharply increase (reaching the maximal value when $\tilde{r}_1 = 0.255$) due to the increase of the real part of the fundamental frequency and the decrease of the diffractional losses. This effect is similar to the behaviour of the q-factors in the vicinity of $\varepsilon_2 = 1.72$ when ε_2' varies.

For $\tilde{r}_1 > 0.255$ the diffractional losses considerably increase, and, hence, the q-factors fall up to the order 10^3 at $\tilde{r}_1 = 0.23$. If the medium in the second layer has the small losses (for example, $\varepsilon_2'' = 0.001$), there are no splashes of the q-factors as functions of ε_2' in the frequency ranges considered above, and they grow insignificantly with respect to the increasing inner radius.

Two situations are characteristic for the intertype oscillations: for symmetrical H_{03} and H_{41} oscillations the curves $\operatorname{Re}\kappa_{1,2}$ form the Wien graphs. When the permittivity or the parameter $\tilde{r}_2$ vary, the mutual transformation of oscillations occurs. The spectral curves $\operatorname{Re}\kappa_{1,2}$ do not form the Wien graphs for non-symmetrical H_{02} and H_{21} oscillations, and there is no transformation of oscillations when the dielectric permeability of the second layer varies.

In addition to this, the q-factors grow when H_{41} oscillations are transformed to H_{03} oscillations, and they fall when H_{03} oscillations are transformed to H_{41} oscillations. The q-factors of H_{02} and H_{21} oscillations always increase. The emergence of these intertype oscillations may be explained by the presence of the Morse critical points in the intervals of variation of the non-spectral parameters, for example, of the permittivities ε_1' or ε_2'.

§ 4. Theory of the Morse critical points of empty open and closed resonators

It is shown in section 3 that for symmetrical confocal empty resonators there exists a sequence of non-spectral geometrical parameters such that their small variations lead to sharp changes of spectral parameters (fundamental frequencies), cause considerable growth or fall of diffractional losses, appearance of intertype (hybrid) field configurations, etc. These phenomena were called "communication" (or "interaction") of eigenoscillations (modes) of different types. Let us define all these phenomena in what follows as intertype oscillations. The model presented in Refs. [57, 58] for closed resonators describes them as intertype oscillations of an "ideal" resonator perturbed by a small inhomogeneity characterized by a double-degeneration frequency point in the dispersion relation and accompanied by the communication of modes of an ideal resonator caused by the presence of this inhomogeneity. In Refs. [57, 58] the theory of Morse critical points is constructed and the nature of intertype oscillations in open resonators is cleared up by means of the known dispersion relations for closed cylindrical resonators.

This section is devoted to the construction of a new mathematical model describing the phenomenon of intertype oscillations in empty open and closed resonators on the basis of the theory of functions of several complex variables and the theory of singularities.

As has been shown in section 1, the spectral theory allows to determine for open cylindrical resonators the characteristic values of the spectral parameter $\kappa = ka$ for which there exist nontrivial solutions to the homogeneous system of the Maxwell equations, satisfying boundary conditions on the resonator surfaces, the Meixner edge conditions and the Reichardt conditions in infinity. The mathematically rigorous solution of this problem has been obtained in section 1 on the basis of the spectral theory of non-self-adjoint holomorphic Fredholm operator-valued functions. The operator-valued function of the problem is determined by expression (1.33). It is holomorphic with respect to κ on the Riemann surfaces Λ and has the form $I - B(\kappa) : H \to H$ where $H = \bigotimes_{n=1}^{2} l_2$ is a Cartesian product of two pieces of the space l_2 having the natural structure of a Hilbert space.

Holomorphic and kernel properties of the operator-valued function $A(\kappa)$ with respect to κ and non-spectral parameters allow us to prove that the spectrum (set) of its characteristic numbers is discrete and has a finite multiplicity. The problem on characteristic numbers may be reduced to the dispersion relation

$$F(\kappa) \equiv \det\{I - B(\kappa)\} = 0. \qquad (1.39)$$

This equation contains the infinite characteristic determinant of the kernel operator $B(\kappa)$ and one can solve it only numerically. Various examples of the corresponding computer simulations are presented in section 3 in the form of dispersion curves obtained, in particular, for intertype oscillations.

It is important that $B(\kappa)$ is also considered as a function of the non-spectral parameters of the resonator (as an example, one can consider the symmetrical open resonator presented in Fig. 9 where l is the distance between the mirrors, a is the radius and b is the aperture of the mirrors, $l_0 = 2[a - (a^2 - b^2)^{1/2}]$, and for $l = l_0$ this open resonator becomes a closed one). Let us take only one of these parameters, namely, $\chi = b/a$ and rewrite the operator-valued function in the form $B(\kappa, \chi)$ assuming that χ varies in the domain $D_\chi \subset C$ where $B(\kappa, \chi)$ is holomorphic (here C is the complex plane). We note also that $B(\kappa, \chi)$ is a kernel operator holomorphic with respect to two complex variables κ, χ for $(\kappa, \chi) \in \Lambda \times D_\chi$. The parameter χ physically corresponds to the adjustment of the resonator. We will assume also that the eigenvalues $\kappa \subset \Lambda_0$, where $\Lambda_0 = \{\kappa \in C : -\pi < \arg \kappa < \pi; \ \kappa \neq 0\}$ is the zero sheet of Λ.

Since $B(\kappa, \chi)$ is a kernel and holomorphic operator-valued function, the set of zeros of the dispersion relation $F(\kappa, \chi) = 0$ coincides with the set of characteristic numbers $\sigma(\kappa)$ of $I - B(\kappa, \chi)$ ($\sigma(\kappa)$ is defined on the subset $D = \Lambda_0 \times D_\kappa \subset C^2$). It is necessary to establish what happens to $\sigma(\kappa)$ when the eigenvalues κ vary in D_κ.

Let us introduce the analytical set $\sigma_0 = \{(\kappa, \chi) \subset D : F(\kappa, \chi) = 0\}$ and consider $F(\kappa, \chi)$ as a mapping $F : C^2 \to C$ defined in the domain D (here we naturally assume that χ is a complex value). As follows from the general results of the theory of functions of several complex variables and the theory of the smooth mapping singularities [64, 65], close to an isolated singularity (κ_0, χ_0) of the mapping F, the local structure of σ_0

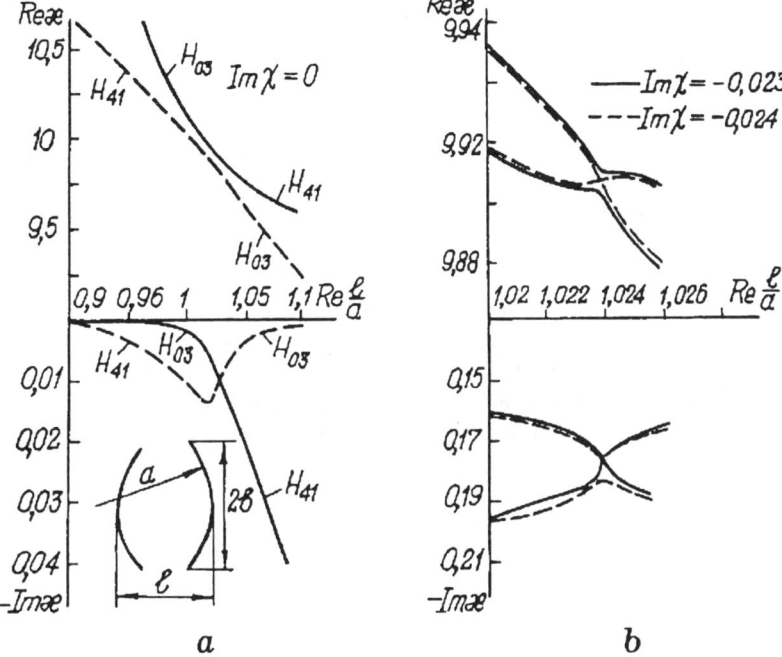

Figure 9.

is defined by the type and position of (κ_0, χ_0), and small changes of χ in the vicinity of (κ_0, χ_0) lead to small changes of $\sigma(\kappa)$.

The phenomenon of intertype oscillations has well-known and commonly accepted features, and the first among them is the dispersion law which regulates the special behaviour of two fundamental frequencies as functions of χ when $\operatorname{Im}\chi = 0$ (the Wien graph). The natural problem appears to describe the conditions, which correspond to the dispersion law for intertype oscillations in terms of regular and singular points of the mapping F.

The local structure of σ_0 (as a hyper-plane) does not correspond to the dispersion law and it is convenient to consider the set of critical points

$$\sigma_K = \left\{ (\kappa, \chi) \in D : \operatorname{grad} F = \left(\frac{\partial F}{\partial \kappa}, \frac{\partial F}{\partial \chi} \right) = 0 \right\}$$

as well as the set of isolated Morse critical points

$$\sigma_{MK} = \left\{ (\kappa, \chi) \in \sigma_K : \frac{\partial^2 F}{\partial \kappa^2} \frac{\partial^2 F}{\partial \chi^2} - \left(\frac{\partial^2 F}{\partial \kappa \partial \chi} \right)^2 \neq 0 \right\}.$$

Let (κ_0, χ_0) be an isolated Morse critical point situated close to σ_0 (i.e., the value of $F(\kappa_0, \chi_0) = \delta$ is small enough). Then one can write the dispersion relation in the vicinity of (κ_0, χ_0) in the form

$$\delta + \frac{1}{2} \frac{\partial^2 F}{\partial \kappa^2}\bigg|_{\substack{\kappa=\kappa_0 \\ \chi=\chi_0}} (\kappa - \kappa_0)^2 + \frac{\partial^2 F}{\partial \kappa \partial \chi}\bigg|_{\substack{\kappa=\kappa_0 \\ \chi=\chi_0}} (\kappa - \kappa_0)(\chi - \chi_0) + \frac{1}{2} \frac{\partial^2 F}{\partial \chi^2}\bigg|_{\substack{\kappa=\kappa_0 \\ \chi=\chi_0}} (\chi - \chi_0)^2 + O^3 = 0,$$

$$(1.40)$$

where O^3 are small terms of the third order and higher. Since $(\kappa_0, \chi_0) \in \sigma_{MK}$, there exists [66] the linear change of variables $\tilde{\kappa} = \psi_1(\kappa, \chi)$, $\tilde{\chi} = \psi_2(\kappa, \chi)$ which transforms (1.40) into

$$\tilde{\kappa}^2 + \tilde{\chi}^2 + \delta = 0, \tag{1.41}$$

and, consequently, $\tilde{\kappa} = \pm\sqrt{-\tilde{\chi}^2 - \delta}$. For $\delta = 0$ we get $\tilde{\kappa} = \pm i\tilde{\chi}$. Qualitative properties of the solutions to (1.40) and (1.41) coincide. Equation (1.41) contains the dispersion law (i.e., it determines the spectral parameter as a function of non-spectral parameters l or a), which is characteristic for the intertype communication, and the value of $\delta = F(\kappa_0, \chi_0)$ determines the degree of this communication. There is a double degeneration of $F(\kappa, \chi)$ when $\delta = 0$ at the point (κ_0, χ_0) with respect to both κ and χ.

It is sufficient to study $\mathrm{Re}\,\kappa(\chi)$ and $\mathrm{Im}\,\kappa(\chi)$ as solutions of (1.40) in a physical domain of variation of the non-spectral parameter (when $\mathrm{Re}\,\chi = 0$) that defines the resonator adjustment, and consider the variation of $\mathrm{Re}\,\tilde{\kappa}(\tilde{\chi})$ and $\mathrm{Im}\,\tilde{\kappa}(\tilde{\chi})$ given by (1.41) along the straight lines $\mathrm{Re}\,\tilde{\chi} = \xi$, $\mathrm{Im}\,\tilde{\chi} = \alpha\xi + \beta$ where ξ and α are the real parameters of the line and $\beta \neq 0$ for $\mathrm{Im}\,\chi_0 \neq 0$. Finally we get

$$\tilde{\kappa}(\xi) = \pm\left[\left(\frac{\sqrt{a^2+b^2}+a}{2}\right)^{1/2} + i\,\mathrm{sign}\,b\left(\frac{\sqrt{a^2+b^2}-a}{2}\right)^{1/2}\right],$$

where

$$a = (\alpha\xi + \beta) - \xi^2 - \delta_1; \quad b = -\delta_2 - 2\xi(\alpha\xi + \beta); \quad \delta = \delta_1 + i\delta_2; \quad \delta_1, \delta_2 \in \mathbb{R}^1.$$

Two-parameter families of the curves $\mathrm{Re}\,\tilde{\kappa}(\xi)$, $\mathrm{Im}\,\tilde{\kappa}(\xi)$ contain for certain values α and β the relations which are analogous to the Wien graph for the fundamental frequencies of coupled oscillation contours. These curves qualitatively repeat the behaviour of the dispersion curves presented in Fig.3,a,b (see section 3). Intertype oscillations that have been found for the symmetrical confocal open resonator with varying apertures ($\chi = b/a$, see Fig.3,a,b) is characteristic for other types of the resonator deformations.

For H_{03} and H_{41} oscillations it is shown in Fig.9 how their fundamental frequencies and diffractional losses depend on the distance between the mirrors (which is described by the non-spectral parameter $\theta = l/a$). One can observe for $\theta = 1 \div 1.05$ a considerable increase of the diffractional losses for the H_{03} oscillation caused by its communication with the H_{41} oscillation. Fig.9b shows a similar dependence obtained for the complex values of the parameter χ. It is important that there exist other types of the curves presented in Fig.10,a–e, which have been obtained earlier in section 3 where the spectral problems have been rigorously solved.

The dispersion curves in Fig.10,a–e obtained from (1.40) and (1.41) describe all possible qualitative situations occurring in the vicinities of the Morse critical points for different values of α and β. We will call these families of curves the communication diagrams. The functions $\tilde{\kappa}(\xi)$ corresponding to $\beta = 0$ are shown in Fig.10,a–e by bold lines, and their perturbations with respect to the small values of $\beta \neq 0$ are shown by bar lines. One can also produce the dependence obtained after the change of variables $\mathrm{Re}\,\tilde{\kappa} \leftrightarrow \mathrm{Im}\,\tilde{\kappa}$.

Hence, the existence of the Morse critical point (κ_0, χ_0) of the characteristic determinant $F(\kappa, \chi)$ of an open resonator is connected with the existence of two solutions $\kappa_+(\xi)$ and $\kappa_-(\xi)$ of the equation $F(\kappa, \chi) = 0$ in the vicinity of (κ_0, χ_0). Behaviour

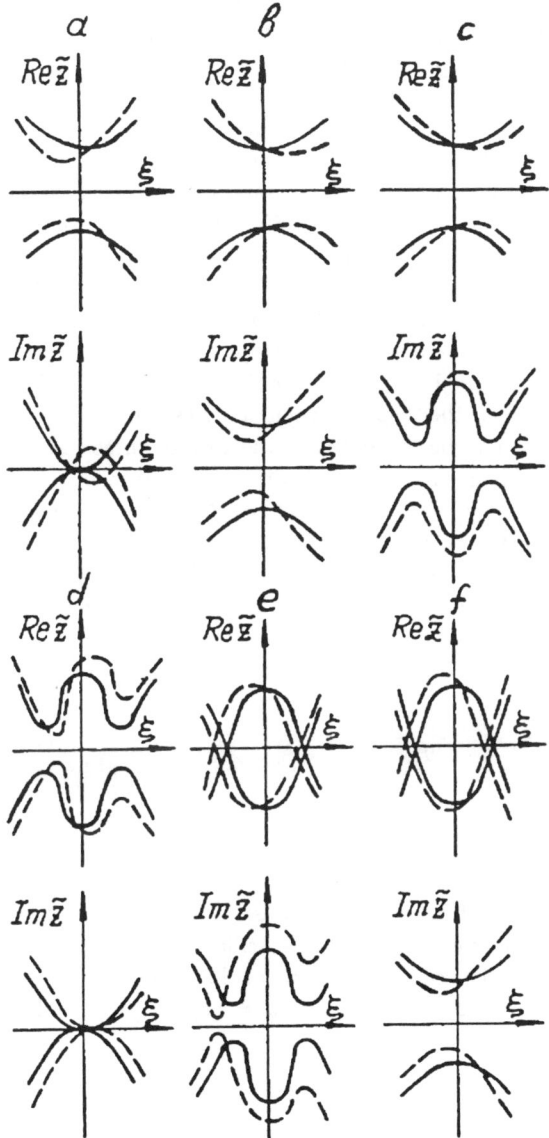

Figure 10.

of these solutions with respect to the varying resonator adjustment χ is completely determined by (1.41) and it describes the intertype oscillations of an open resonator. In particular, it determines the sharp change of the q-factor (and diffractional loss) due to small variations of geometrical parameters (all the results given in this section may be also applied for an ideal closed resonator).

Now one can expect that all the dispersion relations obtained in electrodynamics (as well as in fusion, plasma, solid state physics, etc.) may be considered from the viewpoint of the presence (or the absence) of the Morse critical points and, hence, one can verify the existence of intertype oscillations. Here the frequency is chosen as a spectral parameter. Note that every other parameter of the problem may be taken as a non-spectral variable. It is important for the dispersion relation to have the form $F(\kappa, \chi)$ where κ is a frequency and χ is one of the non-spectral parameters (κ and χ are complex). The procedure of the Morse critical point determination is not complicated. One solves the equations $\frac{\partial F}{\partial \kappa} = 0$ and $\frac{\partial F}{\partial \chi} = 0$ (in the general case, numerically) with respect to $\kappa = \kappa_0$ and $\chi = \chi_0$, determines $\delta = F(\kappa_0, \chi_0)$ and calculates the second derivatives at the point (κ_0, χ_0) to prove that the Hessian is not equal to zero. Then the Taylor series for the function $F(\kappa, \chi)$ in the vicinity of the point (κ_0, χ_0) must be obtained and after the change of variables $\kappa \to \tilde{\kappa}$, $\chi \to \tilde{\chi}$, one finally gets the quadratic equation with respect to $\tilde{\kappa}$ and $\tilde{\chi}$. Its solutions give the set of dispersion curves determining the characteristics of the physical "intertype process" under study.

One has to point out that the proposed method of the Morse critical point determination, where the function $F(\kappa, \chi)$ in the dispersion relation $F(\kappa, \chi) = 0$ depends on two complex variables κ and χ, may be expanded to the case of three independent variables κ, Φ and χ. Such problems are connected with a simultaneous study of the oscillation (with respect to κ) and wave (with respect to Φ) processes when the geometric and other non-spectral parameters (denoted by χ) vary.

Let us show now that the approach based on the analysis of the Morse critical points and developed above for an open resonator may be applied for the analysis of closed resonators. For this purpose, we consider the characteristic equation from Refs. [57] for two sufficiently close fundamental frequencies of a closed resonator excited by a connecting slot (hole)

$$z_\alpha z_\beta - z_{\alpha\beta}^2 = 0, \tag{1.42}$$

where

$$z_\alpha = d_\alpha + i\frac{2(\omega - \omega'_\alpha)}{\omega_{cp}}, \quad z_\beta = d_\beta + i\frac{2(\omega - \omega'_\beta)}{\omega_{cp}}, \quad z_{\alpha\beta} = i\gamma_{\alpha\beta}$$

is the intertype communication coefficient introduced in Refs. [57, 58], d_α and d_β are the attenuation factors (inverse to q-factors) for α- and β-oscillations and ω_α and ω_β are the eigenfrequencies perturbed by the reactive property of the connecting hole.

Let us introduce the notation $z = (\omega - \omega'_\alpha)/\omega_{cp}$ and $w = (\omega'_\beta - \omega'_\alpha)/\omega_{cp}$. Then (1.42) will take the form

$$F(z, w) \equiv \gamma_{\alpha\beta}^2 - 4\left(z - i\frac{d_\alpha}{2}\right)\left(z - w - i\frac{d_\beta}{2}\right) = 0. \tag{1.43}$$

In equation (1.43) the normalized fundamental frequency z and the adjustment factor w are taken as the spectral and non-spectral parameters, respectively. It is clear that the mapping $F(z, w) : C^2 \to C$ is an analytical function of the variables z and w. We will determine isolated Morse critical points of this mapping by solving the system of equations

$$\begin{cases} \dfrac{\partial F}{\partial z} = -4\left(-i\dfrac{d_\beta}{2} + z - w\right) - 4\left(-i\dfrac{d_\alpha}{2} + z\right) = 0 \\[3mm] \dfrac{\partial F}{\partial w} = -4\left(-i\dfrac{d_\alpha}{2} + z - w\right) = 0 \end{cases} \tag{1.44}$$

It follows from (1.44) that there exists the unique isolated Morse critical point (z_0, w_0): $z_0 = i(d_\alpha/2)$, $w_0 = i(d_\alpha - d_\beta)/2$, and (1.43) may be rewritten in the form

$$F(z, w) \equiv \gamma_{\alpha\beta}^2 - 4(z - z_0)^2 + 4(w - w_0)(z - z_0) = 0. \tag{1.45}$$

It follows from this expression that $F(z_0, w_0) = \gamma_{\alpha\beta}^2$ and the value of the characteristic determinant calculated at the Morse critical point coincides with the squared value of the intertype communication coefficient. After the change of variables

$$\begin{aligned} \tilde{z} &= i[2(z - z_0) - w + w_0], \\ \tilde{w} &= w - w_0, \end{aligned} \tag{1.46}$$

we get

$$\tilde{z}^2 + \tilde{w}^2 + \gamma_{\alpha\beta}^2 = 0.$$

Taking into account (1.46) one can show that here $\alpha = 0$ and $\beta = i(d_\beta - d_\alpha)/2$. Hence, $\operatorname{Re} \tilde{z}(\xi)$ and $\operatorname{Im} \tilde{z}(\xi)$ correspond to the first dispersion curve in Fig.10a.

It is convenient to make a precise analysis of the properties of the Morse critical points considering closed cylindrical resonators with small connecting holes. For this purpose, one has to carry out a more detailed study of an operator-valued function and its resolvent. Let us rewrite equation (1.42) in the form

$$A(z, w) = \left\| \begin{matrix} 2i(z - z_0) & i\gamma \\ i\gamma & 2i(z - w + w_0 - z_0) \end{matrix} \right\| : C^2 \to C,$$

where $A(z, w)$ is a matrix-valued operator depending on the parameters z, w, z_0, w_0 and γ and (z_0, w_0) is the Morse point of its determinant:

$$\det A(z, w) = -4(z - z_0)(z - w + w_0 - z_0) + \gamma^2, \tag{1.47}$$

where

$$z_0 = i\frac{d_\alpha}{2}, \quad w_0 = i\frac{d_\alpha - d_\beta}{2}.$$

The non-spectral parameter w characterizes the given family of volume cylindrical resonators with varying length and we will call it the control parameter.

The example considered is in some sense trivial because the operator-valued function of the resonator family has (in a canonical basis) the form of a square 2×2 matrix. In addition to this, the characteristic determinant has only one critical point, namely, the Morse point. However, this problem has sufficiently general features that are characteristic for more complicated electrodynamical structures.

Let us represent the operator-valued function in the form

$$A(z, w) = 2izI + T(w),$$

where

$$T(w) = \left\| \begin{matrix} -2iz_0 & i\gamma \\ i\gamma & 2i(w_0 - w - z_0) \end{matrix} \right\|, \quad I = \left\| \begin{matrix} 1 & 0 \\ 0 & 1 \end{matrix} \right\|. \tag{1.48}$$

The initial problem is equivalent to determination of eigenvalues and eigenvectors of the operator $T(w)$ which is holomorphic with respect to the parameter w. It is typical

for the perturbation theory in a finite-dimensional space to study the behaviour of
eigenvalues (fundamental frequencies) and eigenvectors of the family of operators $T(w)$
when w varies. One has to point out one important question. In fact, the domain of
"physical" values of the parameter w coincides with the set of real numbers. However,
we construct an analytical continuation of the operator-valued function $T(w)$ into the
complex domain. Such a continuation allows us to simplify the analysis of the problem
and to understand the qualitative nature of the behaviour of fundamental frequencies
and eigenvectors when w varies in the physical domain.

It follows from (1.48) that eigenvectors of the family $T(w)$ coincide with eigenvectors of (1.42), and the eigenvalues are given by the relation $z = i\lambda/2$. Eigenvalues λ
of the family $T(w)$ satisfy the equation

$$\det\{T(w) - \lambda\} = 0,$$

or

$$\lambda^2 + 2i(2z_0 + w - w_0)\lambda + 4z_0(w_0 - w - z_0) + \gamma^2 = 0. \tag{1.49}$$

Equation (1.49) implies that eigenvalue $\lambda(w)$ is a branch of the two-valued analytical
function

$$\lambda_\pm(w) = -i\{2z_0 + w - w_0 \pm \sqrt{(w - w_0)^2 + \gamma^2}\}. \tag{1.50}$$

The eigenvalues z of the operator-valued function $A(z, w)$ obey similar relations

$$z_\pm(w) = \frac{i}{2}\lambda_\pm(w) = z_0 + \frac{1}{2}\{w - w_0 \pm \sqrt{(w - w_0)^2 + \gamma^2}\}. \tag{1.51}$$

Let us point out that this property, which holds for a model example, is valid
in the general case too. The characteristic determinant is considered in the vicinity
of the Morse point, i.e., this property holds locally. In other words, fundamental
frequencies as functions of the control parameter are the branches of a two-valued
analytical function in the vicinity of the Morse point of the characteristic determinant

It follows from (1.50) that $\lambda(w)$ has branch-type singularities $w_\pm = w_0 \pm i\gamma$. For
sufficiently small values of $|\gamma| \ll 1$ these branch-points lie in the vicinity of the Morse
point w_0 and in the particular case $\gamma = 0$ they coincide with w_0. For all values of
$w \neq w_\pm$ the family $T(w)$ has two different eigenvalues which coincide for $w = w_\pm$.
Such a qualitative behaviour of eigenvalues is also characteristic for the general case.

It is interesting to point out that $w = w_\pm$ are not singularities of the operator-
valued function $T(w)$. Analysis of (1.50) allows us to show that the behaviour of $\lambda_\pm(w)$
has the form of a typical Wien graphs. Namely, when w varies in a physical domain,
one of the branch points may find itself in a physical domain of the parameter w for
certain values of the parameters γ, d_α and d_β. The situations are possible when the
boundary of the domain of variation of w goes around one or two branch points or
this domain contains none of them. The possible positions of the branch points for
the model problem with respect to the Morse point w_0 and the physical domain of
variation of w are presented in Fig.11.

From the physical viewpoint, c) and f) are the most interesting cases. Here the
line of variation of w goes around one of the branch points only once and hence one has
the transference from one branch to another (if to assume that the analytical property
holds). in addition to this, the eigenvalues (the frequencies) behave as a single object,

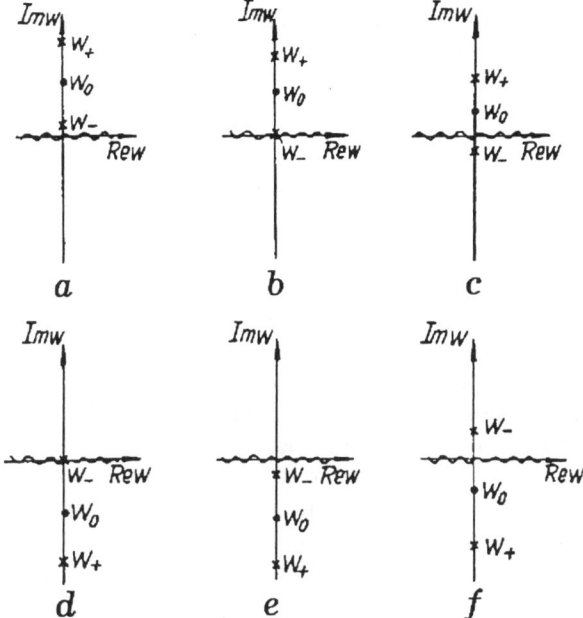

Figure 11.

namely, as a two-valued analytical function. If to introduce the Riemann surface for the parameter w, then $\lambda(w)$ becomes a usual analytical function. The corresponding eigenvalues may continuously transform one into another when w varies and hence they represent the same object.

Let us consider the behaviour of eigenprojectors $P(w)$ corresponding to the eigenvalues $\lambda(w)$ of the operator. The study of this problem allows one to understand how fundamental frequencies change when the parameter w (the resonator "disjustment") varies. We introduce the operator integrals

$$P(w) = -\frac{1}{2\pi i} \int_\gamma R(\lambda, w) d\lambda, \tag{1.52}$$

where

$$R(\lambda, w) = \{T(w) - \lambda\}^{-1}.$$

One can show that

$$R(\lambda, w) = \{[\lambda - \lambda_+(w)][\lambda - \lambda_-(w)]\}^{-1}[-\lambda I + B(w)]. \tag{1.53}$$

Here

$$I = \begin{Vmatrix} 1 & 0 \\ 0 & 1 \end{Vmatrix}, \quad B(w) = \begin{Vmatrix} 2i(w_0 - w - z_0) & -i\gamma \\ -i\gamma & -2iz_0 \end{Vmatrix},$$

and $\lambda_\pm(w)$ are determined by (1.50). Let γ be the circle containing only $\lambda_+(w)$ and $w \neq w_\pm$ where $w_\pm = w_0 \pm i\gamma$. One can see that at $w = w_\pm$ the characteristic equation has the form

$$\lambda - \lambda_\pm^0 = 0, \quad \lambda_\pm^0 = -i(2z_0 \pm i\gamma).$$

Hence for these values of w there exists a double-degenerated eigenvalue. Substituting (1.53) in (1.52) and calculating the integrals we obtain

$$P_+(w) = \frac{1}{+\sqrt{(w-w_0)^2+\gamma^2}}\left\{\frac{1}{2}\left[2z_0+w-w_0+\sqrt{(w-w_0)^2+\gamma^2}\right]I - 2iB(w)\right\},$$

$$P_-(w) = \frac{1}{-\sqrt{(w-w_0)^2+\gamma^2}}\left\{\frac{1}{2}\left[2z_0+w-w_0-\sqrt{(w-w_0)^2+\gamma^2}\right]I - 2iB(w)\right\}.$$

$$(1.54)$$

It follows from (1.54) that eigenprojectors $P_\pm(w)$ are the branches of one analytical operator-valued function with the same branch points as the eigenvalues $\lambda(w)$ have if to consider them as functions of the parameter w. Thus, when the parameter varies in a physical domain, there exists a continuous transformation of the eigenprojector (and, hence, of the eigenoscillation) corresponding, for example, to the eigenvalue $\lambda_+(w)$ to the eigenprojector (eigenoscillation) corresponding to $\lambda_-(w)$.

We may conclude that these eigenoscillations demonstrate the behaviour of the same object. However, there is a considerable difference in the behaviour of eigenoscillations and eigenvalues in the vicinity of branch points. The analysis of equation (1.54) enables us to assert that eigenprojectors always have a pole and $\lambda(w)$ is continuous in a branch point. This influences eigenoscillations and especially when the branch points are sufficiently close to the physical domain of variation of the parameter w. Here one observes sharp changes in the structure of eigenoscillations subject to relatively small variations of w in the vicinity of the Morse critical point. S Situations are possible when there is no mutual transformation of eigenoscillations (for example, like the pictures shown in Figs.11,a,e). However, one of the branch points is sufficiently close to the physical domain of w, and such a position of the branch points yields a considerable modification of the structure of a certain oscillation and does not influence another oscillation. It means that one of the oscillations dominates in the vicinity of the Morse point. Formulas (1.54) hold in the vicinity of branch points, but they are not valid for degenerated values of $w = w_\pm$. In this case one can also calculate the projectors

$$\{T(w_-) - \lambda\}^{-1} = -\left\{\frac{I}{\lambda - \lambda_0^-} + \frac{\gamma\left\|\begin{smallmatrix} 1 & i \\ i & -1 \end{smallmatrix}\right\|}{(\lambda - \lambda_0^-)^2}\right\},$$

$$\{T(w_+) - \lambda\}^{-1} = -\left\{\frac{I}{\lambda - \lambda_0^+} + \frac{\gamma\left\|\begin{smallmatrix} -1 & i \\ i & 1 \end{smallmatrix}\right\|}{(\lambda - \lambda_0^+)^2}\right\},$$

$$(1.55)$$

where $\lambda_0^\pm = -i(2z_0\pm i\gamma)$ and these projectors coincide with the unit operator $I = \left\|\begin{smallmatrix} 1 & 0 \\ 0 & 1 \end{smallmatrix}\right\|$. Such a behaviour of the resolvent $\{T(w - \lambda)\}^{-1}$ in the vicinity of degenerated eigenvalues leads to appearance of associated oscillations. One gets a qualitatively different situation as compared with the one considered above, and it requires additional considerations.

Hence the model example allows us to outline basic qualitative features of the eigenvalues and eigenoscillations in the vicinity of a Morse critical point. Namely, the latter are considered together with their behaviour with respect to the control parameter. The fact that both fundamental frequencies and eigenoscillations display the same properties is the most essential feature and this very phenomenon may be called the intertype oscillations.

Intertype oscillations have been experimentally observed for the first time in open resonators with spherical mirrors. For the confocal geometry the appearance of intertype oscillations were explained in Refs. [30, 31] on the basis of the oscillation interaction [57, 58] described by the asymptotical theory [14]. The reason for this interaction was explained in Refs. [30, 31] as a result of imperfect quality of the resonator mirrors and small inhomogeneities that are always present in an experimental plant. However, the results of experimental works [30, 31, 14] contradict this conclusion.

Here we present a theoretical model of these phenomena which is based on a rigorous mathematical approach including both the mathematical theory of the spherical resonator and qualitatively new ideas about the phenomenon of intertype oscillations. Application of this approach allows us to clear up the mechanism of the fall of the resonator q-factor in the "vicinity of a confocal geometry" (where the asymptotical theory prescribes the maximal q-factor) as a result of the interaction of one well-known type of oscillations with another oscillation, which is not described by the asymptotical theory of oscillations [27].

The open spherical resonator formed by two coaxial perfectly conducting spherical segments is presented in Fig.5 as the projection on the plane of symmetry. It is characterized by the (relative) curvature radii ka, opening angles α and the distance kl between the centers of the mirror curvatures. Let us consider the problem of the excitation of this resonator by the vertical electric dipole placed on the axis of symmetry. An exact statement of the diffraction problem and its regularization are analogous to the approach developed in Ref. [7]. We look for the electric Debay potential U determining the scattered field which satisfies the Helmholtz equation, the Sommerfeld radiation condition in infinity, the finiteness of the field energy in every bounded spatial domain and the natural boundary conditions: the tangential components of the electric field vanish on the mirror surface. Here the magnetic potential $V \equiv 0$. The regularization procedure leads to the infinite system of the second kind linear algebraic equations

$$(I - H)x = b; \quad x, b \in l_2^2 \tag{1.56}$$

with respect to the unknown vector-column x, where $l_2^2 = l_2 \otimes l_2$, l_2 is the space of the sequences $\{\xi_n\}$ of complex numbers such that $\sum_n |\xi_n|^2 < \infty$ and H is the kernel operator acting in l_2^2.

The spectral theory of two-dimensional diffraction problems constructed in section 1 can be applied to the three-dimensional problem considered here. The natural domain of the analytical continuation for the Green function in three-dimensional problems is the whole complex plane C^1. The statement of the diffraction problem (symmetrical with respect to the coordinate axes) remains the same with the only one exception: the Sommerfeld radiation condition must be replaced by the condition (equivalent to that for $k > 0$) that the potential U admits expansion in absolutely convergent series in the vicinity of an infinitely far point:

$$U(r, \theta) = \sum_{n=0}^{\infty} C_n(kr)^{-1/2} H_{n+1/2}^{(1)}(kr) P_n(\cos \theta), \quad k \in C^1, \tag{1.57}$$

where r, θ are the coordinates of the point (r, θ, ϕ) in an arbitrary fixed spherical coordinate system with the origin situated on the mirror symmetry axis, $H_{n+1/2}^{(1)}$ is a

Hankel function, $P_n(t)$ are Legendre polynomials, C_n are some coefficients and a time dependency, which is omitted in (1.57), is assumed to be $e^{-i\omega t}$. One can show that this boundary value problem formulated for $k \in C^1$ defines the analytical continuation of the Green function of the initial boundary value problem with the condition (1.57) for $ka \in C^1$. The values of ka, for which there exist non-trivial solutions, are considered as the spectrum of quasi-eigenvalues. One can show that this spectrum lies in the lower half-plane, forms a discrete set of points with finite multiplicity and coincides with the poles of an analytical continuation of the Green function. Note that it follows from (1.57) that the defined quasi-eigenfunctions exponentially increase when $r \to \infty$.

Initial boundary value problem depends not only on ka but on other (non-spectral) parameters as well as on kl. Similar to section 1, we will consider the operator H in (1.56) as an operator-valued function $H = H(z, w)$, where $z = ka$, $w = a/l$, $z \in C^1$, $w \in D \subset C^1$ and D is a domain in C^1. One can show that $H(z, w)$ is a C^2-holomorphic and kernel operator-valued function for $(z, w) \in C^1 \times D \subset C^2$. The spectrum of the boundary value problem coincides for every fixed value of w with the set of zeros of the function

$$F(z, w) = \det\{I - H(z, w)\}, \quad (z, w) \in C^1 \times D; \tag{1.58}$$

To calculate the roots of $F(z, w)$ with a prescribed accuracy it is sufficient to find zeros of the function $F_N(z, w)$. The definition of $F_N(z, w)$ is similar to that given by (1.58), where the operator-valued function $H(z, w)$ should be replaced by a sufficiently close to it (in a local manner, with respect to z and w) finite-dimensional operator-valued function $H_N(z, w)$

The mathematically correct definition of the interaction of eigenoscillations of an open resonator is presented in Refs. [10, 63] and it means that $F(z, w)$ has an isolated Morse critical point (z_0, w_0). The set of roots of $F(z, w) = 0$ from a small C^2-vicinity of (z_0, w_0) is described with the accuracy up to the higher-order small terms as the set of the roots of the corresponding non-degenerated quadratic form

$$\tilde{F}(z, w) = \frac{\partial^2 F}{\partial z^2}(z - z_0)^2 + 2\frac{\partial^2 F}{\partial z \partial w}(z - z_0)(w - w_0) + \frac{\partial^2 F}{\partial w^2}(w - w_0)^2 + 2F(z_0, w_0), \tag{1.59}$$

where the partial derivatives are calculated at the point (z_0, w_0). In particular, if $|\text{Im}\, w_0| \ll 1$, we get the spectral curves $z_k(w)$ ($k = 1, 2$) for $\text{Im}\, w = 0$ and $\text{Re}\, w$ varying in the vicinity of $\text{Re}\, w_0$. These curves are well approximated by the solutions of the equation $\tilde{F}(z, w) = 0$ for $\text{Im}\, w = 0$.

The numerical search of the Morse critical points gives us one of them, $(z_0, w_0) \approx (13.0908 - i0.1799; 1.19132 + i0.028093)$ for the aperture $\alpha = 130°$. Here the approximate values of the coefficients in (1.59) are

$$F(z_0, w_0) = (-1.113665 - i1.05208) \cdot 10^{-3};$$

$$\left.\frac{\partial^2 F}{\partial z^2}\right|_{\substack{z_0 \\ w_0}} = (4.304253 - i3.620324) \cdot 10^{-1};$$

$$\left.\frac{\partial^2 F}{\partial z \partial w}\right|_{\substack{z_0 \\ w_0}} = 2.079782 + i1.618900;$$

$$\left.\frac{\partial^2 F}{\partial w^2}\right|_{\substack{z_0 \\ w_0}} = 8.279679 + i7.666498.$$

The value of z for this Morse point is presented in Fig.12 as a function of w obtained for $\operatorname{Im} w = 0$ from the equations $\tilde{F}(z, w) = 0$ (bar lines) and $F(z, w) = 0$ (bold lines). One can see that the structure of spectral curves is determined locally and with a high accuracy by the presence of the Morse point (z_0, w_0). Its projection (for $\operatorname{Im} w = 0$) is shown in Fig.16 by stars.

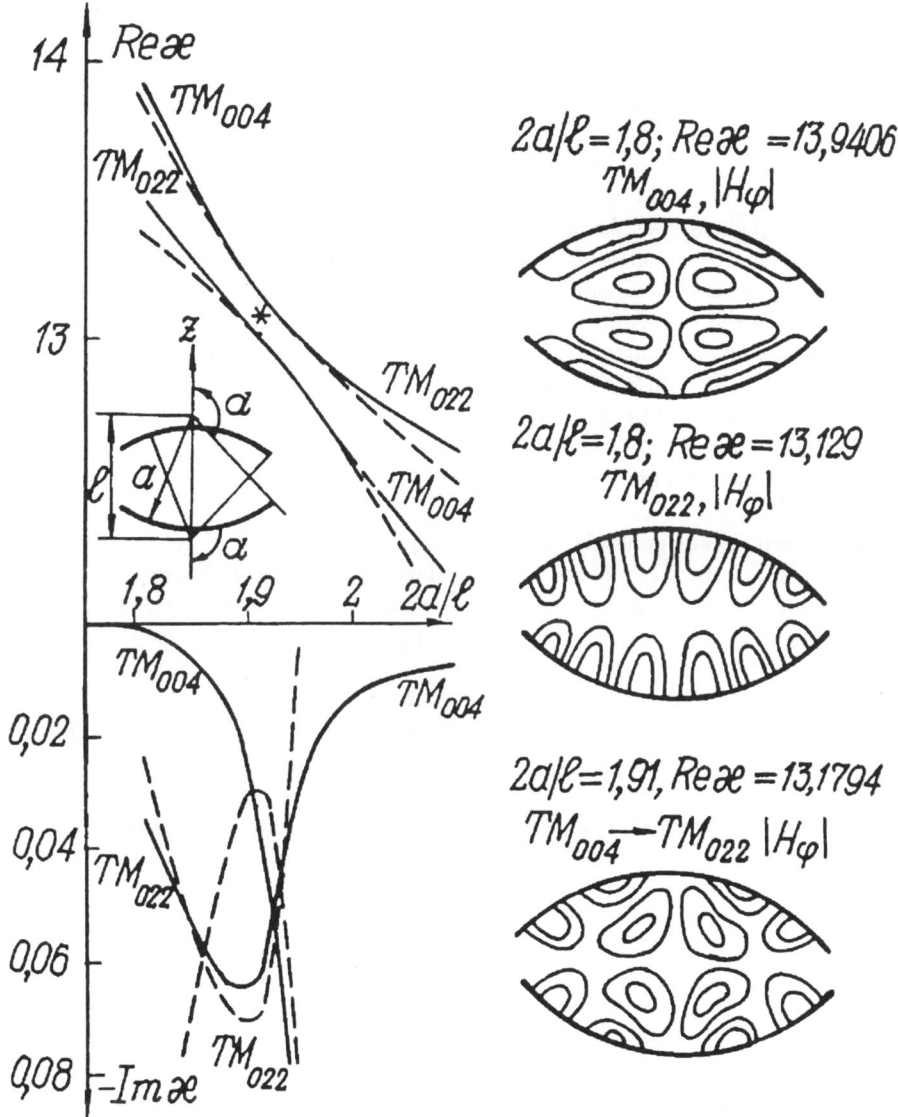

Figure 12.

To identify the resonant fields (quasi-eigenfunctions) and the considered spectral curves, the field distributions have been constructed (see Fig.12) that are excited in the resonator by the vertical electric dipole (for $\operatorname{Im} z = \operatorname{Im} w = 0$). It follows from their analysis that the interaction of TM_{004} and TM_{022} oscillations takes place in the vicinity of the Morse point. This interaction is accompanied by a fall of the q-factor for the TM_{004} oscillation and by the "exchange" of types between the spectral curves when l monotonically varies. We assign, as in Fig.12, the type TM_{0nq} to an eigenoscillation when the magnetic field H_ϕ has q maxima along the axis of the spherical open resonator and n knots (except for the points of the axis) in the direction perpendicular to the axis. There is the hybrid $TM_{004} + TM_{022}$ oscillation in a small vicinity of the point $w = \mathrm{Re}\, w_0$ which has the features of both TM_{004} and TM_{022} types. The isolines $|H\phi|$ of this oscillation are presented in Fig.12 in relative units for two "pure" and one hybrid types. Note that TM_{022} and $TM_{004} + TM_{022}$ oscillations are not described by the asymptotical theory [14].

Thus, the phenomenon of intertype oscillations is observed together with the well-known resonance effects in open resonators with spherical mirrors. This phenomenon results in sharp changes of the spatial field structure and their q-factors as well as in the behaviour of dispersion curves when resonator's parameters have small variations. Its correct description is possible only within the frames of a rigorous model. It is established that the fall of the q-factor observed in experiments for confocal spherical resonators (and which contradicts the results of asymptotical theory [14]) holds not because of the defects in resonator's construction, but, on the contrary, this effect displays the nature of spherical resonators as well as circular cylindrical two-mirror open resonator (described in sections 1–3).

Let us consider the Morse critical points of open resonators with dielectric inclusions.

It has been already pointed out that the fundamental frequencies of the considered open metallic-dielectric resonators are determined from the dispersion relation $F(\kappa) = 0$ where $F(\kappa) = \det\{I - B(\kappa)\}$. Since the characteristic determinant $F(\kappa)$ depends not only on κ but, for example, on $\varepsilon_n = \varepsilon_n' + i\varepsilon_n''$ $(n = 1, 2)$, we will use the notation $F(\kappa, \varepsilon_n)$ instead of $F(\kappa)$ where $\varepsilon_n \in D = \{\varepsilon_n = \varepsilon_n' + i\varepsilon_n'' : \varepsilon_n' \geq 1,\ \varepsilon_n'' \geq 0\}$.

One can show that $F(\kappa, \varepsilon_n)$ is an analytical function of two complex variables $(\kappa, \varepsilon_n) \in \Lambda \times D$. We can limit ourselves to the case when $\kappa \in \Lambda_0$ and introduce the set $\tilde{D} = \Lambda_0 \times D$, $\tilde{D} \subset C^2$ and the analytical set $\sigma_0 = \{(\kappa, \varepsilon_n) \in \tilde{D} : F(\kappa, \varepsilon_n) = 0\}$, $\sigma_0 \subset \tilde{D}$.

Let us go back to expression (1.16), $F(\kappa, \varepsilon_n) = \exp\{\sum_{j=1}^{+\infty} \ln[1 - \lambda_j(\kappa, \varepsilon_n)]\}$, where $\lambda_j(\kappa, \varepsilon_n)$ are eigenvalues of the operator $B(k)$ (for H-oscillations).

One can show that

$$\frac{\partial F}{\partial \kappa} = -F(\kappa, \varepsilon_n)\, \mathrm{sp}\left\{\frac{\partial B(\kappa, \varepsilon_n)}{\partial \kappa} R_B(\kappa, \varepsilon_n)\right\},$$

$$\frac{\partial F}{\partial \varepsilon_n} = -F(\kappa, \varepsilon_n)\, \mathrm{sp}\left\{\frac{\partial B(\kappa, \varepsilon_n)}{\partial \varepsilon_n} R_B(\kappa, \varepsilon_n)\right\},$$

where $\mathrm{sp}\{\dots\}$ is the operator trace and $R_B(\kappa, \varepsilon_n) = \{I - B(\kappa, \varepsilon_n)\}^{-1}$. Here $B(\kappa, \varepsilon_n) = B(\kappa)$. We will introduce the sets

$$\sigma_K = \left\{(\kappa, \varepsilon_n) \in \tilde{D} : \mathrm{grad}\, F = \left(\frac{\partial F}{\partial \kappa}, \frac{\partial F}{\partial \varepsilon_n}\right) = 0\right\}$$

of the critical points and

$$\sigma_{MK} = \left\{(\kappa, \varepsilon_n) \in \sigma_K : \frac{\partial^2 F}{\partial \kappa^2}\frac{\partial^2 F}{\partial \varepsilon_n^2}\left(\frac{\partial^2 F}{\partial \kappa \partial \varepsilon_n}\right)^2 \neq 0\right\}.$$

of the Morse critical points and assume that $(\kappa_0, \varepsilon_n^0)$ is an isolated Morse critical point which is close to σ_0. We will assume that $(\kappa_0, \varepsilon_n^0) \notin \sigma_0$, although the situation is possible when $(\kappa_0, \varepsilon_n^0) \in \sigma_0$. In other words, it follows from the expressions for $\frac{\partial F}{\partial \kappa}$ and $\frac{\partial F}{\partial \varepsilon_n}$ that two situations are possible for $(\kappa_0, \varepsilon_n^0)$:

1.

$$
\begin{cases}
\mathrm{sp}\left\{ \dfrac{\partial B(\kappa_0, \varepsilon_n^0)}{\partial \kappa} R_B(\kappa_0, \varepsilon_n^0) \right\} = 0, \\[2mm]
\mathrm{sp}\left\{ \dfrac{\partial B(\kappa_0, \varepsilon_n^0)}{\partial \varepsilon_n} R_B(\kappa_0, \varepsilon_n^0) \right\} = 0, \\[2mm]
F(\kappa_0, \varepsilon_n^0) \neq 0,
\end{cases}
$$

and

2.

$$
\begin{cases}
\mathrm{sp}\left\{ \dfrac{\partial B(\kappa_0, \varepsilon_n^0)}{\partial \kappa} R_B(\kappa_0, \varepsilon_n^0) \right\} = 0, \\[2mm]
\mathrm{sp}\left\{ \dfrac{\partial B(\kappa_0, \varepsilon_n^0)}{\partial \varepsilon_n} R_B(\kappa_0, \varepsilon_n^0) \right\} = 0, \\[2mm]
F(\kappa_0, \varepsilon_n^0) = 0.
\end{cases}
$$

Taking two first terms of the Taylor expansion for the left-hand side of the equation $F(\kappa, \varepsilon_n) = 0$ in the vicinity of $(\kappa_0, \varepsilon_n^0)$ we get a quadratic equation for κ and ε_n which is similar to (1.40).

The Morse critical point obtained above has been calculated with respect to varying permittivity ε_n $(n = 1, 2)$. For $\varepsilon_1 = 1$ the Morse point is determined by the parameters

$$\mathrm{Re}\,\kappa = 10.0238, \quad \mathrm{Im}\,\kappa = -0.0126, \quad \varepsilon_2' = 1.3877, \quad \varepsilon_2'' = 0.0493.$$

For $\varepsilon_2 = 1.4 + 0.001i$

$$\mathrm{Re}\,\kappa = 10.0207, \quad \mathrm{Im}\,\kappa = -0.0118, \quad \varepsilon_1' = 0.9955, \quad \varepsilon_1'' = 0.0753.$$

Here $\delta = F(\kappa_0, \varepsilon_n^0)$. In the case of non-symmetrical H_{02}- and H_{21}-oscillations the Morse point has the parameters

$$\varepsilon_1 = 1, \quad \mathrm{Re}\,\kappa = 6.7013, \quad \mathrm{Im}\,\kappa = -0.001, \quad \varepsilon_2' = 1.15989, \quad \varepsilon_2'' = 0.0002.$$

Let us point out that the imaginary part of the permittivity ε_2 corresponding to the Morse point changes its sign (becomes negative) and it is important in this case that $\delta = 0$, i.e., in the vicinity of these parameters H_{02} and H_{21} oscillations degenerate with respect to both κ and ε_n.

In the case of the non-degenerated Morse critical point the spectral curves $\mathrm{Re}\,\kappa_{1,2}$ form the Wien graphs, and the curves $\mathrm{Im}\,\kappa_{1,2}$ intersect.

If the Morse critical point is degenerated, $\mathrm{Re}\,\kappa_{1,2}$ intersect and their intersection point coincides with the real part of the fundamental frequency corresponding to the Morse point. Note that the curves $\mathrm{Im}\,\kappa_{1,2}$ intersect to the left of the intersection point of the real parts of the fundamental frequencies.

In addition to the above conditions that provide the vanishing of $\operatorname{grad} F(\kappa, \varepsilon_n)$, two more conditions are possible, for example:

$$\begin{cases} \operatorname{sp}\left\{\dfrac{\partial B(\kappa_0, \varepsilon_n^0)}{\partial \kappa} R_B(\kappa_0, \varepsilon_n^0)\right\} \neq 0, \\[2mm] \operatorname{sp}\left\{\dfrac{\partial B(\kappa_0, \varepsilon_n^0)}{\partial \varepsilon_n} R_B(\kappa_0, \varepsilon_n^0)\right\} = 0, \\[2mm] F(\kappa_0, \varepsilon_n^0) = 0 \end{cases}$$

or

$$\begin{cases} \operatorname{sp}\left\{\dfrac{\partial B(\kappa_0, \varepsilon_n^0)}{\partial \kappa} R_B(\kappa_0, \varepsilon_n^0)\right\} = 0, \\[2mm] \operatorname{sp}\left\{\dfrac{\partial B(\kappa_0, \varepsilon_n^0)}{\partial \varepsilon_n} R_B(\kappa_0, \varepsilon_n^0)\right\} \neq 0, \\[2mm] F(\kappa_0, \varepsilon_n^0) = 0. \end{cases}$$

These equations determine the degeneration points with respect to ε_n ($n = 1, 2$) and κ.

§ 5. Experimental study of the amplitude, phase and polarization characteristics of open resonators

The spectral theory of cylindrical structures and the methods of solution to their excitation problems by cylindrical and plane waves allow us to determine within the frame of the developed mathematical model all the characteristics of open resonators. In real conditions open structures have complicated geometrical configurations of mirrors, the finite conductivity and thickness of the screen metal, connecting holes, etc. Therefore, one can establish the properties of open resonators in the millimeter and submillimeter wavelength ranges, if theoretical results are sufficiently complemented with experimental data. The method of visualization of electromagnetic fields [6, 8, 70] which is worked out by the author and his co-workers, enables one to study the amplitude field distribution in resonators together with their phase and polarization characteristics. A special attention will be paid in this section to the holographic methods developed for the studies of the characteristics of high q-factor open resonators.

With the purpose to increase sensibility in the electromagnetic field measurements, the high q-factor open resonators containing a small scatterer [70] are applied. Scattering characteristics of some objects are measured and the near fields of different radiating apertures are studied. The registered interference pictures are considered as radioholograms with the resonance supporting beam.

Here we will also describe the experimental method of reconstructing the phase front of the wave beam in the given cross-section of a quasi-optical two-mirror open resonator.

The final result of the experiment is the choice of optimal open resonators such that their geometrical and material properties allow one to support oscillations with the required frequency, q-factor and such a field distribution in the resonator volume and on its mirrors that may be excited in the simplest way by concentrated or distributed sources. The frequency, q-factor, and power are usually measured for closed resonators. The field distribution in open resonators of the millimeter and submillimeter wavelength ranges is one of their most important characteristics. Creation of the field with a given configuration is of crucial significance for various applications in the radio engineering and electronics. Here we will consider the method of the field visualization in the volume of an open resonator.

The field visualization method is based on the scattering of resonator's eigenfield by the sufficiently small probe body. One can find out how the resonator inhomogeneity influences the energy of oscillations coming through the connecting slot from the waveguide, when an eigenoscillation with a constant amplitude is excited in the resonator (without inhomogeneity). A small body does not change the oscillation frequency and influences only the field amplitude in the resonator, while the field amplitude inside the inhomogeneity remains constant. An inhomogeneity may have a form of a small perfectly conducting ball with the radius a, for which $\rho \gg a > \lambda$ (we assume that for the dominant oscillation of the open resonator $\rho \approx w_0$, where w_0 is the minimal width of the spot of the resonator field). In the empty resonator $Q^{-1} = Q_0^{-1} + Q_{fr}^{-1}$ where Q, Q_0 and Q_{fr} are the q-factors of the empty resonator oscillations calculated with an account of diffractional and communication losses. If we place the scattering ball inside the resonator, then $Q_l^{-1} = Q_0^{-1} + Q_{fr}^{-1} + Q_{rad}^{-1}$, where Q_l is the q-factor of the loaded resonator that takes into account the radiation losses proportional to Q_{rad}. In the case considered the total energy of the field scattered by the ball is small if to compare with the oscillation energy. For $ka \ll 1$ one can determine the total energy by the Rayleigh formula. The open resonator exit power is given by the formula

$$w_{rad} \approx Q_{fr}^{-1} |AE_0(r_0, \phi_0, z_0)|^2 \frac{4}{3} 10\pi a^2 (ka)^4, \tag{1.60}$$

where $AE_0(r_0, \phi_0, z_0)$ is the amplitude of the electric oscillation of the open resonator in the observation point (r_0, ϕ_0, z_0). One can see from (1.60) that the exit signal is proportional to the field intensity at the point occupied by the sensor, and its value depends on the ball size and the field amplitude of the exciting oscillation.

To obtain a visible picture of the resonator field one has to move the sensor continuously in a certain plane inside the resonator. While moving along a given trajectory, the sensor passes in different moments of time the domains where the field intensity is different: greater or less the intensity, greater or less the sensor reradiates (or absorbs) the energy. Hence the variation of the resonator q-factor (or of its transmission coefficient) depends on the sensor position inside the resonator. The sensor displacements are accompanied by the movement of the beam on the screen of the electron-beam tube (EBT), and its trajectory identically repeats the sensor position. The beam brightness is controlled by the signal taken from the open resonator. Hence, successively, point by point, the electromagnetic field intensity is continuously transformed to the brightness of the picture of the field intensity distribution. The corresponding transformation coefficient considerably depends on the type of the resonator, its q-factor, dimensions and the material of the sensor. The size of the sensor has to satisfy contradictory requirements: it must be sufficiently small to feel the complicated field configurations and, at the same time, comparatively big to provide for a reasonable sensitivity. When visualizing the principal types of oscillations of an open resonator, the shape of the sensor does not play a significant role; but for the higher-order oscillations its shape sufficiently influences the results of measurements. The sensor has to move uniformly and straightforwardly in the domain of the field visualization. The potential-meters connect the sensor movement with the beam on the EBT screen. The EBT with after-luminescence is used to observe the whole picture. A special scanning mechanism [32, 67, 6] controls the sensor movements along the given trajectory and allows one to obtain an exact visible picture of the field distribution in the given plane inside the open resonator.

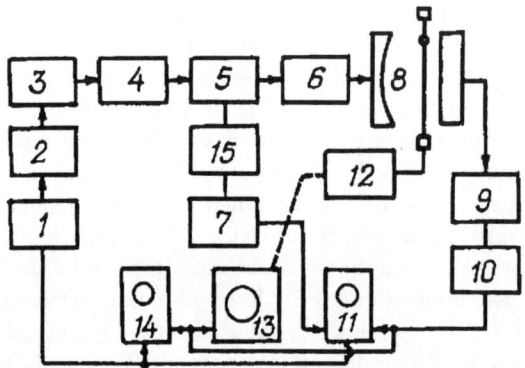

Figure 13.

An experimental plant for the electromagnetic fields visualization in open resonators is depicted in Fig.13. Its elements are fixed on a massive bed, which enables one to carry out precise mirror displacements. The initial impulse produced by generator 1 simultaneously arrives at the saw-type voltage generator 2 and at the schemes of the initiation of electron oscilloscopes 11 and 14. The saw-type voltage goes from generator 2 to reflector 3. It causes the synchronous initiation of the clystron and oscilloscopes. Generator 3 generates frequency-modulated radio-impulses with the frequency equal to that of the initiating impulses. From the generator energy goes through the attenuator to junction 5, which takes a part of it to the control scheme (this energy comes through wave-meter 15 to the amplitude detector connected with one of the entrances of the two-beam oscilloscope where one can observe the picture of an oscillation zone). The main part of the power is directed to resonator 8, then it is registered at the exit by detector 9 and comes through amplifier 10 to the second entrance of two-beam oscilloscope 11. If resonator 8 is adjusted with respect to the frequency in the clystron zone of generation, the characteristic splash appears on the oscilloscope screen repeating the form of the resonant curve. The signal from the resonator comes both to the two-beam oscilloscope and to the impulse oscilloscope with memory 14 as well as to the EBT luminescence control electrode 13, which is necessary to use the plant in different operation modes. The following units are also presented in the scheme in Fig.14: unbinding attenuator 4, detector 7 and scanning mechanism 12. The plant is applied for the measurements of the q-factor, the field amplitude, as well as to perform the visualization of the electromagnetic field distribution in an open resonator and in the free space. An analysis of the errors of the method and of the plant showed that, if the absolute error of the determination of the mirror position is equal to 0.002 mm, the distortion do not exceed 6% for the field configurations observed by sight, 12% for the amplitude and 18% for the q-factor.

Note that various modifications of the electromagnetic field visualization technique have been recently worked out [68] on the basis of the considered method and the constructed plant. All such measurements are automatized and the data is processed by computers, which allows one to study the resonator characteristic in the real time and with a high accuracy.

Consider the discrete reconstruction of the phase front in open two-mirror quasi-optical resonators with the Fresnel number $N \geq 2$. In a quasi-optical approximation the generalized Fresnel number is $N = a_1 a_2 / (\lambda l) \ll l^2 / (a_1 a_2)$, where a_1 and a_2 are the mirrors' radii, l is the maximal distance between them and λ is the working wavelength. In this case the diffractional loss of energy is negligible and the oscillations can be well approximated by Gaussian beams. Depending on the mirror geometry, the resonator eigenoscillations are described by the product of the Gauss function and the Laguerre polynomial for circular apertures [30] or by the Gauss function and the Hermit polynomial for rectangular apertures [69].

Before describing the method of reconstructing the surfaces of equal phases in the resonant Gaussian beams, let us consider how one should understand the phase measurement in the resonance field. Let us consider a two-mirror resonator formed by two rectangular spherical mirrors. Assume that the dominant oscillation is excited in the resonator which is characterized by a monotonous amplitude-phase functions of the magnetic field distribution in the planes perpendicular to the resonator axis. In order to perform the phase measurements, the supporting quasi-plane wave is brought into the resonance volume. The resonator field and the field of the supporting wave have a linear polarization and the electric field vectors are colinear. The superposition of the intensities of the resonance field and the field of the supporting wave yields the stationary interference picture in the working volume of the resonator

$$J(x) = A_s^2(x) + \left[\frac{B}{r(x)}\right]^2 + 2A_s(x)\frac{B}{r(x)} \cos\left[\frac{2\pi}{\lambda}x\cos\alpha + \phi_0 + \psi_s(x)\right], \qquad (1.61)$$

where ϕ_0 is the initial phase of a signal, $r(x)$ is the distance between the source and the observation point, B is the mean value of field amplitude on the source aperture, $A_s(x, y, z_0)$ is the amplitude distribution of the resonator resonance field in the cross-section of the excited s-oscillation and $\psi_s(x)$ is the desired phase distribution function in the same cross-section. This is the equation of the Gaussian beam resonance radio-hologram with the supporting quasi-plane wave having the amplitude $Br^{-1}(x)$. It is easy to see that all the parameters (except for the function $\psi_s(x)$) may be measured experimentally. One can get the analytical expression for the function $\psi_s(x)$ from (1.61) in the form

$$\psi_s(x) = \arccos \frac{J(x) - A_s^2(x) - \left[\frac{B}{r(x)}\right]^2}{2A_s(x)\frac{B}{r(x)}} - \frac{2\pi}{\lambda}x\cos\alpha - \phi_0 + 2\pi n, \qquad (1.62)$$

where $n = 0; \pm 1; \pm 2; \ldots$.

The determination of the intensity distributions $J(x)$ in the interference picture and of $A_s^2(x)$ in the excited oscillation is carried out by the known method of small perturbations [31]. When the values of $J(x)$ and $A_s^2(x)$ are reconstructed from the record of the signal giving the values of the resonance transmission coefficient, one should normalize the data by the factor equal to the maximal signal corresponding to the system that is not perturbed by the sensor. The value of Br^{-1} is uniquely determined by the depth of the resonance Gaussian beam modulated by the supporting electromagnetic field in the regions where the interfering waves have opposite phases (we call them the regions of destructive interference).

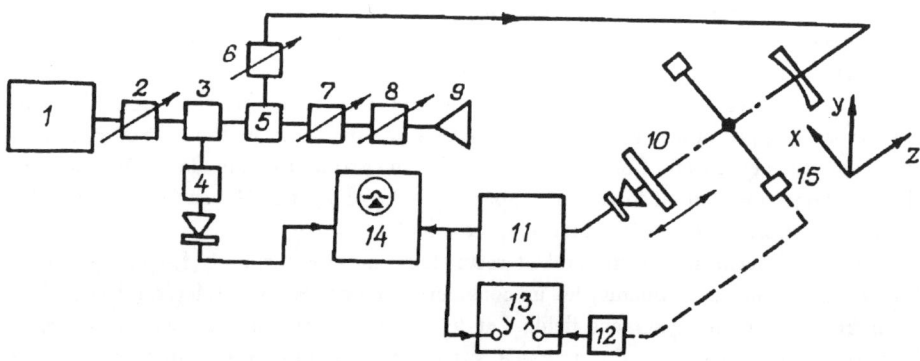

Figure 14.

On the basis of these methods the radio-holographic plant [8, 70] has been constructed. It consists of the two-channel interferometer whose block-scheme is presented in Fig.14. The signal form the thermostat clystron generator 1 (the operating frequency $f = 58$ GHz and the relative instability $\delta = 10^{-5}$) comes simultaneously through the ferrite valve 2 (with the inverse attenuation ≈ 20 dB and the direct attenuation ≈ 1.5 dB) and junction 5 (with the transfer attenuation 8 dB) to the main channel containing resonator 10 and to the supporting channel ending with antenna 9. Attenuators 6 and 7 are used to control the appropriate rate of the amplitudes in both channels. The open resonator is mounted on the measurement bed, along which one can smoothly shift the resonator mirrors in order to adjust them with the accuracy 5 mm. The waveguide slots with the dimensions 3.6×0.1 mm placed in the mirror centers connect the waveguide section and the resonator. In order to bring in the supporting electromagnetic wave at a given angle to the resonator axis, the supporting channel is fixed on the bar rotating around the origin on the resonator axis. The field of the eigenoscillation excited through the connecting slot interferes with the field of the supporting wave. The depths of interference is determined by the ratio of the amplitudes of the subject, $A_s(x)$, and supporting, $Br^{-1}(x)$, fields. The number of the interference strips depends on the angle α between the geometrical axes of the resonator and the antenna. The interferogram is recorded by the attenuating sensor stretching along the coordinate x; its movement is synchronized (by means of pick-up 12) with the position of a pen of the two-coordinate self-recorder 13 modulated with respect to the amplitude by the exit resonator signal. The graphite sphere (with the diameter $= 0.1\lambda$) is used as a sensor. The intensity distribution in the interference picture, which is reconstructed with the help of the method [31] by recording the resonance transmission coefficient, is presented in Fig.15. The intensity distribution $A_s^2(x)$ of the excited oscillation in the resonance beam is registered in a similar manner by attenuator 7 (≈ 50 dB) placed inside the supporting channel.

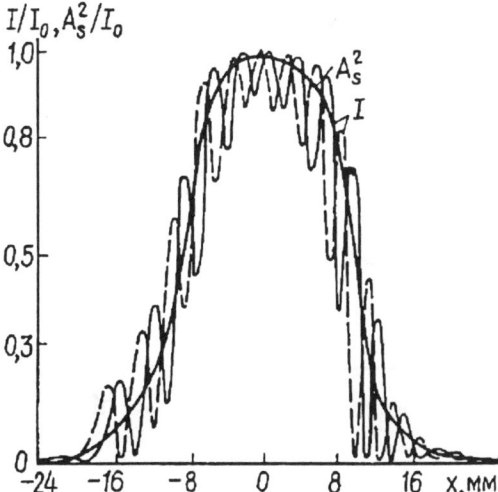

Figure 15.

When the attenuating sensor is used, the parts of the supporting wave energy which are spent for a nonresonant background excitation (due to the scattering by the sensor) and transformed into the energy of the resonator eigenoscillations are negligible. The rest of the energy is absorbed by the sensor material. Hence the registered interferogram (see Fig.18) has a considerable distortion in the regions where the interfering fields have equal phases (i.e., in the regions of constructive interference). Therefore, the reliable phase data will be registered in the following intervals

$$2\pi n + \frac{\pi}{2} < \psi_s(x) + kx\cos\alpha + \phi_0 < \frac{3\pi}{2} + 2\pi n. \tag{1.63}$$

However, as can be seen from (1.62), the direct determination of $\psi_s(x)$ is too difficult even with the help of (1.63), because the inverse to $\cos[\psi_s(x) + kx\cos\alpha + \phi_a]$ is not a single-valued function. Such an obstacle is removed in the active radio-holographic measurements of the amplitude-phase distribution by applying special sensors in the form of multiple receivers with slotted waveguide bridges [71]. In the case considered, in order to reconstruct the phase structure in an open resonator along the directions orthogonal to the electric field vectors of the interfering waves, we have used different artificial methods.

The plot of the relative phase distribution of the dominant eigenoscillation in the open semi-symmetrical spherical resonator is presented in Fig.16 (where $d = 16.7$, $b = 34.75$, $z_0 = 9.5\lambda$ and the mirror diameter is 13.5λ). The calculated curve of the resonance Gaussian beam phase distribution is shown by the bold line; its explicit representation has the form

$$\psi_s(z, z_0) = 2\sqrt{\frac{l}{b-l}}\left(1 - \frac{z_0}{l}\right)\frac{x^2}{\frac{b\lambda}{\pi} - \frac{l}{b-\lambda}\left[1 - \frac{l}{b}\left(1 - \frac{z_0^2}{l^2}\right)\right]}. \tag{1.64}$$

The crosses denote the phase values obtained experimentally; the phase is reconstructed from the interferogram presented in Fig.15. The measured and the calculated . values are in a good agreement.

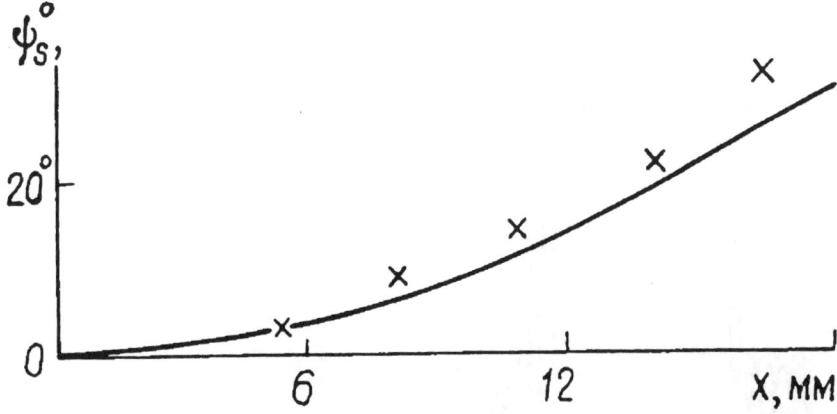

Figure 16.

We may conclude that, within the frames of the developed method of reconstructing the wave front of resonance Gaussian beams, [72], an attenuating sensor should be applied in the regions of destructive interference. In order to obtain a continuous phase distribution, one has either to vary the initial phase ϕ_0 of the supporting wave by phase-rotator 8 (see Fig.14) or to change the supporting wave angle α. The latter method may cause spatial displacements of the interference picture (denoted by the bar line in Fig.15).

The developed phase measurement method allows one to reconstruct the front of the resonance field in the given cross-section of an open resonator with negligible perturbations in the spatial structure of the excited oscillation. The trustworthiness of the results is confirmed by the computational data.

We have considered the principles of the measurement of the field phase structure in the resonance Gaussian beams along the trajectories coinciding with the direction of propagation of the supporting wave. To obtain the phase information in the whole plane, one has to modify the method with the purpose to arrange phase measurements along the trajectories colinear to the vectors of the interfering wave electric field [8, 73]. The resonator considered as well as the supporting quasi-plane wave are excited by the same high-stable millimeter-wave generator. Such a type of excitation provides the required coherent property of the interfering oscillations.

Assume that the dominant oscillation is excited through the connecting slot in one of the mirrors of the semi-symmetrical spherical resonator under study. The transmission coefficient is recorded by the probe attenuating sensor placed in the maximum of the resonator electric field [8, 31] by scanning along the horizontal axis. As a result of the excitation by a supporting inhomogeneous quasi-plane wave, the radio-hologram (1.61) is formed in the domain of the field spot, which looks like a sequence of interference strips oriented in the direction of the scanning coordinate. A visual analysis of the registered radio-hologram shows that in the case considered a possibility appears to record an interferogram by scanning in the strip of destructive interference. Assume that the hologram is placed in the origin and the geometric axis of the resonator and the source of the supporting wave are in the plane corresponding to the angle $\beta = 0$.

Then equation (1.60) may be rewritten in the form

$$\psi_s(y) = \arccos \frac{J(y) - A_s^2(y) - \left[\frac{B}{r(y)}\right]^2}{2A_s(y)\frac{B}{r(y)}} - \phi_0. \tag{1.65}$$

It is clear that the field phase structure in the resonator can be uniquely determined from (1.60), since the sensor moves along the line of equal phases of the supporting wave.

The intensity distributions of $J(y)$ and $A_s^2(y)$ are registered by the method of small perturbations. In order to find the unknown value B characterizing the average field amplitude on the antenna aperture 9, (Fig.14) one has to carry out an independent consideration. We will determine the amplitude of B when describing the interferogram registration with the help of the following procedure: The sensor is placed on the resonator axis in the maximum of the excited mode electric field. It leads to the shift of a resonance frequency providing the maximal value of the transmission coefficient at the resonator exit, which is used to normalize the intensities $J(y)$ and $A_s^2(y)$. First, the signal corresponding to the value of the transmission coefficient characterizing the amplitude field distribution of the $A_s^2(y)$-oscillation is recorded. Then the resonator volume is excited by the supporting quasi-plane electromagnetic wave in such a way that the main axis of the far-field pattern of antenna 9 (Fig.14) goes through the origin where the sensor is placed. By means of the calibrated attenuators 6, 7 and phase rotator 8 (Fig.14) such conditions are created that the exit transmission coefficient vanishes. This balance situation is in one-to-one correspondence with the field distribution in the origin, and it follows from (1.65) that when the transmission coefficient vanishes in the origin, one should set $B = r(0)A_s(0)$ in order to determine the required amplitude of the supporting wave.

The interferogram is recorded when the sensor moves in the strip of destructive interference, and the detected signal of the transmission coefficient taken from the open resonator comes to the selective amplifier, which separates a videosignal at the frequency 1 KHz. The signal is simultaneously sent from the amplifier exit to the EBT-screen and to the "u"-entrance of the two-coordinate recorder. The obtained dependence of the transmission coefficient are normalized with respect to the relative intensity distributions $J(y)$ and $A_s^2(y)$ using the known formula [31] $E_y^2/E_0^2 = C^2(1 - \sqrt{\xi})/\sqrt{\xi}$, where C is a normalizing factor; $\xi = K(y)/K_0$ is the relative value of the transmission coefficient at the point with coordinate y and K_0 is the value of the transmission coefficient recorded when the sensor is taken out from the resonator.

The intensity distributions in the interferogram $J(y)$ and in the resonance mode $A_s^2(y)$ obtained with the help of the last expression are shown in Fig.17a for a semi-symmetrical open resonator with the parameters $\lambda = 4.28$ mm, $\alpha = 25.5\lambda$, the mirror diameter is 16.35λ, the mirror curvature radius is 42λ and $z_0 = 6.5\lambda$. Curve 2 in Fig.20b describes the phase distribution of the resonance Gaussian beam reconstructed by means of relation (1.65). The bold line corresponds to the phase wave front calculated from the geometry relation (1.64). A sufficiently good coincidence of the experimental and theoretical data confirms the quality of the methods. We explain the divergence of the curves outside the field spot by the fact that in the expressions describing the theoretical behaviour of the resonance field phase, we have assumed a monotonous

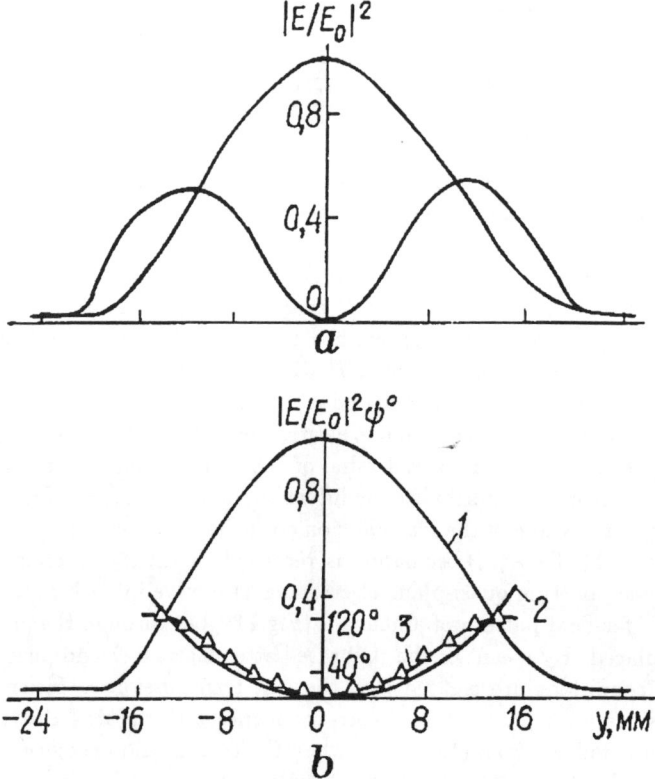

Figure 17.

increase of the phase raid up to the boundary of the cavity. It is difficult to determine the phase outside this boundary. In practice this threshold boundary is determined first of all by the sensibility of the registration system.

The methods described may be also applied when a resonator operates at higher-order modes with an account of the following remarks: for the fixed phase ϕ_0 of the supporting wave the regions of constructive and destructive interference may alternate when the the scanning by a sensor is performed, for example, in case of even symmetrical oscillations. Therefore, one has to provide the fulfillment of the counter-phase condition for interfering waves along the sensor trajectory in the domain of every field spot before to record the interferogram.

Note that the balance scheme for the registration of the phase field distribution in the cross-section of the resonance wave beam allows us to carry out quantitative estimates of the relative phase raid by means of a usual phase-rotator placed in one of the channels of the interferometer. We thus hasten the process of receiving the necessary information by means of balancing the system with respect to the minimal value of the transmission coefficient at every point of the sensor position. However, the accuracy of the phase measurement decreases because of inaccurate determination of the domains of the destructive interference and additional errors caused by properties of the phase-rotator.

Let us compare the techniques presented in this section. They differ by the methods of selecting the scanning domains (characterized by different types of interference) and of the phase front reconstruction from the registered interferogram. In both cases one manages to determine the character of the resonance beam surfaces of equal phases by comparing the phase with a certain constant value or with a certain known dependence. One can carry out this comparison with the help of the balance scheme, because the constant scanning by the sensor is performed in the domain where the supporting wave has a homogeneous phase. We use the phase periodicity of the propagating electromagnetic wave for the discrete reconstruction of the wave front curvature. In fact, the sensor finds itself in the domains of equal phases of the supporting quasi-plane wave, which are situated along its trajectory with a certain period, and the period is determined by the wavelength and by the angle between this trajectory and the direction perpendicular to the wave surface of the supporting wave.

It is clear that these methods provide the possibility to obtain the field phase structure in the whole analyzed plane. Note also that the present methods of the reconstruction of the phase fronts in the resonance Gaussian beams may be applied both for the transfer and reflection resonators. In the latter case, when there is only one communication element, additional coordination of the wave resistances of the scheme elements is necessary in order to provide the minimal value of the standing wave ratio (SWR).

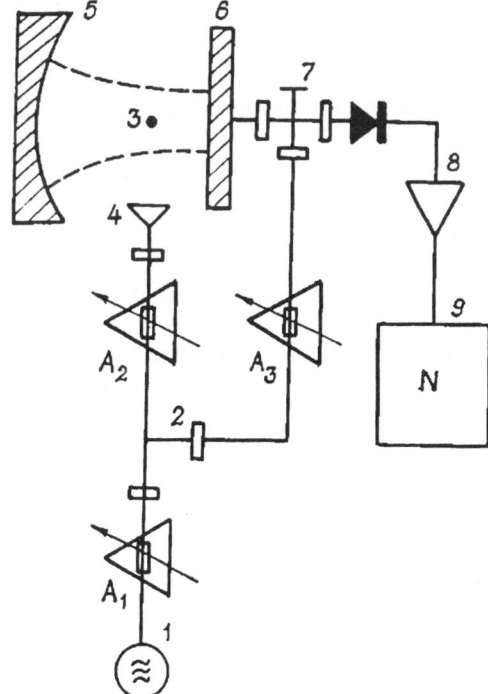

Figure 18.

One can use an artificial way of generating the supporting wave for the phase measurements. It is sufficient to take the scattering sensor in the form of a small conducting sphere in place of the attenuating sensor. The corresponding scheme of the experimental plant for an open resonator operating in the "excitation" mode is presented in Fig.18. It is known [14] that when an elementary electric dipole with the moment $\mathbf{P}_l$ is placed in the resonator volume, the oscillations with the amplitude $E_0 = -(q/N_s)\mathbf{P}_l\mathbf{E}_s$ is excited, where Q_s is the q-factor of the excited oscillation, N_s is the normalizing factor proportional to the energy stored in the field of excited oscillation and $\mathbf{E}_s$ is the function describing the field distribution inside the resonator. Consider now the excitation of the electric oscillation. When the radius r of the ball is sufficiently small, one may assume that $\mathbf{P}_l = \kappa_l\mathbf{E}$, where $\kappa_l = [(\varepsilon - 1)/(\varepsilon + 2)]r^3$ ($\mathbf{E}$ are the components of the electric field calculated at the point where the sensor is placed and ε is the permittivity of the ball. The quasi-plane wave scattered by the ball in the resonator excites the resonance oscillation with the amplitude

$$\mathbf{E}_l = -\frac{Q_s}{N_s}\kappa_l\mathbf{B}e^{-i\phi}A_se^{i\psi_s}. \tag{1.66}$$

Here the coordinates x, y, z determine the position of dipole 3, which is the communication element moving between the resonator and free space. The far field of antenna 4 serves as an exciting field. The energy goes out of the resonator volume through the waveguide communication slot in the center of plane mirror 6. In order to take into account the latter property we introduce the coefficient into (1.66) (denoting it by γ). To measure the phase distribution described by the function ψ_s, one has to bring a coherent signal directly into the waveguide, and this signal will play the role of the artificially created supporting wave [71]. Here the usual double waveguide T-junction 7 works like a mixer. If to add up the exit signal and the supporting signal with the constant amplitude B_0, one can write the resulting voltage on the quadratic detector connected with the mixer as

$$u_g \sim J(x,y,z) = |E_0(x,y,z)|^2 + B_0^2 + 2\gamma|E_0|B_0\cos[\psi_s(x,y,z) + \phi(x,z) + \phi_0]. \tag{1.67}$$

This expression contains information about the amplitude and phase field structure of the excited wave beam. Note that if to apply a confocal resonator and perform the scanning by the dipole in the focal plane where the phase front of the resonance beam is planar, one can obtain a qualitative information about the phase structure of the exciting field.

The explicit expression for the function ψ_s does not differ from (1.65). The determination of the field phase distribution in a resonator is described below.

Let the principal oscillation with the intensity $|E_0|_{max}^2$ be excited as a result of the electromagnetic wave diffraction by the sensor placed on the resonator axis. Assume that the scanning is performed by the sensor along the line of the homogeneous wave front of the exciting field generated by the quadratic detector loaded with the waveguide communication element. Then the resulting signal will be proportional to the intensity distribution of the resonance field in the cross-section of excited oscillation. In order to measure the phase distribution, the coherent supporting signal with the amplitude $B_0 = |E_0|_{max}^2$ and the phase opposite to that of the resonance signal generated by the sensor placed on the resonator axis, is sent to mixer 7. In the process of scanning the interference signal is registered by the recorder in the chosen cross-section of the wave beam. Taking into account the phase raid of the supporting wave $\phi(x) = kx\cos\alpha$ up to the line of scanning, the obtained relations for $J(y)$ and $|E_0(y)|^2$ are substituted to equation (1.65). Finally we determine the sought-for phase field distribution $\psi_s(y)$.

The analysis showed that a relative error of the discrete measurement of the field phase distribution has the order of 16%. The errors of phase measurements by means of the balance scheme are equal to 11% for the resonator with the q-factor $Q_s = 27000$ and $r = 4.67\lambda$ at $\lambda = 4.28$.

It is important to point out that the prescribed possibility of the antenna rotation around the resonator enables one to solve some essential problems concerning the design of open resonators. It is shown in Refs. [74, 75], where the diffractional fields in open resonators are considered, that for small values of the Fresnel numbers ($N \leq 2$) there is an interaction of a resonator with the free space by means of the diffractional field. That is why, when the described methods are applied, one has to choose a rotation angle with an account of the maximal difference of interfering fields. From the other side, the system open resonator + scattering sensor is the high q-factor resonance directed antenna [76]. This fact explains the necessity to use a rotating antenna in the method of the exciting sensor.

The homodyne interference phase method with the suppressed bearing frequency and with a single resonance mixer in the form of a quasi-optical open resonator connected with one of the interferometer shoulders is also applied for the phase measurements.

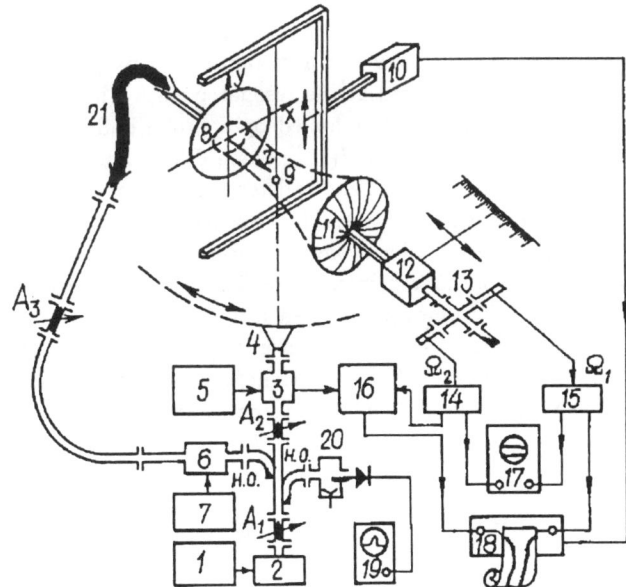

Figure 19.

Here the way to get the information about the wave front of the resonator resonance beam may be cleared up if to consider the modified scheme of the experimental plant presented on Fig.19 (compare with Fig.14).

The frequency-modulated signal of the microwave generator 2 comes through the unbinding attenuator (with the attenuation 5-7 dB) or the ferrite valve (with the direct attenuation -1.5 dB and inverse attenuation -20 dB) and junction 8 connected by means of the attenuator (the maximal attenuation is 50 dB) of the supporting channel and the working power unit 1 (with the internal modulation) to the feeding channel of the considered open resonator. The required conditions of the resonator excitation at one of its eigenoscillations are attained by the smooth tuning of the resonator by measuring device 12 with the accuracy 0.5 μm. The probe sensor 9 is placed on the resonator axis in the electric field maximum. The excited oscillation is controlled by the method described in Refs. [31, 8]. Then the supporting quasi-plane wave formed by antenna 4 is brought in the resonator volume and the power block 1 starts to perform the continuous generation. The phase modulator 3 fed by the sound generator 5 carries out a linear phase modulation of the microwave signal in the supporting channel. Finally the supporting wave field in the far-field region of antenna 4 (in the cross-section z_0) takes the form

$$E_2(x, y, z_0; t) = B(x, y, z_0)e^{-i[\phi(x,y,z_0)+\phi_0]} \tag{1.68}$$

where $B(x, y, z_0)$ and $\phi(x, y, z_0)$ are the spatial amplitude and phase distributions of the supporting wave in the analyzing cross-section of the resonance beam and $\Omega_2 = 136$ Hz is the frequency of the phase modulation. Note that here it is necessary to find the way of selecting the modulation frequency in order to eliminate the net component (50 Hz) which hampers the registration and leads to additional errors of measurements.

The intensity of the interference electromagnetic field becomes the function of the time and coordinates. However, not the intensity but a certain signal proportional to the transmission coefficient [31, 8] is registered at the resonator exit. This signal is connected with the field intensity at the point occupied by a probe body in the resonance volume by the formula $k_p(t) = C^4/[C^2 + J(t)]^2$, where C is a normalizing factor which takes a specific value for a given sensor and is determined for every series of measurements from the condition $\max |E/E_{\max}| = 1$. The presence of a sensor in the resonator leads to a nonlinearity of the registration system and causes the appearance of the signal corresponding to the value of the transmission coefficient and the spectral components proportional to Ω_2 (at the intermediate frequencies).

In the scheme described above the narrow-band resonance amplifier 14 is adjusted with respect to the phase-modulated frequency. Then the signal at the amplifier exit, which is proportional to the signal of the intermediate frequency transmission coefficient, will have the form

$$u(t) \sim k_p(t) = \frac{C^4}{2A_s^2B^2} \frac{\cos \kappa_2}{(\alpha_2 - 1)^{3/2}} \tag{1.69}$$

This signal is sent (as the basic component) to the measurement entrance of the frequency phase-meter 16. The supporting signal comes directly from the exit of the supporting voltage generator 5 to the second entrance of the phase-meter, which measures the difference between these two sinusoidal signals having the same frequency. The origin is chosen under the assumption that the sensor occupies a fixed position in the resonator (for example, on its axis). The latter assumption is provided by the manual tuning of the initial phase ϕ_0 of the microwave signal with the help of the phase-rotator placed directly in the phase-shifter. The tuning is performed until these two signals coincide on the screen of the two-beam oscillograph 17. Such a position corresponds to the initial point of the phase-meter indicator. The phase-meter transforms the measured phase difference to a time interval. This transformation is accompanied by the measurement (and the output to the registering device) of the constant component of the current impulses with the duration equal to this time interval. In the case of inhomogeneous phase distribution the variation of the sensor position leads to the temporal shift of the resonator signal (1.69) with respect to the supporting signal. This time interval determines the phase difference $\theta = \psi_s - \phi$ or, in other words, the phase raid of the resonance beam with respect to the constant phase of the supporting beam along the sensor trajectory. The synchronization of the sensor movement with the recorder ribbon 18 is provided by the potential-meter electrically connected with the scanning device 10. As a result, the phase distribution of the resonance field ψ_s is directly registered by the recorder. The amplitude-phase distribution of the principal oscillation in the open resonator obtained by this method is presented in Fig.20.

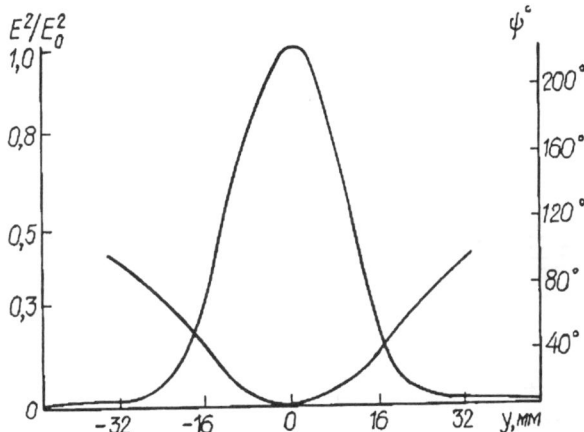

Figure 20.

We can simultaneously register the amplitude and phase field structures in an open resonator by the considered experimental plant, which considerably decreases the duration of the experiment. The double-modulation method [77] allows us to increase the efficiency of the system. With this purpose we put the semi-conductor amplitude modulator 6 (see Fig.19) fed by the sound generator 7 operating at the frequency $\Omega = 1$ KHz to the main channel of the interferometer. It is obvious that here, when the sensor moves inside the resonator, the exit signal is a function of two parameters: $\kappa_1 = \sin \Omega_1 t$ and κ_2. From the amplifier exit the signals come to registering device 18 (for example, a two-channel recorder). The signal parameters are controlled by the two-channel oscillograph 17. Hence, the methods developed for a two-channel interferometer allow one to carry out a synchronous recording of the phase distribution and the signal of the transmission coefficient corresponding to the amplitude field distribution [8, 31] in two-mirror open resonators. An automation of measurements enable us to obtain the necessary information concerning the amplitude-phase distribution on the whole plane of the beam (see Fig.21,a,b).

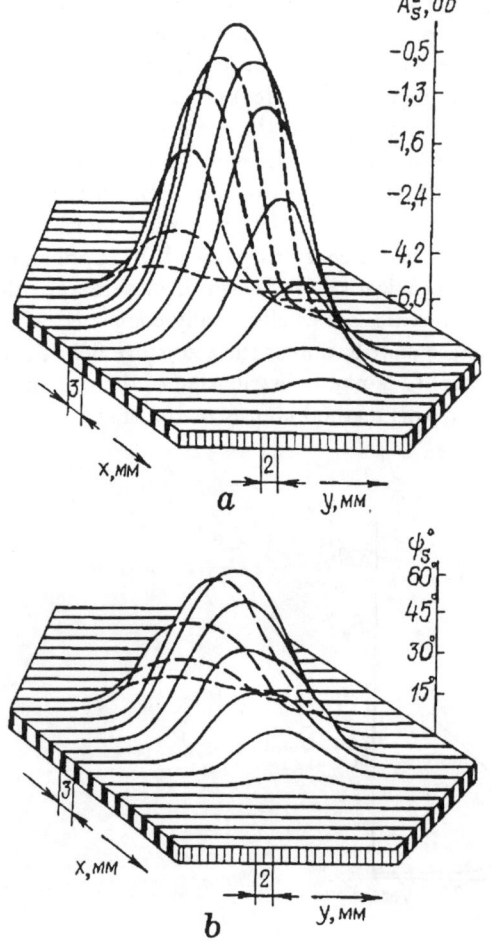

Figure 21.

Let us consider one more spectral method of analyzing the amplitude-phase prop-erties of open resonators. We construct the field representation using the eigenmode spectrum of a high q-factor open resonator together with recording the complex value (amplitude and phase) of the resonance transmission coefficient at every oscillation. This value characterizes the coefficient of the field transformation into the given oscilla-tion. In other words, it determines the field expansion by its components (see sections 1 and 2). The amplitude and phase spectra are the measured values. The complete information (the complex expansion coefficients) allows one to reconstruct the spatial structure of the wave beam.

The method of the radio-structural analysis is realized with the help of the uni-versal experimental plant operating in the two-millimeter wavelength range [78, 79, 8]. The scheme of the plant is presented in Fig.22. A two-channel scheme with the subject and supporting channels is used for the amplitude and phase measurements.

The plant consist of the following units. The transmission line contains the waveg-

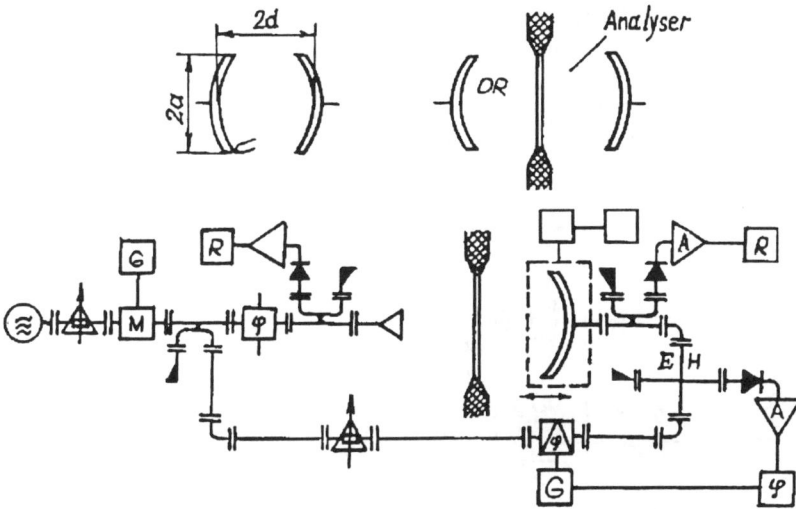

Figure 22.

uides with the cross-section of 0.8×1.6 mm and the beam-guides with the diameter 20 mm. The generator of diffractional radiation (GDR) is used as a feeder. A signal generated by the GDR goes through the attenuator with the attenuation ≈ 6 dB and is modulated by the semi-conductor modulator at the frequency 1 KHz. The phase-rotator and the feeding element in the form of a waveguide slot or an antenna (in the cases of concentrated or distributed feeding, respectively) is placed in the subject channel. The two-mirror unit may operate either as the "transfer-type" or the "reflection-type" resonator. In the first case, the exit signal is divided by the junction and comes to the supporting channel and to the device performing the measurement of the transmission coefficient. There the signal is detected and amplified by the selected amplifier (U) at the amplitude modulation frequency and then goes to the registering device (R). The supporting channel contains the measuring attenuator, the polarization phase modulator fed by the low-frequency generator (G) and the double waveguide bridge where the supporting and exit signals are mixed. The entrance mirror is fixed on the powerful bed and the exit mirror is placed on the moving carriage connected with the longitudinal displacement mechanism (with the distance measurement accuracy 0.005). The exit mirror movements are synchronized with the oscillograph beam and the recorder ribbon. The mechanical scanning device 8 with the analyzing field frame 70×70 mm and the frame formation time ≈ 10 sec is used for registering the amplitude field distribution. The copper or aluminium communication elements that are placed in the mirror centers, have the form of narrowing waveguides with the slot 1.6×0.03 mm. Hence one can excite the principal oscillations in the resonator. The aluminium quartz plates and lavsan films are used as the semi-transparent entrance mirrors.

In order to completely describe the resonator wave field, it is necessary to obtain the information concerning the beam polarization [81, 82, 8]. In this way the problem arises to organize a selective control of the electric field orientation using a resonance wave beam cross-section. A synchronous registration of the amplitude-phase charac-

teristics formed the basis for the resonance polarimetry development in the studies of
two-mirror resonators of the millimeter and submillimeter wavelength ranges.

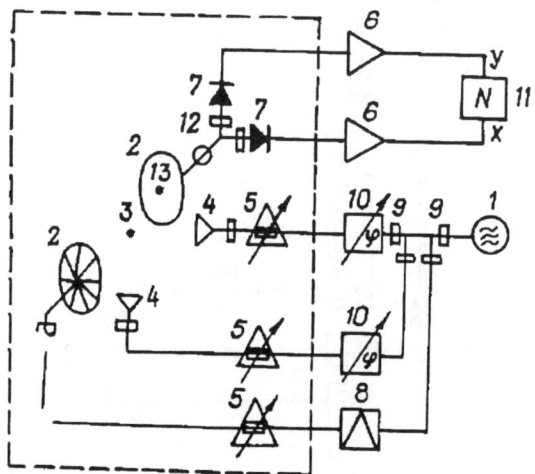

Figure 23.

The electromagnetic field polarization may be described by the four Stokes parameters [83] where three of them are linearly independent. Different polarization
states may be geometrically described with the help of these parameters (like a Poincare
sphere [83]). Hence, to obtain the polarization picture of the field in the cross-section
of the wave beam considered, one has to register the amplitude-phase distribution
along two orthogonal projections of the vector **E**. If to assume that all the Stokes
parameters are known, one can reconstruct the polarization (coordinate) structure of
the field considered applying a discrete sequence of the sought-for values. Here the
difficulties are caused by the necessity to rotate the resonator around its axis (coordinated with the direction of the electric field vector of the supporting wave) in order
to find two fixed positions differing by the azimuth angle equal to 90°. This process
becomes more complicated for a resonator with inclusions that change the polarization of the resonance wave beam. In addition to this, strict requirements are laid on
the accuracy of the spatial combination of the scanning planes in both channels. A
possible solution to this problem is in creating the automatic device in the form of
the three-channel quasi-optical interferometer having the second supporting channel,
which is cross-polarized with respect to the first channel [84]. The block-scheme of the
plant is presented in Fig.23. The microwave signal from the high-stable generator 1
is divided by two couplers 9 into three channels. The resonator formed by mirrors 2
and placed in the main channel of the interferometer is excited through the waveguide
communication slot in the mirror center at one of the resonant eigenmodes. The angle
between the vector of the exciting electric field and each of the electric field vectors
in the supporting channels ending with the identical antennas 4 is equal to 45°. As
a result of initiating the supporting channels, the complicated interference picture is
formed inside the resonator, where only the co-linear field components interfere. The
small scattering sphere 3 placed in the maximum of the excited electric field is used as
a sensor.

In order to perform the measurements with the help of this method, one has to provide the phase synchronization and the equality of the supporting electromagnetic field amplitudes. Here we calibrate the preliminary signals in the supporting channels by phase-rotators 10 and attenuators 5 with respect to the linearly polarized field directed along the bisector dividing the propagation directions of the supporting wave beams. The scanning by the sensor in these two domains is characterized by the following differences: For one polarization the sensor moves back and forth relative to the supporting wave source, while for another polarization it moves parallel to the wave front of the quasi-plane wave. Hence we must correctly determine the law of the intensity variation as well as the initial phase of the supporting electromagnetic wave when calculating the Stockes parameters. This problem may be solved in two ways. The first possibility is to place the electrically controlled phase-rotator in the supporting channel to compensate the supporting wave phase raid occurring in the process of the sensor movement. Here we must provide the synchronization with the speed of the varying phase. The second possibility is to scan by the sensor and the supporting wave source. Such a procedure is associated with technical difficulties and complicates the measurements. The analysis of the polarization measurements yields the necessity to construct the hard unit containing sensor 3 and antennas 4. The design of the scanning system is changed. In this way we have proposed and created the modified device where the open resonator is scanned in two orthogonal directions with respect to an immovable sensor and is connected with the supporting channels.

The information about a microwave signal is transferred to the intermediate frequency in the three-channel interferometer at two polarizations by the phase modulator 8 standing in the main channel. The output of the energy from the resonator volume is made by the isotropic (with respect to the polarization) communication element placed in the center of the second mirror. The resonator's exit is connected with the waveguide or the quasi-optical polarization divisor 12. The narrow-band resonance amplifiers 6 (with the band width of about 0.7 for the power and 4–8% for the selection frequency) separate the signals of two orthogonal components of the electric field at the phase modulation frequency Ω. The exit amplifier signals may be represented in the form

$$u_1(t) \sim K_\Omega^y = a^y \cos[(\psi_s^y + \phi^y + \phi_0) + \Omega t],$$
$$u_2(t) \sim K_\Omega^x = a^x \cos[(\psi_s^x + \phi^x + \phi_0) + \Omega t],$$

$$(1.70)$$

where a^y and a^x are the amplitude coefficients, ψ_s^y and ψ_s^x are the resonator field phase distributions for the E^y and E^x components and ϕ^y and ϕ^x are the phase values of the corresponding components in the supporting channel calculated at the point occupied by the sensor. These signals are delivered to both entrances of the phase detector which controls the phase shift between the orthogonal field components in the process of scanning. The phase resolution is $\pm 2°$. The phase-meter exit signal proportional to the phase shift $\psi_s^x - \psi_s^y$ and the exit signals of amplifiers 6 proportional to the field amplitudes of the corresponding components measured at the frequency Ω, are simultaneously registered by the device 11. To perform an express-analysis of the polarization of the resonance beam cross-section, both signals given by relations (1.70) are compared by sending to the x and y planes of the oscillograph. The Lissagoue figure observed on the screen is the ellipse of the resonator field polarization. The rotation direction of the electric field hodograph is determined by the brightness corresponding to the signal coming from the supporting channel of the phase modulator 8. As a result, the polarization appears on the screen in the form of luminous point, which allows us to watch the dynamics of the polarization changes directly in the process of scanning.

One has to mention one more advantage of the express-analysis described. Namely, it is not necessary to register a phase shift between the orthogonal components of the electric field by the recorder when the measured values vary in a wide dynamical range (for example, when one must take into account the phase leaps occurring at the higher order resonator oscillations).

Note also that if the waveguide communication element is situated only on one of the resonator mirrors, the interferometer itself must provide the fulfillment of the excitation conditions. In the given situation the resonator is excited at two polarizations by the scattering sensor placed on the line of the wave beam propagation (generated by antennas 4). This interferometer version requires an artificial creation of the supporting wave in both shoulders of the polarization divisor 12 by the phase-modulated signal that imitates the supporting wave. The waveguide T-junctions in both shoulders of the polarization divisor are used in the two-channel interferometer as the mixers to separate the signals of the intermediate part of the frequency range. These coherent signals are compared by sending to the x and y planes of the screen or to the low-frequency phase detector.

Various properties of empty resonators and resonators with inclusions are studied by means of the experimental technique described in this section. Let us begin with the empty resonators applied in the millimeter and submillimeter wavelength ranges. Among them there are usual resonators formed by spherical mirrors (or semi-spherical resonators where one of the mirrors is a metallic plane) such that $ka \approx 10 \div 60$ (a is the radius of the mirror aperture). If $a^2/(e\lambda) \ll (l/a)^2$ (l is the distance between the mirrors), then the mirror dimensions and the distance between them weakly influence the resonator characteristics, as has been shown by our measurements. Here the Fresnel number is one of the most important parameters. We have established experimentally (with the help of the plant depicted in Fig.14) that in the 4-millimeter wavelength range the small diffractional losses of the low-order eigenoscillations are caused by the presence of the caustic surfaces. Their form can be completely reconstructed by the visualization method when the shapes of the field spots are analyzed in the different resonator cross-sections. As l increases the diffractional losses grow and attains a maximal value in the case of a semi-concentric resonator. One can determine the resonator spectral characteristics by considering the q-factors and the field configurations. It turns out that in the 4-millimeter wavelength range the spectrum varies from one to eleven oscillations (from a concentric to a confocal resonator) and their q-factors lie in the interval from 300 to 15000. The analysis of the obtained photos of the field distributions and their amplitudes shows that the use of the slot (as a communication element) instead of a circular hole does not cause a strong distortion of the field distribution. The influence of the mirror asymmetry on the field distortion is also clearly exposed. The detailed information concerning these phenomena (examined in the presence of inhomogeneities on the mirrors in the form of diffractional gratings) is presented in Ref. [6]. Some experimental results for these resonators are presented below.

Here we will consider in details the experimental data for the resonators with $ka \lesssim 10$ and $b/a \lesssim 10$ The high q-factor oscillations with the H-polarized field having distinct cavities appear in cylindrical resonators with circular mirrors even for $ka \sim 7$ and $b/a \sim 0.8$. Hence, even for $a/\lambda \approx 1$ there exist resonance open structures that may be applied in the millimeter and submillimeter wavelength technology. We will call them small resonators if $a/\lambda \lesssim 1 \div 3$. One could study the properties of such open resonators only after elaborating the rigorous spectral theory of such structures and different methods of their excitation. The development of this branch of radiophysics and electronics is connected with the solution to new theoretical and experimental problems. Traditionally the quasi-optical resonators with $a/\lambda \lesssim 10 \div 15$ are used in the millimeter and submillimeter technology. For the laser open resonators this ratio reaches $500 \div 1000$.

Let us consider the results of the measurements of the transmission coefficient, q-factors and field distributions in small open resonators formed by spherical and plane mirrors. The experimental methods have been described above. We assume that the resonator is excited through the communication slot in the plane mirror and the energy goes out through the same hole in the center of the spherical mirror. The measurements are made for $\lambda = 3 \div 5$ mm, $b = 9, 15, 20$ mm, $a = 6, 8, 9, 17$ mm. The spectrum of the semi-spherical resonator with $\lambda = 4$ mm and the Fresnel number $N_0 = 3.27$ consists of eight types of axial-symmetrical oscillations. The number of oscillations (spectral components) decreases together with the Fresnel number. For the resonator with $N_0 = 1.4$ there are four oscillations in the spectrum. Note that the short-focus open resonators have several features. In particular, when the Fresnel number increases and $b = $ const the depth of the spherical mirror increases what leads to the considerable field concentration in the resonator volume (because of the diffraction on its edges). The resonance oscillations with the Fresnel numbers $N_0 = 3$ corresponds to an over-concentric resonator. The resonance transmission coefficient in this case is less than that of the principal oscillation and the difference is 26 dB. When the distance between the mirrors increases, one observes at $l/b = 1.3$ an increase of the resonant transmission coefficient. The change of this coefficient is accompanied by the distortion of the resonance field distribution. The field reconstruction in the open resonator by the method of visualization with respect to varying $l/b = 0.9 \div 1.7$ gives the complete picture of the principal oscillation destruction. For $l/b = 1.7$ the field is concentrated close to the edges of the spherical mirror. The oscillation q-factors are of the order $300 \div 500$ (the diffractional loss here is comparatively high) and their maximal value for short-focus open resonators, which has been registered for the principal oscillation at $N_0 = 3$, is equal to $5 \div 10^3$.

It is important to know how the spherical property of the mirror and the position of the communication hole influence the oscillations and field distributions in small resonators. With this purpose the mirrors are made on a lathe and copper plates are squeezed out on spherical arbors. The measurements of the resonators formed by such mirrors show that a violation of the spherical shape leads to elimination of the frequency degeneration of the higher-order oscillations with respect to transversal indices. The oscillation spectrum of the resonator with such squeezed mirrors is considerably condensed; at the same time, there are only three types of oscillations in the spectrum.

In different circuits the mirrors are placed in a cylindrical body. The presence of an outside screen in short-focus resonators yields the appearance of oscillations typical for cylindrical resonators. For example, when $N_0 = 3.27$ the oscillation spectrum in the cylindrical body with the radius of the inner surface 14.5 mm ($b/a = 1.03$) has three types of oscillations of a cylindrical resonator. The amplitude resonance transmission coefficient is comparable with that of the corresponding empty open resonator. The number of oscillations in cylindrical resonators as well as their amplitudes increase together with N_0.

It is of a practical interest to study the q-factors and field distributions in small open resonators with the plane mirror partially covered by a diffractional grating. We will consider the resonators for which this grating has a form of a comb with the width 10 mm, period $d = 0.4$ mm, height $h = 0.92$ mm. The spherical mirror apertures are 14, 12 and 9 mm. The grating leads to the elimination of the frequency degeneration with respect to transversal indices. In addition to higher-order oscillations, the principal oscillation appears in the spectrum with the q-factor reaching $4.5 \cdot 10^3$. Some of the higher-order oscillations are also characterized by comparable q-factors. The oscillations appear because the field spots situated along the diffractional grating have low q-factors (the latter is connected with a big ohmic loss on the grating). The q-factor maximal value considered as a function of the distance between the mirrors shifts towards greater distances (with the increasing mirror aperture) due to the decrease of the diffractional losses in these open resonators.

Hence, our experimental investigations of small open resonators show that they have sufficiently high q-factors and the rarefied oscillation spectra. A decrease of the field spot on the plane mirror plays a positive role for some applications in the millimeter and submillimeter technology.

Further experimental studies of open resonators with the geometrical dimensions that are close to or greater than 10λ is carried out by the methods described above.

The study of the field distribution of the resonance oscillations in the resonator cross-section was performed by the following methods: reactive sensing [3], measurements of the phase distribution with the help of the technique developed in Refs. [72, 77, 85, 8] and measurements of the loaded q-factor Q_l by registering the values along the resonance curve. Usually $Q_l \approx 10^4 \div 10^6$ and it is determined with the accuracy $10 \div 15\%$.

We considered the resonators with the spherical mirror curvature radii $R = 20 \div 40\lambda$, the diameter $2a = 15 \div 30\lambda$ and the distance between the mirrors l varying from 10λ to the values corresponding to the before-concentric geometry. The behaviour of the amplitude spectra of the symmetrical ($R_1 = R_2 = 30\lambda$) and semi-symmetrical ($R_1 \to \infty$, $R_2 = 30\lambda$) resonators measured for l/R varying within the limits $0.4 \div 0.5$ shows that the spectra can be divided (for increasing l/R) into characteristic groups containing the oscillations disposed with respect to the decreasing radial index. The relative phase shifts between the neighboring oscillations are 180° (in the case of the concentrated excitation). For a distributed excitation the phase shifts are determined by the beam structure on the entrance reflector. There exists a part of the spectrum of the semi-symmetrical almost confocal $l/R \approx 0.4$, resonator where there are two groups of oscillations with $\approx \lambda/2$ containing separate oscillations with even and odd indices. In the case considered ($R = 2a = 30\lambda$, $l \approx a$) the diffractional losses connected with the value of the Fresnel number $N_s = N_0\sqrt{(R/l) - 1}$ $N_0 = a^2/(2l\lambda)$ characteristic for the given resonator geometry is small for all oscillations up to the twelfth (the amplitudes vary in the range ≈ 5 dB). The diffractional losses start to increase beginning from the thirteenth oscillation.

The cross-sectional field structure is measured by the probe sensor moving on the surface of the beam wave front. This cross-section corresponds to the beam mouth in the case of the rectilinear transference. The pictures of the field brightness obtained by means of visualization [67, 6] and two-dimensional scanning (see section 3) display a satisfactory coincidence of the theoretical and experimental results. The increase of the phase steepness in the central part of the resonance curve causes (as the q-factor increases) the enhancement of the requirements related to the generator stability and the accuracy of the resonator adjustment with respect to the maximum value of the transmission coefficient. For $Q_l = 10^4$ the difference of 0.005% between the generator and resonant frequencies leads to the 6% and 20% changes of the transmission coefficient and the phase, respectively. In our experiments we applied the GDR-2 with the relative frequency instability $\sim 10^{-6}$ which is a high-stable source of submillimeter waves. The absolute error of the phase measurement was about 5% and the experimental error of the resonance adjustment along the resonance amplitude curve was not greater than 4%.

In our experiments we have established the optimal parameters of the open resonators under study. For $\lambda = 2.21$ mm, $R_1 = \infty$, $R_2 = 60$ mm, $2a = 60$ mm and $l_{00} = 26.48$ mm there are 13 non-degenerated and non-interacting modes within the limits of the working part of the spectrum. These oscillations are characterized by the axial symmetry, and their parameters are in agreement with the calculated and experimental field distributions.

If the fields have a more complicated structure than in the axial-symmetrical case, an additional theoretical analysis should be performed. In this connection the problem appears to carry out a specific control of the spectrum of fundamental frequencies and the field parameters in the beam cross-section. In order to solve this problem in the optical wavelength range the main attention is paid to eliminating the higher-order oscillations. The latter procedure is performed, for example, to provide the stability of the quantum light single-mode generators or operating at close-to-axis modes. Such an elimination is achieved by diaphragming the beam with the help of open resonators formed by conducting disks, washers [86] and absorbing films [87], or by varying the mirror geometry [88].

In some cases the perturbation of the eigenoscillation field is small and its cross-sectional distribution may be described by various asymptotical approximations where the spherical [89] or the orthogonal Hermit-Raus functions may be applied. The use of these functions for the analysis of the fields that are not axial-symmetrical, is caused by the necessity to work out the effective ways to eliminate the degeneration of the fundamental frequencies without their selection in the resonator with the mirror parameters satisfying the condition $2a > \lambda$, $a \approx \lambda$ (note that a mathematical description is possible only for the small beam deformations). It is important to optimize the parameters of this resonator in order to use it as an element of the analyzer of the field spatial structure.

Spherical resonators with the big-aperture rectangular mirrors [79] have been also experimentally studied (using the plant whose block-scheme is presented in Fig.22). In these resonators the series of axial-symmetrical oscillations is excited for semi-confocal distances between the mirrors: most of the higher-order oscillations have an approximately equal transmission coefficient and hence the equal total loss. The difference is caused by the increase of optical losses in the resonator operating at higher-order oscillations because the field spots increase on the mirrors. For definite values of the radial index (close to 12) the diffractional losses sharply increase, the transmission coefficients fall (and the elimination of degeneration takes place) and the q-factors grow because at equal diffractional losses the field volume of the higher-order modes is greater than that of the principal mode. Note that here the quality of the mirror surfaces and their proper adjustment plays an essential role.

When the resonator is close to the concentric one $(l = R)$, there are three oscillations in the spectrum having non-zero amplitudes. The higher-order oscillation remains degenerated but the field symmetry is violated in the transversal direction. The rectangular mirror aperture starts to influence the oscillation types. In fact, the diffractional losses in the resonator formed by rectangular mirrors and with the field spots placed along one of the coordinate axes are greater than the losses of the oscillations with the field spots situated close to the axis. The interacting oscillations in such a rectangular resonator contribute to the total field and destroy the axial symmetry. In the resonators with the square big-aperture mirrors the axial-symmetrical oscillations are excited except for the almost semi-concentric resonators, where they are a superposition of the "rectangular" oscillations. Here one can also eliminate the degeneration in different ways and, in particular, by inserting local perturbing inhomogeneities, like a spherical absorbing sensor, $\emptyset \approx \lambda/2$. The elimination of degeneration may also occur due to small mirror deformations [79]. When a spherical mirror is squeezed along one of the coordinate axis, the curvature radii change in the perpendicular planes. If the deformation is symmetrical with respect to the mirror axis, then the field perturbation is similar to the one caused by a phase filter placed in the resonator or by the use of astigmatic mirrors.

The polarization properties of open resonators are studied by means of the plant depicted in Fig.23. Note that the data concerning the field polarization structure in open resonators is of principal importance for applications in radio-spectroscopy and in the systems of estimating the rate of nuclear polarization. In fact, in the latter systems the magnetic component of the microwave field must have a strictly determined direction with respect to the polarization of the external constant magnetic field.

In order to illustrate the influence of different types of inhomogeneities on the field polarization structure, we present the results of the systematic treatment of the resonance field polarization characteristics of the dominant and the first higher-order oscillations in the semi-symmetrical resonator. The measurements are carried out with the help of the three-channel scheme (Fig.23). at the frequency 75 GHz for the resonator formed by spherical mirrors with the curvature radius $R = 45\lambda$, the aperture $2a = 20\lambda$, the distance between the mirrors $l = 18.5 \div 42\lambda$ and the distance between the plane mirror and the scanning plane 7.75λ. The resonator is excited by a linearly polarized wave generated from the circular communication element situated on the spherical mirror. The inhomogeneous distribution of the polarization in the resonator cross-section is the general property of the field polarization structure. Here one registers the alteration of ellipticity and the orientation angle of the big ellipse axis. Note that the inhomogeneous distribution of the field vector cannot be found by the known methods of the amplitude measurements and is extremely undesirable in experiments where, for example, the magneto-optical phenomena in open resonators are investigated.

Consider the distortion of the field polarization structure in resonators caused by the influence of communication elements in the form of a communication slot in the mirror and the wire vibrator placed in the resonator [90]. The experimental study of the effect of communication slots for a semi-spherical resonator is carried out by the specially constructed polarimeter (its description is given in the second part of the monograph). Let us describe the mechanism determining the variation of the field polarization when a communication slot turns with respect to the electric field vector. We will assume that the slot in the mirror is equivalent to the same slot cut in an infinite metallic screen (this assumption holds because $2a \gg \lambda$ and the relative width of the slot $\sigma \approx 0.025\lambda$). It is known [90] that the total scattering cross-section of the slot in the case of normal incidence is different for the H- and E-polarizations and it is equal to $\sigma^H = 0.168$ and $\sigma^E = 0.149 \cdot 10^{-4}$, respectively. Hence, the main contribution to the change of the electric field is made by the slot oriented in the orthogonal direction with respect to the electric field vector. Our calculations and experiments show that when the slot turns at an angle varying in the interval from $-45°$ to $90°$, one can observe the rotation of the polarization plane and the change of the resonance field ellipticity: the maximal phase shift between the orthogonal field components for two perpendicular positions of the slot is the same and is equal to $14°$. A rectangular communication slot in a resonator is the polarization inhomogeneity, and its different orientation with respect to the exciting field vector leads to the change of polarization state caused by elimination of the polarization degeneration. The latter may be explained by the phase difference between the reflection coefficients of the waves with orthogonal polarizations.

Let us clear up the influence of a load in the transmission line connected with a resonator on the resonance field polarization. With this purpose, we will put a short-circuit piston into the waveguide, which is connected with a resonator. When the piston moves, the phase ratio of the reflected waves changes together with the polarization ellipse parameters. Hence one can control the field polarization by varying the type and position of the load in the waveguide section. In order to construct the polarimetric schemes based on the use of open resonators, it is necessary to provide the coordination between a resonator and transmission lines

We have examined the resonance curve of the principal oscillation as a function of the angle of the slot orientation for the generator operating at the 75 GHz frequency modulation, the resonator's q-factor is $35 \cdot 10^3$ and it is excited through the circular hole in the spherical mirror. The signal is taken out of the resonator through the slot in the center of the plane mirror. The results show that there is a fall in the resonance curves when the slot angle varies from $-90°$ to $-70°$ and $-45°$. For the H-polarized slot the resonance curve has a regular character. A small turn of the E-polarized slot at an angle close to $-45°$ also leads to a sharp fall of the resonance curve, and the maximal perturbations of polarizations are observed for this angle of the slot orientation. It is explained by the fact that when the slot turns, a part of the exciting field energy is spent to generate a resonance oscillation with the electromagnetic field vector oriented parallel to the communication slot. Hence all the field energy is redistributed practically between two oscillations, and this process leads to the fall of the resonance curve. The q-factor of the oscillation with an electric field parallel to the slot is much greater than the q-factor of the registering oscillation. This effect takes place for all oscillations in the resonator considered. Consequently, the communication slot leads not only to phase shifts but also to the strong deformation of the resonance curve. A slotted communication element with big dimensions causes the complete elimination of the polarization degeneration. It is confirmed by the measurement data obtained for the resonator containing the thread (the diameter is 0.05 mm) saturated with soot: its turn causes the effect which is similar to the above-considered phenomena.

Let us consider now the influence of a wire vibrator on the resonance field polarization properties. Its position inside the resonator also substantially influences the field polarization. This fact may be explained with the help of the reciprocity principle [90]: a slot in the screen is equivalent to a metallic strip in the free space. Consider, for example, the vibrator (with the length $l \approx \lambda$ and the diameter 0.03 mm) placed on the resonator's axis. We have established that here the orientation angle and the ellipticity angle of the polarization ellipse are not constant. Namely, by measuring the field polarization in different planes, we have demonstrated that the field is not linearly polarized. The influence of the wire vibrator placed at an angle of 45° to the electric field vector causes the high ellipticity in the center and its decreasing up to zero on the periphery of the wave beam. Hence if the vibrator or the slot are directed either perpendicular or parallel to the electric vector of the exciting field, they weakly influence the polarization of the field in the resonator.

We have also analyzed the effect of the spherical scatterer (in the form of the metallic ball with the diameters 1.5 mm and 1.0 mm) on the field polarization. It turns out that small spheres ($a \ll \lambda$) do not distort the polarization properties and may be applied in resonance polarimetric schemes. The situation sharply changes when the form deviates from the spherical.

It has been shown that a strong focusing which is applied to hold the field in resonators formed by small mirrors (compared with the wavelength λ), may be one of the reasons leading to the complication of the polarization structure in open resonators. Such open resonators are applied in the millimeter and submillimeter wavelength ranges.

The spatial field structure of the electric components of the resonance electromagnetic field is studied experimentally in the semi-symmetrical resonator with the following parameters: the radius of the spherical mirror is $R_{\text{sp}} = 20$ mm, the aperture is $2a = 28$ mm, the aperture of the plane mirror is $2a \approx 70$ mm nd the wavelength is $\lambda = 4.2$ mm. The pictures are obtained by the visualization method for the dominant E_y-components. We have registered the cross-sectional distribution of the field cross-components of the TEM_{007}-oscillation in the resonator with a communication slot on the plane mirror placed parallel (or perpendicular) to the mirror's displacement and with the exciting slot on the spherical mirror placed perpendicular (or parallel) to this displacement. The results show that only the central part is left in the cross-component signal. Hence two identical oscillations are simultaneously excited in the resonator, differing by the polarization and amplitude. We explain this effect by the polarization degeneration in the resonator.

It follows from the analysis of the dispersion relation of the resonator considered that for the semi-concentric geometry TEM_{116}- and TEM_{007}-oscillations are frequency-degenerated. There may also exist intertype oscillations. The cross-sectional field distributions obtained for TEM_{00q}-oscillations with different values of q show that as the index q increases (that is, the resonator is transformed from the semi-confocal to semi-concentric geometry) the connection between the TEM_{00}-oscillation and the field with the orthogonal polarization becomes more pronounced. One can show that higher-order intertype oscillations are also excited in semi-concentric resonators. The bounded mirror apertures are the additional sources of a cross-polarized field in empty

resonators (note that the edge wave diffraction occurs in the vicinities of the apertures). For example, spherical mirrors with small curvature radii become the main
source of edge waves. If to determine the currents on these mirrors, one can show
that their inhomogeneous components influence the resonance field polarization structure. These processes are considered in Ref. [90], where the resonators are investigated
which are perturbed transparent anisotropic plates. Such inhomogeneities eliminate
the polarization degeneration in resonators and are applied for recording the edge wave
distribution. We have measured the edge wave field structure and the polarization of
the principal oscillation in a semi-symmetrical resonator with respect to the aperture
a and the distance between the mirrors l (at the frequency 75 GHz). It turns out
that contribution of the edge effect to the formation of the field polarization increases
together with the decrease of the plane mirror. One can observe this effect if to examine the variation of the resonance field polarization when transferring from the plane
mirror in the form of a disk with the radius $a = 15\lambda$ to the mirror with $a = 4.5\lambda$. Here
the decrease of the plane mirror aperture influences the q-factor and yields a decrease
of the field amplitude.

Table 2.

Materials	h, μm	ε'	$\varepsilon_{\parallel}$	$\varepsilon_{\perp}$	$|\Delta\varepsilon|$
Polyethylene	32	2.52 ± 0.05			
isotropic					
$c/p = 2$	190		2.33	2.25	0.08
$c/p = 4$	130		2.32	2.23	0.09
$c/p = 4$	68		2.53	2.39	0.13
$c/p = 3,62$	54		2.47	2.37	0.1
oriented	120				0.012
Fluoric plastic FT-4	45	2.04 ± 0.07			
Lavsan					
isotropic	23	3.2 ± 0.1			
anisotropic	48		3.33	3.18	0.15
Polyvinyl alcohol					
non-thermo-processed	53	3.41 ± 0.05			
thermo-processed	53	3.62 ± 0.05			
150%, oriented	28		3.6	3.4	0.2
300%, oriented	20		4.40	3.9	0.2
Polyurethane	94	2.92 ± 0.04			

It is of a considerable interest to perform the experimental studies of open resonators with anisotropic inclusions and, in particular, with thin dielectric films. The
results of our systematic treatment allow us to make the following conclusions. First of
all, an anisotropic plate strongly deforms the principal oscillation. We have not indicated the violation of the amplitude-phase structure for the plate with thickness 0.835λ
at $\lambda = 4$ mm. The spectrum splits in the resonator with the anisotropic inclusion having an arbitrary orientation of axes with respect to the exciting electric field vector.
The value of this split $\Delta l = l_{\perp} - l_{\parallel}$ (where $l_{\perp}$ and $l_{\parallel}$ define the resonant states for the
perpendicular and parallel components of the principal oscillation) is determined by
the anisotropy parameter $\Delta\varepsilon = (\varepsilon_{\perp} - \varepsilon_{\parallel})$ where $\varepsilon_{\perp}$ and $\varepsilon_{\parallel}$) are the permittivities of
the inclusion along the axes of anisotropy. When the film rotates around the exciting

electric field vector, the perpendicular and parallel components of the resonator's principal oscillation are equal to each other when the angle between the axis of anisotropy and the electric field vector is 45°. In the case of weak anisotropy ($\Delta\varepsilon < 0.015$) one can measure the spectrum split with the help of the resonance polarimeter by registering the field polarization along the resonance curve [90].

Application of the described experimental techniques in the millimeter and sub-millimeter wavelength ranges allow us to measure the characteristics of dielectrics with the help of open resonators and, in particular, the properties of anisotropic films. The results of these measurements are presented in Table 2. The comparison of the obtained and known data confirms the efficiency of these methods.

Chapter 2

Propagation of electromagnetic waves in open waveguides

As has been pointed out above, the open waveguide having the form of a cylindrical slotted waveguide (CSW) is a perspective type of the transmission line of the millimeter and submillimeter wavelength ranges. The fields in CSW are quasi-static. Elaboration of the mathematical model describing the wave propagation in CSW is connected with the necessity of the correct statement of the corresponding boundary value problem. The mathematically rigorous approach to the analysis of spectral properties of open structures is developed in chapter 1. The results of this analysis enables one to perform, with an account of errors, simple engineering calculations of the CSW parameters by means of an equivalent scheme with definite values of capacitances, inductances, resistances, active and passive elements, etc.

A quasi-static character of electrodynamic fields in CSWs gives unique possibilities to construct the element base and various radio-systems based on the application of this line. Since the electric and magnetic field components are concentrated close to the slot and inside the line cavity, respectively, one can place the controlled microwave devices inside the CSW in such a way that the elements whose properties are determined by the vector of electric field (mixers, detectors, $p - i - n$ diodes, etc.) are situated in the domain of the slot. The non-reciprocal elements (ferrites, etc.), whose operation parameters are determined by the magnetic field are placed inside the CSW close to its wall and in front of the slot. It is essential that for small values of the CSW radii (less than one millimeter in the millimeter wavelength range) these elements (and CSW itself) may be constructed using the well developed film-drifting technology applied for creating cylindrical dielectric surfaces of transmission lines.

The CSW is a wide-band line. Its transmission band reaches two octaves. In real CSWs the attenuation is about fractions of dB/m due to the losses in dielectric and non-ideal conductivity of the metal. When the width of the slot increases, the slot wave transfers to one of the waves of a dielectric waveguide and for the fixed value of ε it may radiate to the free space. Hence a CSW may be gradually transferred from the state of the device operating as a guiding structure to the radiating antenna system. The appropriate choice of the permittivity ε, the slot dimensions θ and the wavelength λ enables one to provide the conditions under which the CSW wave resistance coincides with that of the free space or even with the wave resistance of a planar slot line (PSL).

The cylindrical strip waveguide (CStW) or the image slot waveguide (ISW) formed by a curvilinear metallic strip enveloping a dielectric pivot (with circular, rectangular

85

or other types of the transversal cross-section) is also a variety of an open waveguide. The wave band broadening and the decrease of losses is provided here by the waveguide design: the longitudinal slot is formed by the screen edge and the coating and a part of the metal screen is made of conducting materials. A smaller size of the CStW cross-section allows to broaden the transmission band for the principal slot H_{00} wave up to 20%, if to compare CStW with a usual CSW, and to decrease losses up to 1.5 times. It is possible, in addition, to elaborate a rather simple manufacturing technology for the ISW, similar to that applied for PSLs.

The strip transmission lines are the basic devices of the integrated schemes applied in the centimeter wavelength range. We have been elaborating the methods of creating analogous schemes with small dimensions and low losses caused by functional elements for the application in the millimeter wavelength range. Such integrated schemes are made on the basis of CSWs or ISWs. Unlike strip transmission lines, these open waveguides have the definite shape of the cross-section and, therefore, the dimensions of their elements play an important role.

In this chapter we develop the approach for solving the problem of the wave propagation in a slotted cylinder with the finite thickness of the walls based on the results of A.I. Nosich and A.N. Svezhentsev. This technique enables one to calculate the propagation constants and attenuation factors of CSWs and CStWs. The properties of ISW was also studied by G.I. Komar'. The experimental data obtained by Yu.I. Glotov, V.V. Kryzhanovskij, G.I. Komar', A.G. Kovalenko, and G.I. Khlopov allowed us to establish the limits of applications of theoretical results and to verify simple engineering formulas for calculating the parameters of CSWs and ISWs.

§ 6. Spectral theory of cylindrical slot and strip waveguides

The spectral theory of the CSW and CStW is constructed within the framework of the mathematical model developed in chapter 1. In order to study the problem of the wave propagation, it is sufficient to consider one perfectly conducting circular waveguide of the radius a with a longitudinal slot of the angle width 2θ (Fig. 24). The internal ($r < a$) and external ($r > a$) media consist of the radial dielectric layers (with the unit magnetic permeability). For $\theta < \pi/2$ such a waveguide serves as a model of the multi-layered CSW and for $\delta < \pi/2$ ($\delta = \pi - \theta$) it models the CStW. The circle $r = a$ will be referred to as the main boundary. Let the external medium contains $N + 1$ dielectric layers where in the jth layer $r_j^+ < r < r_{j+1}^+$ and $\varepsilon = \varepsilon_j^+$ ($j = 1, 2, \ldots, N$), and the inside medium contains $M + 1$ layers where in the lth layer $r_{l+1}^- < r < r_l^-$ and $\varepsilon = \varepsilon_l^-$ ($l = 1, 2, \ldots, N$). The case of perfectly conducting boundaries of the layers is included into consideration. Hence the considered CSW and CStW may be both open and shielded (closed) waveguides. Since the structure is regular with respect to the axis Oz, we will represent the guided field in the form of a normal wave [91] which depends on the coordinate z and time t as $e^{i(hz - \omega t)}$ where h is the propagation constant and ω is the frequency parameter.

The spectral problem is to determine the set of the values h (h is the spectral parameter) for which there exist nontrivial solutions

$$
\begin{aligned}
E_z &= {}^{+(-)}E^{j(l)}(r, \phi; h)e^{i(hz - \omega t)}, \\
H_z &= {}^{+(-)}H^{j(l)}(r, \phi; h)e^{i(hz - \omega t)}
\end{aligned}
\quad (j = 1, 2, \ldots, N, \ l = 1, 2, \ldots, M)
$$

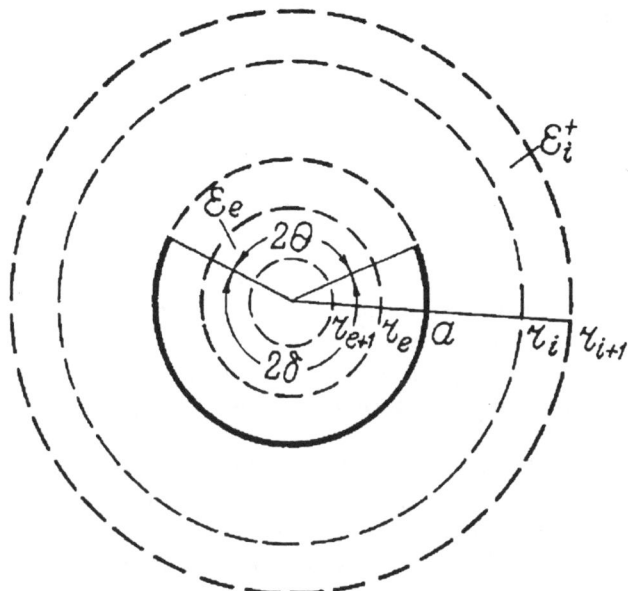

Figure 24.

of the system of Helmholtz equations

$$(\Delta_{r,\phi} + \tilde{k}_{j(l)}^2)\{{}^{+(-)}E^{j(l)}, {}^{+(-)}H^{j(l)}\} = 0 \tag{2.1}$$

(here $\Delta_{r,\phi}$ is the two-dimensional Laplace operator in the polar coordinate system, $\tilde{k}_{j(l)} = \pm\sqrt{k^2\varepsilon_{j(l)}^{+(-)} - h^2}$ are the transversal wavenumbers in the layers and $k = \omega/c$ is the vacuum wavenumber) satisfying the following conditions:

$$
\begin{aligned}
{}^{+}E_z^1 &= {}^{-}E_z^1, \ \forall\phi, & {}^{+}E_\phi^1 &= {}^{-}E_\phi^1, \ \forall\phi, \\
{}^{+}H_\phi^1 &= {}^{-}H_\phi^1, \ |\phi| < \theta, & {}^{+}H_z^1 &= {}^{-}H_z^1, \ |\phi| < \theta, \\
{}^{+}E_z^1 &= {}^{-}E_z^1, \ \theta < |\phi| \le \pi, & {}^{+}E_\phi^1 &= {}^{-}E_\phi^1, \ \theta < |\phi| \le \pi,
\end{aligned}
\tag{2.2}
$$

on the main boundary $r = a$;

$$
\begin{aligned}
{}^{+(-)}E_z^{j(l)} &= {}^{+(-)}E_z^{j+1(l+1)}, & {}^{+(-)}E_\phi^{j(l)} &= {}^{+(-)}E_\phi^{j+1(l+1)}, \\
{}^{+(-)}H_z^{j(l)} &= {}^{+(-)}H_z^{j+1(l+1)}, & {}^{+(-)}H_\phi^{j(l)} &= {}^{+(-)}H_\phi^{j+1(l+1)},
\end{aligned}
\quad r = r_{j(l)}^{+(-)}
\tag{2.3}
$$

for $r = r_j^+$ $(j = 1, 2, \ldots, N)$ and $r = r_l^-$ $(l = 1, 2, \ldots, M)$;

$$\iint_D (\operatorname{Re}\varepsilon|\mathbf{E}|^2 + |\mathbf{H}|^2)r\,dr\,d\phi < \infty \tag{2.4}$$

in an arbitrary bounded domain D (this condition corresponds to the requirement concerning the finiteness of energy) and the generalized radiation condition [40] (for open structures) determining the choice of the solution in the open domain $r > r_{N+1}^+$ in the form

$$\{^+E^{N+1}, {}^+H^{N+1}\} = \sum_{n=-\infty}^{\infty} C_{E,H}^{N+1} H_n^{(1)}(\tilde{k}_{N+1}r)e^{in\phi}. \tag{2.5}$$

It is assumed in (2.5) that the domain of variation of the spectral parameter h is the Riemann surface R of the analytical continuation of the Hankel function $H_n^{(1)}(\tilde{k}_{N+1}r)$ from the real interval $|h| < k\sqrt{\varepsilon_{N+1}^+}$ (this surface R will be considered below).

Note that the demand specifying the exponential decay at infinity holds for surface waves [92], and condition (2.5) allows us to expand the spectral theory for all complex waves including the leaky waves of an open waveguide.

The electromagnetic fields which may exist in the considered structure are hybrid, i.e., they have all six components and four of them are expressed through E_z and H_z (see [92]).

In order to solve the spectral problem (2.1)-(2.5) by the semi-inversion method (by the method of the Riemann-Hilbert problem) we will obtain representations for the z-components in the radial domains adjoining the main boundary, i.e., for $r_2^- < r < a$ and $a < r < r_2^+$ ($r_1^+ = r_1^- = a$). Such representations for H_ϕ and H_r that take into account the layered-radial character of the external and internal waveguide media, may be derived with the help of consecutive relations (similar to the procedure described in section 1).

We will write the Fourier transforms of the z-components in the domains adjoining the main boundary $r = a$ in the form

$$\begin{bmatrix} {}^+H_z^1 \\ {}^+E_z^1 \end{bmatrix} = [\alpha^{(1)}(r)]\begin{bmatrix} C_H^1 \\ C_E^1 \end{bmatrix}, \quad \begin{bmatrix} {}^-H_z^1 \\ {}^-E_z^1 \end{bmatrix} = [\beta^{(1)}(r)]\begin{bmatrix} A_H^1 \\ A_E^1 \end{bmatrix}, \tag{2.6}$$

where $\alpha^{(1)}(r)$ and $\beta^{(1)}(r)$ are the coefficients determining the data about all layers situated outside the first one, as well as about all internal layers.

One can get two systems of the coupled summation equations of the first kind with the kernel in the form of trigonometrical functions after satisfying the boundary conditions (2.2) on the curve $r = a$. Applying the semi-inversion procedure (based on the solution to the Riemann-Hilbert problem [5]), we will reduce the problem to the coupled homogeneous infinite system of linear algebraic equations (SLAE) of the second kind

$$\begin{cases} \rho_m = \sum_{n=-\infty}^{\infty} (\Delta_n^E \rho_n + \alpha n \varepsilon_n^E \mu_n) T_{mn}(u), \\[4mm] \mu_m = \sum_{n=-\infty}^{\infty} (-1)^{m+n}(\beta n \varepsilon_n^H \rho_n + \Delta_n^H \mu_n) T_{mn}(-u), \end{cases} \tag{2.7}$$

where

$$u = \cos\theta, \quad \Delta_n^E = |n| - \frac{\Psi_n(x^2+y^2)}{2(\varepsilon_1^+ + \varepsilon_1^-)}, \quad \Psi_n = \frac{AD - BC}{\Delta_2 - \frac{\Delta_\Gamma}{y\Delta}},$$

$$\varepsilon_N^H = 1 + \left[\frac{(w_4\Delta - w_2\Delta_\Gamma)\tilde{\Delta}}{\Delta q(\Delta_\Gamma - y\tilde{\Delta}\Delta_2)} + \frac{\Delta_\Gamma}{y\tilde{\Delta}\Delta_2 - \Delta_\Gamma}\right]\frac{x^2+y^2}{x^2},$$

$$\varepsilon_N^E = 1 + \left[\frac{(\varepsilon_1^+ w_0 \Delta_\Gamma - \varepsilon_1^- w_2 \Delta)\tilde{\Delta}}{\Delta q(y\tilde{\Delta}\Delta_2 - \Delta_\Gamma)} - \frac{\Delta_2 y \tilde{\Delta}}{y\tilde{\Delta}\Delta_2 - \Delta_\Gamma}\right]\frac{x^2 + y^2}{y^2},$$

$$\Delta_n^H = |n| + \frac{(x^2 + y^2)\Delta_\Gamma \tilde{\Delta}}{\Delta_\Gamma - y\tilde{\Delta}\Delta_2}, \quad \beta = -\frac{xq}{n}, \quad \alpha = -\frac{qxy^2}{2n(\varepsilon_1^+ + \varepsilon_1^-)},$$

$$A = \frac{\varepsilon_1^- \beta_4'}{y} - \frac{\varepsilon_1^+ \beta_4 w_1}{x\Delta} - \frac{\varepsilon_2^+ \beta_2' w_0}{y\Delta} + q\beta_2 - \frac{\varepsilon_1^+ q\beta_4 w_0}{\Delta},$$

$$B = -\frac{\varepsilon_1^- \beta_3}{y} + \frac{\varepsilon_1^+ \beta_3 w_1}{x\Delta} + \frac{\varepsilon_1^- \beta_1' w_0}{y\Delta} - q\beta_1 - \frac{\varepsilon_1^+ q\beta_3 w_0}{\Delta},$$

$$C = -\beta_2 + \frac{w_2 \beta_4}{\Delta} + \frac{x\Delta_1 \beta_2'}{y\Delta} - \frac{xq\beta_4 \Delta_1}{\Delta}, \quad \tilde{\Delta} = \frac{\Delta}{x\Delta_1},$$

$$D = \beta_1 - \frac{\beta_3 w_2}{\Delta} - \frac{x\Delta_1 \beta_1'}{y\Delta} + \frac{xq\beta_3 \Delta_1}{\Delta},$$

$$\Delta = \alpha_1' \alpha_4 - \alpha_2' \alpha_3, \quad \Delta_\Gamma = \beta_1' \beta_4 - \beta_2' \beta_3, \quad \Delta_1 = \alpha_1 \alpha_4 - \alpha_2 \alpha_3,$$

$$\Delta_2 = \beta_1 \beta_4 - \beta_2 \beta_3, \quad w_4 = \alpha_3' \alpha_4 - \alpha_4' \alpha_3, \quad w_3 = \beta_3' \beta_4 - \beta_4 \beta_3,$$

$$w_1 = \alpha_1' \alpha_4' - \alpha_2' \alpha_3', \quad w_0 = \beta_2 \beta_2' - \beta_1' \beta_2, \quad w_2 = \alpha_1' \alpha_2 - \alpha_2' \alpha_2,$$

$$\alpha_s = \alpha_s^1, \quad \beta_s = \beta_s^1 \quad (s = 1, 2, 3, 4),$$

$$q = \frac{inh}{k}(y^{-2} - x^{-2}), \quad x^2 = (ka)^2\left[\varepsilon_1^+ - \left(\frac{h}{k}\right)^2\right], \quad y^2 = (ka)^2\left[\varepsilon_1^- - \left(\frac{h}{k}\right)^2\right],$$

$$T_{mn}(u) = \frac{1}{m}V_{m-1}^{n-1}(u) \ (m \neq 0), \quad T_{0n}(u) = \frac{1}{n}V_{n-1}^{-1}(u) \ (n \neq 0),$$

$$T_{00} = -\ln\frac{1 + u}{2},$$

and functions $V_{m-1}^{n-1}(u)$ are determined in [5].

Using (2.4) we can show that

$$\sum_{n=-\infty}^{\infty} |\rho_n|^2 |n| < \infty, \quad \sum_{n=-\infty}^{\infty} |\mu_n|^2 |n| < \infty.$$

Hence it is convenient to introduce, as well as in [93], the Hilbert space $l^2 = l_2 \times l_2 = \{x = \binom{\rho}{\mu} \in l_2^2, \ \rho = (\rho_n)_{n=-\infty}^{\infty} \in l_2, \ \mu = (\mu_n)_{n=-\infty}^{\infty} \in l_2\}$ with the inner product $(xx') = \sum_{n=-\infty}^{\infty}(\rho_n \rho_n^* + \mu_n \mu_n^*)$ where $*$ denotes the complex conjugation.

The matrix of system (2.7) defines a matrix block-type operator-valued function $I - A(h)$, and the spectral problem (2.1)-(2.5) is reduced to the determination of the characteristic numbers of $I - A(h)$. The operator-valued function $A(h)$ has the form

$$A(h) = \begin{bmatrix} A_1 & A_2 \\ A_3 & A_4 \end{bmatrix}.$$

$A_j(h) : l_2^2 \to l_2^2 \ (j = 1, 2, 3, 4)$ are the kernel operator-valued functions considered in the domain $G(A) = R \setminus C_h$ where C_h is the set of poles $A_j(h)$ [93] and R is the surface of the analytical continuation of the Hankel function $H_m^{(1)}(\tilde{k}_{N+1}r)$ from the interval $|h| < k\sqrt{\varepsilon_{N+1}}$. Here one has to take into account the double-valued character of the function $\tilde{k}_{N+1} = \pm\sqrt{k^2 \varepsilon_{N+1} - h^2}$. $A_j(h)$ is an even function of $\tilde{k}_{j(l)}$ $(j \neq N+1)$ [94]. It means that the spectral values situated on different sheets of the function $\tilde{k}_{j(l)}(h)$ $(j \neq N+1)$ are indistinguishable.

The Riemann surface R is "double infinite-sheet": each of the logarithmic sheets $\left\{ -(\pi/2) + 2\pi n < \arg \tilde{k}_{N+1} < (3/2)\pi + 2\pi n \right\}$ of the Hankel function consists, in turn, of two more sheets $R_n^{1,2}$ because of the branch points $h = \pm k\sqrt{\varepsilon_{N+1}}$ of the function $\tilde{k}_{N+1}(h)$. Therefore, we take $R = \cup_{n=-\infty}^{\infty} R_n = \cup_{n=-\infty}^{\infty}(R_n^1 \cup R_n^2)$ as the domain of definition of the spectral problem for the considered open structure. For shielded (closed) CSWs and CStWs the domain of definition coincides with the whole complex plane. The results of Ref. [93] enables us to formulate and prove the equivalence theorem: the propagation constants of the multi-layered CSW and CStW coincides with the characteristic numbers of the operator-valued function $I - A(h)$.

The equivalence theorem and fundamental results of the spectral theory of finite-meromorphic operator-valued function [95] imply the discreteness and finite multiplicity of the spectrum of propagation constants on the Riemann surface R proved in Ref. [93]. In other words, the characteristic numbers σ of the operator-valued function $I - A(h)$ form a countable set of points in every bounded subset of R and not more than a finite number of eigenfunctions corresponds to every characteristic number.

Estimates of the matrix elements and the results of Refs. [49, 95, 96] lead us to the conclusion that the sequence of the roots of the reduced finite-dimensional equations

$$\det\{I - A_{NR}(h)\} = 0, \tag{2.9}$$

where NR is the order of the block reduction, converges to the eigenvalues of the initial spectral problem (2.1)-(2.5). Since $A_{NR}(h)$ is an analytical function of h, one can apply the algorithm based on the Newton's method to calculate the roots of (2.9). The results of our systematic calculations have shown that it is sufficient to take the reduction order of every block $A(h)$ as $NR =$ entier$[\max(\tilde{k}_j r_j) + 3]$ to find the roots with the accuracy 10^{-4}.

The properties of the analytical continuation of $H_m^{(1)}(\tilde{k}_{N+1}r)$ on R and the form of $A_j(h)$ yield the symmetry of the eigenvalue positions on the surface R for the considered multi-layered waveguides without losses (Im $\varepsilon_{j(l)} = 0$). Namely, if $h = h_0$ is the eigenvalue, then $-h_0$ and h_0^* are also the solutions of the spectral problem. If to take into account the structure of the Riemann surface of the function $\tilde{k}_{N+1}(h) = \pm\sqrt{k^2 \varepsilon_{N+1} - h^2}$ and to paste together the sheets along the lines starting from the points $h = \pm k\sqrt{\varepsilon_{N+1}}$ parallel to the imaginary axis (Re h Im $h > 0$) and if h_0 is on the sheet R_n^1, then

$$-h_0 \in R_{-n}^1, \quad \pm h_0^* \in \begin{cases} R_{-n}^1, & |\text{Re } h_0| > k, \\ R_{-n}^2, & |\text{Re } h_0| < k. \end{cases} \tag{2.10}$$

Note that the symmetry of eigenvalues on the sheet R_0 for H_{0m} waves has been established in [94] for the three-layered Goubau line.

Hence the wave spectrum of multi-layered CSWs and CStWs without losses always contains the fours of points $\pm h_0$, $\pm h_0^*$ situated on the sheets R according to relation (2.10). The waves corresponding to the eigenvalues positioned on the main physical sheet R_0 are of special interest. Two sub-sheets $R_0^{1,2}$ in the complex plane h are depicted in Fig.25a. The cut line for the Hankel function goes along the real axis of

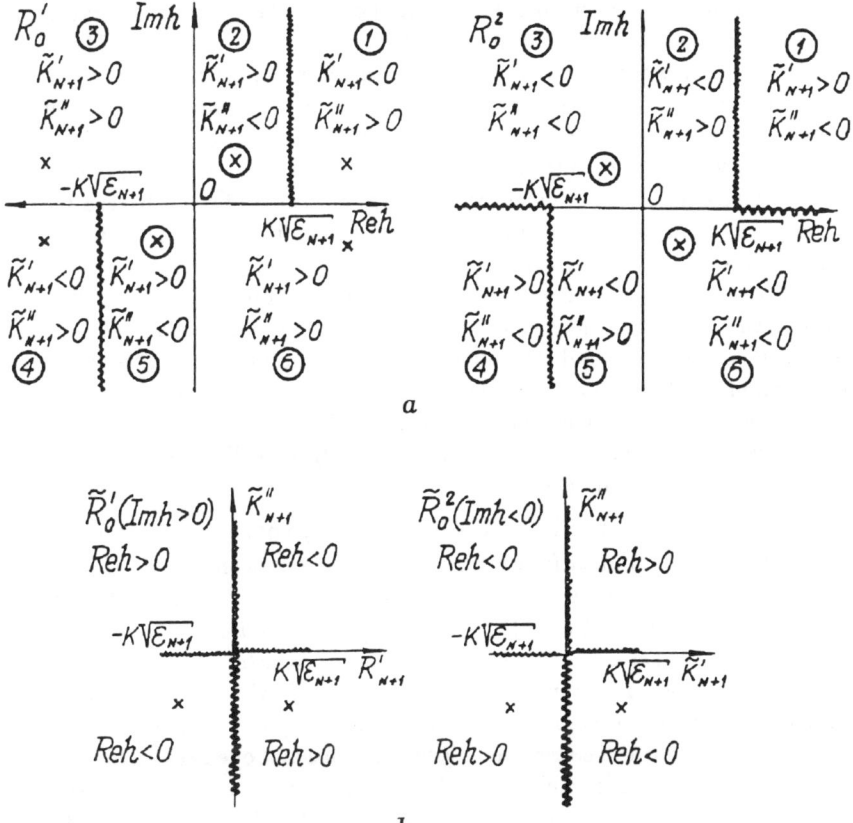

Figure 25.

the sheet R_0^2 for $|\operatorname{Re} h| > k\sqrt{\varepsilon_{N+1}}$ with $\operatorname{Im} \tilde{k}_{N+1} < 0$, $\operatorname{Re} \tilde{k}_{N+1} = 0$. The classification of waves in open waveguides is given in Ref. [97] on the complex plane of the transversal wavenumber of the external (open) domain (here this wavenumber is denoted by $\tilde{k}_{N+1}$) and is illustrated by the diagram in Fig.25. The sheets are chosen with respect to the sign $\operatorname{Im} h > 0$ and $\operatorname{Im} h < 0$ (the cut line is $\operatorname{Im} h = 0$). The sheets $\tilde{R}_0^{1,2}$ are shown in Fig.25b. In this case the cut line for the Hankel function is situated only on the sheet $\tilde{R}_0^2$. The symmetry of the solutions described by (2.10) is clearly seen in Fig.25b. Let us note that (2.10) allows one to consider the spectral points in the first quadrant of every sheet R_0^1 and R_0^2. However, the symmetry is violated in the presence of losses ($\operatorname{Im} \varepsilon_{j(l)} \neq 0$). The correct physical definition of the obtained solutions is the very important problem. For the fours of the spectral points from the sheet R_0 shown in Fig.25a (we consider the solutions indicated by h_0 and $-h_0$) the corresponding waves propagate in the directions $\pm z$ going off the open waveguide ($\tilde{k}'_{N+1} > 0$). Their amplitudes increase with respect to r and decrease in the direction of propagation. They are called the leaky waves. Such waves describe the radiation processes [98]. The waves that propagate in the directions $\pm z$ and come to the waveguide correspond to the solutions h_0^* and $-h_0^*$ taken from the sheet R_0^2. Their amplitudes decrease with respect to r and increase in the direction of propagation. These solutions may describe the generation processes in the waveguides filled with an active medium. The physical sense of the solutions situated on other sheets of the Riemann surface R_n $(n \neq 0)$ is still not clear. However, one has to point out that the results presented in Ref. [99] show that one can reconstruct scattering matrices by the spectrum of complex frequencies situated on these "nonphysical" sheets of the Riemann surface. The importance of taking into consideration the higher-order sheets is also confirmed by the fact that eigenvalues may transfer from one sheet to another when nonspectral parameters vary.

The spectral theory constructed here for open waveguides has much in common with the corresponding results obtained in chapter 1 for open resonators. The main difference is that the waveguide spectral problem is three-dimensional, although it is reduced to the system of two-dimensional problems. A more complicated dependence of the radiation conditions on the spectral parameter occurs In the spectral problem for open waveguides the longitudinal wavenumber h is taken as a complex spectral parameter, while the frequency is one of the nonspectral (real) parameters (as well as complex transversal wavenumbers $\tilde{k}_{j(l)}$). The spectral parameter h is a double-valued function of $\tilde{k}_{j(l)}$ $(j = N+1)$ with two algebraic branch points $h = \pm\sqrt{k^2 \varepsilon_{N+1} - \tilde{k}_{N+1}^2}$. Consequently, the Riemann surface R for the eigenwave problems in open waveguides has a more complicated structure than that for the eigenoscillation problems in open resonators. It is shown in Refs. [19, 96] that one can state the spectral problem to determine the frequency-dependent k-spectrum $(k = \omega/c)$ of an open waveguide assuming that h is given parameter. If to put $h = 0$, then the determination of the complex k-spectrum of an open waveguide is reduced to the determination of the complex k-spectrum for the corresponding two-dimensional open resonator.

The spectral problem for multi-layered structures is reduced to the analysis of two independent dispersion relations corresponding to the families of $H_z^+ E_z^-$ and $H_z^- E_z^+$ waves:

$$\left\| \begin{matrix} A_{mn}^{11} & A_{mn}^{12} \\ A_{mn}^{21} & A_{mn}^{22} \end{matrix} \right\| = 0, \tag{2.11}$$

where

$$A_{mn}^{11} = \left\{ \frac{1}{2} \chi_n \Delta_n^E T_{mn}^{-(+)}(u) - \delta_{mn} \right\}_{m,n=1(0)}^{\infty},$$

$$A_{mn}^{12} = \left\{ \alpha n \varepsilon_n^E T_{mn}^{-(+)}(u) \right\}_{m,n=1(0)}^{\infty},$$

$$A_{mn}^{21} = \left\{ (-1)^{m+n} \beta n \varepsilon_n^E T_{mn}^{-(+)}(u) \right\}_{m,n=0(1)}^{\infty},$$

$$A_{mn}^{22} = \left\{ (-1)^{m+n} \frac{1}{2} \chi_n \Delta_n^H T_{mn}^{-(+)}(-u) - \delta_{mn} \right\}_{m,n=0(1)}^{\infty},$$

$$T_{mn}^{-(+)} = T_{mn}(\pm u) \mp T_{-mn}(\pm u), \quad \chi_n = \begin{cases} 1, & n = 0, \\ 2, & n \neq 0. \end{cases}$$

All the values and parameters in (2.11) are given above.

We use dispersion relations (2.11) to calculate the wave propagation constants in closed and open CSWs and CStWs with arbitrary finite number of layers. In every specific case the coefficients $\hat{\alpha}_s^{(1)}$ and $\hat{\beta}_s^{(1)}$ $(s = 1, 2, 3, 4)$ corresponding to the radial domains adjoining the partially shielded surface are determined by consecutive formulas. For the layers belonging to the external domain $(r > a)$ the consecutive procedure is started from the external layer $(j = N + 1)$ and continued to the internal layer, and in the internal domain we begin from the internal layer $(l = M + 1)$ and proceed to the external layer $(l = 1)$.

The spectra of normal waves in CSWs and CStWs filled with homogeneous media are studied in details in Refs. [19, 100-102]. The approximate expressions for the propagation constants are derived that are valid in the limited range of the parameter variation, namely, for $\theta \to 0$, $\delta \to 0$ and $\kappa = ka\sqrt{\varepsilon - 1} < 1$. For the CSW and CStW with a layered-radial dielectric filling it is rather difficult to obtain such expressions explicitly even in the case when the external and interior media have only two layers. However, the analysis of equation (2.11) in the limiting cases $\theta \to 0$ and $\delta \to 0$ enables one to derive approximate explicit dispersion equations and to estimate the degree of perturbation with respect to parameters θ and δ.

Let us consider the waves of the family $H_z^+ E_z^-$ in the CSW when $\theta \to 0$. Using the approximate expressions for $T_{mn}(u)$ from Ref. [5] we reduce the dispersion relation (2.11) to the quasi-diagonal form

$$
\left\|
\begin{array}{cc}
\left\{ B_{mn} + \delta_{mn} \right\}_{m,n=1}^{\infty} & \left\{ C_{mn} \right\}_{m,n=1}^{\infty} \\[2mm]
\left\{ D_{mn} \right\}_{m,n=0}^{\infty} & \left\{ E_{mn} + \dfrac{(x^2+y^2)\tilde{\Delta}_n \Delta_n^{(r)}/(y\Delta_n^{(2)})}{\tilde{\Delta}_n - \Delta_n} \delta_{mn} \right\}_{m,n=0}^{\infty}
\end{array}
\right\| = 0, \qquad (2.12)
$$

where

$$
B_{mn} = O(\theta^4), \quad C_{mn} = O(\theta^4),
$$

$$
D_{mn} = \beta \varepsilon_n^H + O(\theta^4) \ (n \neq 0), \quad D_{mn} = O(\theta^4) \ (m \neq n, \ m, n \neq 0),
$$

$$
D_{m0} = 0, \quad D_{0n} = -\beta \varepsilon_n^H \ln^{-1} \sin \frac{\theta}{2}, \ (n \neq 0)
$$

$$
E_{m0} = \Delta_0^H m^{-1} + O(\theta^4) \ (m \neq 0), \quad E_{mn} = O(\theta^2) \ (m, n \neq 0),
$$

$$
E_{0n} = -\Delta_n^H \ln^{-1} \sin \frac{\theta}{2} \ (n \neq 0), \quad E_{00} = \ln^{-1} \sin \frac{\theta}{2}.
$$

The analysis of (2.12) shows that the spectrum of the multi-layered CSW contains three subsets of waves from the family $H_z^+ E_z^-$. The first is the spectrum of waves of the external domain $(r > a)$. Their propagation constants satisfy the equation

$$
\tilde{\Delta}_m = O\left(\ln^{-1} \sin \frac{\theta}{2} \right), \qquad (2.13)
$$

or

$$
\alpha_1^{(2)'}(a)\alpha_4^{(1)}(a) - \alpha_2^{(1)'}(a)\alpha_3^{(1)}(a) = O\left(\ln^{-1} \sin \frac{\theta}{2} \right).
$$

The roots of (2.13) for $\theta = 0$ are the eigenvalues of $H^+ E_{mn}^-$ $(m \geq 1)$, $E^- H_{mn}^+$ $(m \geq 1)$ and H_{0n}^+ waves of the multi-layered Goubau line. It is known [103] that this line has three principal waves that have no cut-off frequencies: the symmetrical E_{00} wave and double-degenerated with respect to orientation of vectors $\mathbf{E}$, $\mathbf{H}$, $H^{\pm} E_{11}^{\mp}$ wave. Hence, (2.13) describes the principal quasi-$H^+ E_{11}^-$ together with higher order quasi-σ_1^+ waves, having cut-off frequencies.

The second spectral family of the multi-layered CSW contains the waves of the internal domain ($r < a$) which are close to $E^- H_{mn}^+$ ($m \geq 1$) and $H^+ E_{mn}^-$ ($m \geq 1$) H_{0n} waves of the circular cylindrical multi-layered waveguide. Their spectral values satisfy the equation

$$\Delta_m^{(r)} = O\left(\ln^{-1} \sin \frac{\theta}{2} \right),$$

where

$$\beta_1^{(1)\prime}(a)\beta_4^{(1)}(a) - \beta_1^{(2)\prime}(a)\beta_3^{(1)}(a) = O\left(\ln^{-1} \sin \frac{\theta}{2} \right).$$

It is known that the propagation constants of the circular multi-layered waveguide may be real (for propagating modes), imaginary (for attenuating modes) and complex [104, 105]. Let us note that inverse waves may exist in circular metallic waveguide. The group and phase velocities of these waves have opposite directions.

Third, there is the slot quasi-H_{00}' wave in the spectrum of $H_z^+ E_z^-$ waves of the multi-layered CSW. The left-hand side of the corresponding dispersion relation belongs to the zero diagonal cell of (2.12) and for this wave $x^2 + y^2 = o[\ln^{-1} \sin(\theta/2)]$ when $\theta \to 0$ and $h/k \to \sqrt{(\varepsilon_1^+ + \varepsilon_1^-)/2}$.

The spectral composition of the family of $H_z^- E_z^+$ waves in the multi-layered CSW is considered for $\theta \to 0$ in the same way. It is shown that the spectrum contains the waves perturbed by the narrow slot, namely, the waves of the N-layered Goubau line whose eigenvalues h satisfy the equation

$$\alpha_1^{(1)\prime}(a)\alpha_4^{(1)}(a) - \alpha_2^{(1)\prime}(a)\alpha_3^{(1)}(a) = O(\theta^2), \tag{2.14}$$

There are also the waves that are close to the eigenmodes of the circular metallic M-layered waveguide. The propagation constants of these waves are the roots of the dispersion relation

$$\beta_1^{(1)\prime}(a)\beta_4^{(1)}(a) - \beta_2^{(1)\prime}(a)\beta_3^{(1)}(a) = O(\theta^2), \tag{2.15}$$

E_{00} and $H^- E_{11}^+$ waves of the multi-layered Goubau line perturbed by the narrow slot are the dominant waves in this family. In the case of a CSW filled with a homogeneous dielectric, quasi-σ_0 and quasi-σ_2 waves correspond to these principal waves.

Let us consider now how the waves in the multi-layered dielectric waveguide are perturbed by the narrow metallic strip with the angle width δ placed on one of the boundary of the layers. We will call this waveguide the multi-layered CStW. One can show that its dispersion relation for the waves of the family $H_z^+ E_z^+$ has the form $\Psi_n = O(\delta^2)$ (for $\delta \to 0$) and describes perturbation of the waves of the circular layered waveguide caused by the strip. For the family $H_z^- E_z^+$ whose eigenvalues satisfy the equation $\Psi_n = O(\ln^{-1} \sin \delta/2)$, the spectrum contains the waves for which $x^2 + y^2 = O(\ln^{-1} \sin \delta/2)$ when $\delta \to 0$. Among them there is the principal quasi-σ_0 wave which is present in the spectrum only if the waveguide has a strip.

Hence, the principal waves of the multi-layered CStW are quasi-σ_0 and $H^- E_{11}^+$ waves in the family $H_z^- E_z^-$ and quasi-$H^- E_{11}^+$ waves in the family $H_z^- E_z^-$. A narrow metallic strip perturbs the waves from the family $H_z^- E_z^+$ of the multi-layered circular dielectric waveguide stronger than the waves from the family $H_z^+ E_z^-$.

We considered open CSWs and CStWs. It is obvious that a part of the electromagnetic energy in these structures is always radiated to the external domain. If the transmission line must be completely isolated from external factors (for example, inhomogeneities and fields), one should make it shielded (such lines are also called screened or closed). Consider the corresponding mathematical model of the shielded CStW. Here the external $(r > a)$ and internal $(r < a)$ domains have one layer $(N, M = 0)$ and each of the coefficients α_s^1 of the dispersion relation may be determined as a result of satisfying the boundary conditions on the metallic surfaces at $r = b$. Finally we obtain

$$\alpha_1^{(1)}(a) = \frac{H_n^{(1)\prime}(x_1)}{J_n(x_1)} J_n(x) + H_n^{(1)}(x), \quad \alpha_{23}^{(1)}(a) = 0,$$

$$\alpha_4^{(1)}(a) = -\frac{H_n^{(1)}(x_1)}{J_n(x_1)} J_n(x) + H_n^{(1)}(x),$$

where $x_1^2 = (kb)\left[\varepsilon_2 - (h/k)^2\right]$ and $x^2 = (ka)\left[\varepsilon_1 - (h/k)^2\right]$.

Taking into account that $\beta_s^1(a) = 0$ one can show that in this case of two-layered shielded structures the following values enter the dispersion relations:

$$\Delta_n^E = |n| - \frac{x^2 + y^2}{2(\varepsilon_1 + \varepsilon_2)} \left\{ \varepsilon_1 f_n - \varepsilon_2 \Phi_n - \frac{n^2 h^2 k^2 a^4 (\varepsilon_1 - \varepsilon_2)^2}{x^4 y^4 (f_n - F_n)} \right\},$$

$$F_n = \frac{\alpha_1^{(1)\prime}(a)}{x \alpha_1^{(1)}(a)}, \quad \Phi_n = \frac{\alpha_4^{(1)\prime}(a)}{x \alpha_n^{(1)}(a)},$$

(2.16)

where

$$\alpha_1^{(1)\prime}(a) = -\frac{H_n^{(1)\prime}(x_1)}{J_n'(x_1)} J_n'(x) + H_n^{(1)\prime}(x), \quad \alpha_4^{(1)\prime}(a) = -\frac{H_n^{(1)}(x_1)}{J_n(x_1)} J_n'(x) + H_n^{(1)\prime}(x).$$

One can determine a static limit for the deceleration coefficient h/k in the shielded cylindrical strip line [106] when $|x| < 1$, $|y| < 1$ and $\delta \to 0$:

$$\frac{h}{k} = U_0 \sqrt{\frac{(\ln d - \ln \tau)\varepsilon_2}{U_0^2 \ln d - \varepsilon_2 \ln \tau}},$$

$$\tau = \sin \frac{\delta}{2}, \quad U_0^2 = \frac{\varepsilon_1 + \varepsilon_2}{2}, \quad d = \frac{b}{a}.$$

§ 7. Isolated critical points and the wave transformation in the multi-layered CSW and CStW

In the previous section the spectral problem for the waves in the multi-layered CSW and CStW is reduced to the determination of the roots of the dispersion relation

$$F(h, s_n) = \det\{I - A(h, s_n)\} = 0, \tag{2.17}$$

where s_n denotes a finite number of non-spectral parameters. Traditionally the structure of the set of roots σ_n is studied for different s_n. In particular, the evolution of σ_n

with respect to the frequency, slot angle, radii of dielectric layers and other parameters was considered in Refs. [100-102, 106, 107]. Physical situations when the eigenvalues h corresponding to two different waves draw together and the waves "exchange" their fields may be defined as interaction or "intertype coupling" ("communication") of eigenwaves (see Refs. [57, 58]).

Using the approach presented in [63], one can study the phenomena of interaction in open resonators and waveguides considering the so-called isolated Morse critical points.

The isolated singular points at which function $F(h, w)$ has the double multiplicity zero

$$F(h_0, w_0) = 0; \quad \frac{\partial F}{\partial h} = 0; \quad \frac{\partial^2 F}{\partial h^2}\frac{\partial F}{\partial w} \neq 0 \tag{2.18}$$

are of a special interest in the theory of the CSW and CStW. The critical points defined by (2.18) will be called the degeneration points. We will show that eigenwaves may interact in the vicinity of these points.

Let us consider the situations which may be described by means of the isolated singular points. The methods of the theory of singularities and singular points form the mathematical basis to determine the laws of the eigenwave interaction. Before to start the analysis it is important to point out that function $F(h, s_n)$ is analytical with respect to all its variables except for the slot angle. θ.

Slot and strip waveguides have two characteristic features. First, the spectrum in the limiting cases $\theta = 0°$ and $\theta = 180°$ differs from the case $0 < \theta < 180°$ because the operator-valued function $A(h, \theta)$ is singular with respect to θ [108]. Consequently, there are singular waves in the spectrum: the slot quasi-H_{00}^+ wave and strip quasi-T_0^- wave. They exist for $0 < \theta < 180°$ but there are neither slot waves for $\theta = 0°$ no strip waves for $\theta = 180°$. The second basic feature is the communication of their external and internal domains through the slot in the metallic surface. Each domain contains different dielectric layers, and a redistribution of electromagnetic energy between them occurs.

The qualitative differences of the CSW spectrum for $\theta = 0$ and $\theta \neq 0$ are shown in Fig.26 where the dispersion characteristics of $H_z^+ E_z^-$ waves are presented. The slot waves belong to this family. One can see that the dispersion curves of the circular metallic waveguide ($\theta = 0$) intersect with the line $h/k = \sqrt{(\varepsilon^+ + \varepsilon^-)/2}$ corresponding to the slot wave for $\theta \to 0$. However, the CSW dispersion curves for $\theta \to 0$ ($\theta \neq 0$) have no degeneration points and the so-called Wien graph is observed which is characteristic for the intertype coupling (communication). There is a sharp change of the dispersion curve behaviour close to the intersection point where the eigenwaves "exchange" their fields. The surface waves (corresponding to real eigenvalues h) interact in the vicinities of these points [109]. The coordinates of the isolated Morse critical points coincide for the small θ with the coordinates of the intersection of tangents (dot-dashed lines in Fig.26) and expansion (2.18) approximately describes the interaction of H_{00}^+ waves with $H_{11}^+(\theta)$ and $H_{21}^+(\theta)$ waves.

Hence the wave transformation

$$H_{00}^+(\theta) \to H_{mn}^+(\theta)(E_{mn}^-(\theta)); \quad H_{mn}^+(\theta)(E_{mn}^-(\theta)) \to H_{00}^+(\theta)$$

occurs in CSW for $\theta \to 1$ with increasing frequency parameter ka in the vicinity of $(ka)_{0i} = \nu_{mn}\sqrt{2(\varepsilon_1^+ - \varepsilon_1^-)}$, where ν_{mn} is the root of Bessel function or of its derivative.

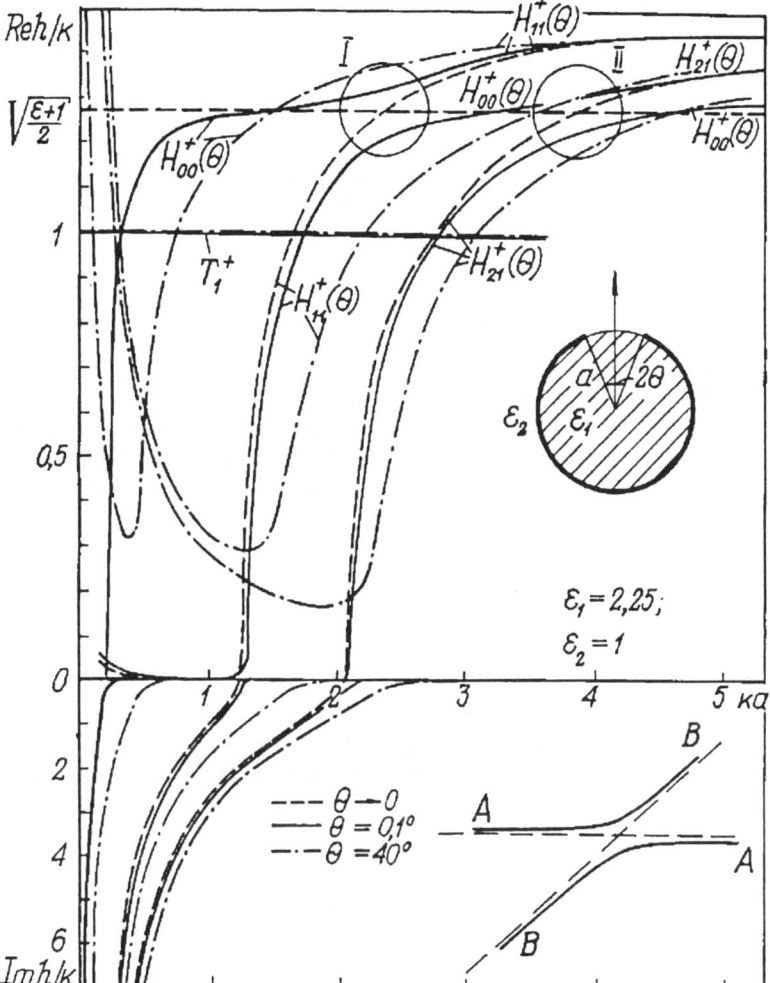

Figure 26.

There exist the Morse critical points in the transformation domains coinciding with $(h_0, (ka)_{0i})$ for $\theta \to 0$ where $h_0 = k\sqrt{(\varepsilon_1^+ + \varepsilon_1^-)/2}$.

A similar physical picture is observed in the CStW. Here for $\delta \to 0$ ($\theta \to 180°$) the principal quasi-T_0 strip wave exists which is an analogue of the E_{00} wave in the Goubau line [94]. For $\delta = 0$ this wave is absent and the CStW spectrum consists only of the waves of a circular dielectric waveguide (containing, in general, several layers). The analysis of the dispersion curves of the homogeneous CStW for $H_z^- E_z^+$ and $H^- E_{11}^+$ waves (corresponding to the waves of the circular dielectric pivot) and the limiting value of the deceleration coefficient of the quasi-T_0 wave $h/k \to \sqrt{(\varepsilon_1^+ + \varepsilon_1^-)/2} = h_0/k$ ($\varepsilon_1^- = 1$, $\delta \to 0$) show that for the CStW and CSW the dispersion curves have no degeneration points and the wave transformation

$$T_0^-(\theta) \to H^- E_{mn}^+(\theta)(E^+ H_{mn}^-(\theta)); \quad H^- E_{mn}^+(\theta)(E^+ H_{mn}^-(\theta)) \to T_0^-(\theta)$$

occurs close to the points $(h_0, (ka)_{0i})$. Here $(ka)_{0i}$ is the value of ka at which the deceleration coefficients of the pivot waves coincide with h_0/k. The Morse points coinciding with $(h_0, (ka)_{0i})$ for $\delta \to 0$ correspond to the domains of interaction. The behavior of the lines of the transversal electric field equal power for quasi-T_0 and $H^- E_{11}^+(\theta)$ waves confirm the phenomenon of the mutual transformation of these waves.

The wave interaction are observed when the slot angle θ varies. However, as it has been pointed out above, the function $F(h, \theta)$ is not analytical with respect to θ and it is not possible to search for the isolated critical points of complex waves if to assume that θ is a complex parameter, although such points exist. When the surface waves interact, the Morse critical points may be found for real values of θ. The transformation of the waves of a circular metallic waveguide to the waves of a circular dielectric pivot caused by the varying parameter θ for the fixed ka is shown in Fig.27. The $H_{00}^+(\theta) \to H^+ E_{11}^-(\theta)$ is the only slow wave that exists for all values of θ and the $H_{11}^+(\theta) \to H_{01}^+(\theta)$ is the slow wave for the angles $\theta < 12°$ and $85° \lesssim \theta \lesssim 152°$. Here the surface wave interaction is observed in the vicinities of the Morse critical points $h/k = 1.11$ and $\theta = 129.5°$. One can reconstruct the spectral curves describing this interaction after calculating all the necessary coefficients in equation (2.18). The relations for $\mathrm{Re}(h/k)$ and $\mathrm{Im}(h/k)$ defining the leaky waves as functions of θ enable one to separate the domains where the curves are similar to the Wien graphs. These domains are denoted by circles in Fig.27: the domains 1, 2 and 3 correspond to the interaction of $H^+ E_{11}^-(\theta)-E^- H_{12}^+(\theta)$, $E^- H_{12}^+(\theta)-H^+ E_{32}^-(\theta)$ and $H^+ E_{32}^-(\theta)-E^- H_{21}^+(\theta)$ waves, respectively. The isolated critical points of the complex (leaky) waves may be found if to consider the analytical continuation of the function $F(h, s_n)$ with respect to ka when the (fixed) value of θ belongs to the interval corresponding to the transformation domains. The dispersion characteristics of $H_z^+ E_z^-$ waves in the CSW with the slot angle $\theta = 115°$ describing the surface wave interaction enable us to assert that the Morse critical point $h_0/k = 0.698 + i0.692$, $ka_0 = 1.73 + i0.1$ corresponds to domain 1. The Morse critical points $h_0/k = 0.263 + i0.235$, $ka_0 = 1.88 + i0.282$ and $h_0/k = 0.053 + i1.264$, $ka_0 = 2.29 + i0.178$ correspond to domain 2. The degeneration point $h_0/k = 0.323 + i0.89$, $ka_0 = 2.43 + i0.0083$ and the Morse point $h_0/k = 0.047 + i0.596$, $ka_0 = 2.62 + i0.223$ are found in domain 3. The answer to the question, which of these two points better describes the given physical situation, depends on the value of $\mathrm{Im}(ka_0)$ and on the distance between the point $\left(\mathrm{Re}(h_0/k), \mathrm{Im}(h_0/k)\right)$ and the values $\mathrm{Re}(h/k)$, $\mathrm{Im}(h/k)$ in the interaction domain. In the given case $E^- H_{21}^+$ and $H^+ E_{11}^-$ waves are better described with the help of the second Morse critical point. On the other hand, the same situation for $E^- H_{21}^+$ and $H^+ E_{22}^-$ waves is better reconstructed by the third degeneration point.

The characteristic example of the degeneration point is presented in Fig.28, where the graphs of h/k for $H_{11}^+(\theta)$ and $H_{21}^+(\theta)$ waves calculated for two close values of ka are depicted. For $ka = 1.35$ the $H_{11}^+(0) - H^+ E_{11}^-(180°)$, $H_{21}^+(0) - E^- H_{12}^+(180°)$ transformations take place. For $ka = 1.4$ one has the Morse critical point $h/k = 0.229 + i0.702$, $ka = 1.705 - i0.179$ and the $H_{11}^+(0) - E^- H_{12}^+(180°)$, $H_{21}^+(0) - H^+ E_{11}^-(180°)$ evolution. The interaction domain is denoted by the index 1, i.e., it has the same number as in Fig.27. Here, as well as in the previous case, an isolated critical point

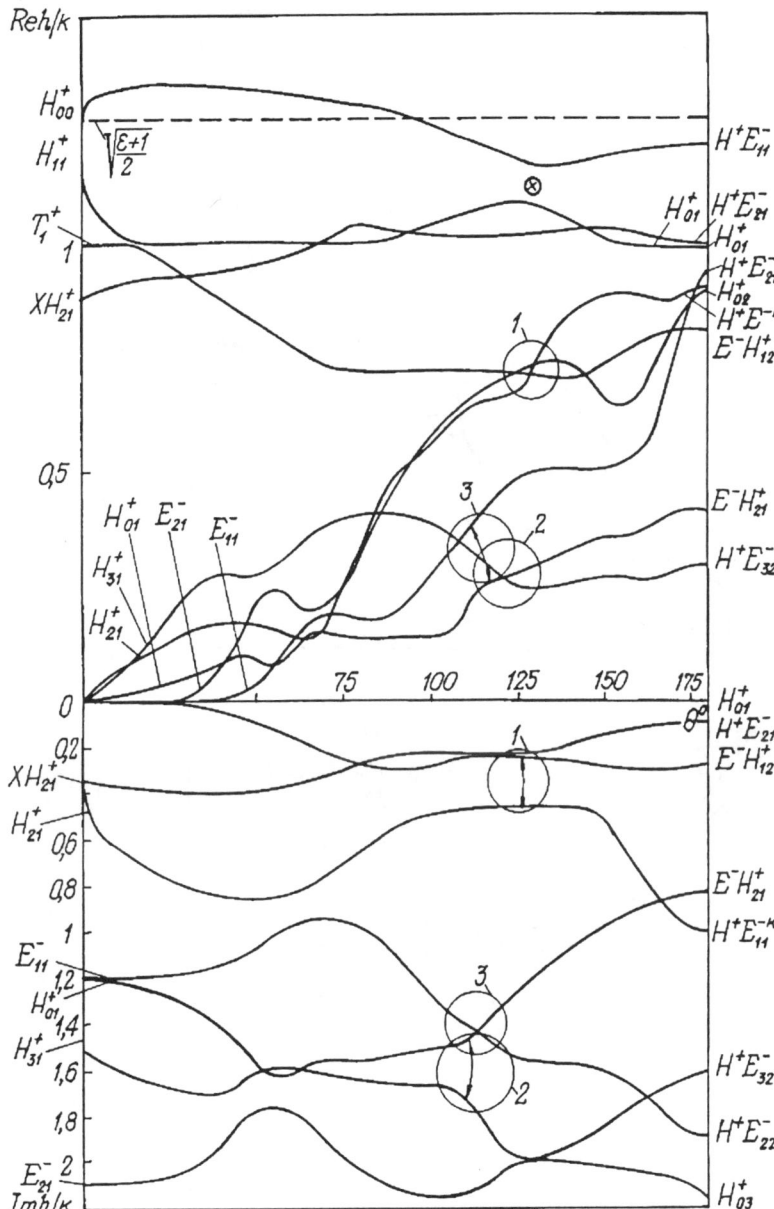

Figure 27.

cannot be found by the analytical continuation with respect to θ. Therefore, we will consider the continuation with respect to the frequency parameter ka to obtain the degeneration point $h_0/k = 0.549 + i1.277$, $ka_0 = 1.39 + i0.0005$ for the CSW with the parameters $\varepsilon = 2.25$ and $\theta = 103.1°$. The dispersion curves calculated for two close values of θ are depicted in Fig.29. It is obvious that Im ka_0 vanishes at the degeneration point for a certain intermediate value of θ. In addition to this, there exists the point

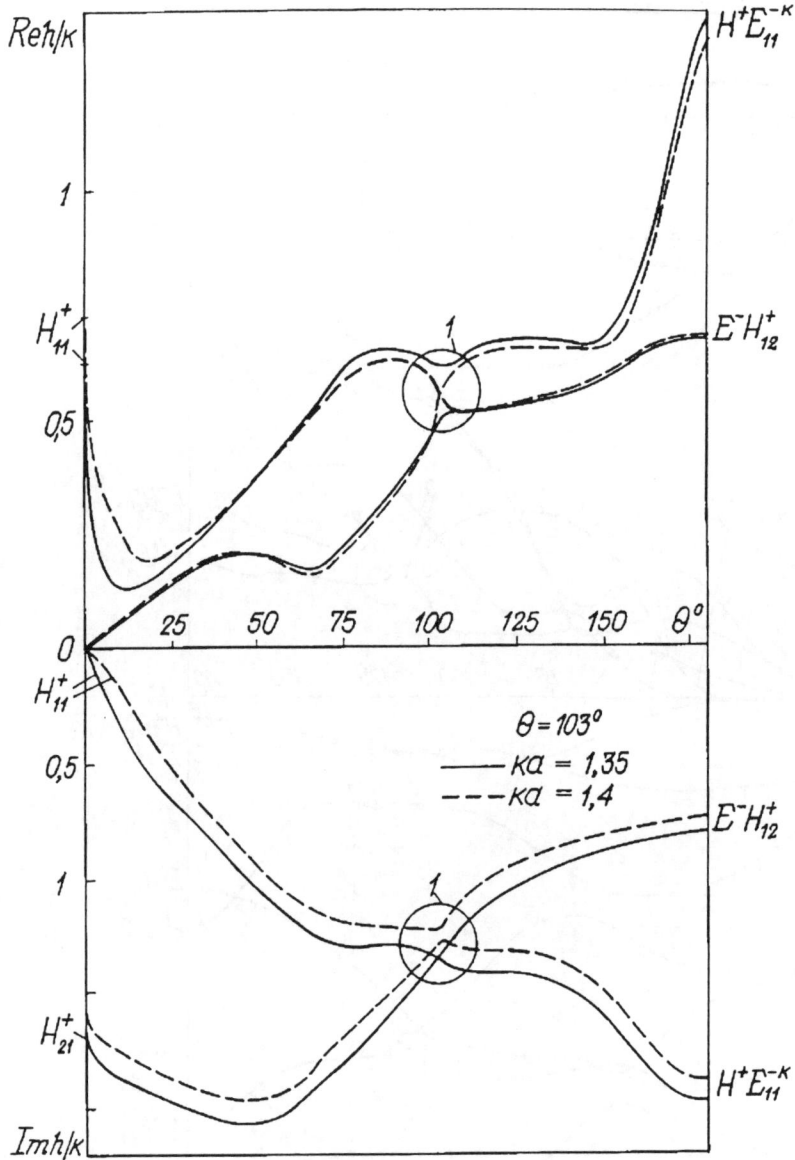

Figure 28.

at which two different eigenvalues coincide for certain real values of ka and θ.

The interaction phenomena have been also indicated in the multi-layered CSW and CStW. One could expect the display of this interaction because of a greater number of the non-spectral parameters, namely, $kr_{j(l)}^{+(-)}$ and $\varepsilon_{j(l)}^{+(-)}$ $(j = 2, \ldots, N, l = 1, 2, \ldots, M)$. If $\theta = 0$, the waves of the external and internal domains (having the forms of a closed multi-layered waveguide or the multi-layered Goubau line) exist independently. When $\theta \neq 0$, these waves may interact. One can observe this phenomenon in the shielded

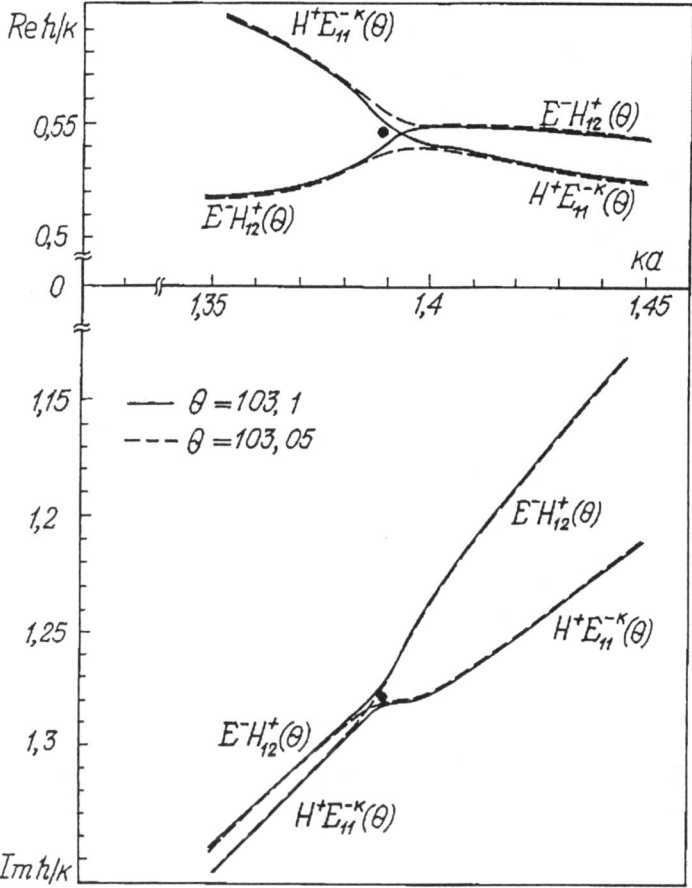

Figure 29.

CSW with a double-layered dielectric filling ($\varepsilon_1 = 4$, $\varepsilon_2 = 2.25$ and $a/b = 0.5$). The dispersion curves of the family $H_z^+ E_z^-$ are shown in Fig.30 where the dispersion curves of the waves of the circular metallic waveguide and the coaxial cable are depicted by bar lines. If these dispersion curves intersect for $\theta = 0$ and the waves of these domains do not interact, the slot causes the wave transformation. The slot quasi-H_{00} wave interact with the waves of a circular metallic waveguide filled with the homogeneous dielectric ($\varepsilon = 4$) and perturbed by a narrow slot close to the points where the dispersion curves of the waves of the internal waveguide intersect the line $h/k = \sqrt{(\varepsilon_1 + \varepsilon_2)/2}$. The Morse critical points coinciding for $\theta \to 0$ with the points of the "asymptotical" tangents constructed in the vicinities of the intersection points (see Fig. 29) lie in the interaction domains.

We may conclude that the wave interaction in open and closed waveguides may be described within the framework of the mathematical model based on the study of isolated singularities of a mapping $F : C^2 \to C$ defined in the domain D, where $F(h, w)$ is an analytical function with respect to two complex variables. In order to analyze the spectral characteristics of complicated waveguide structures, one should consider the

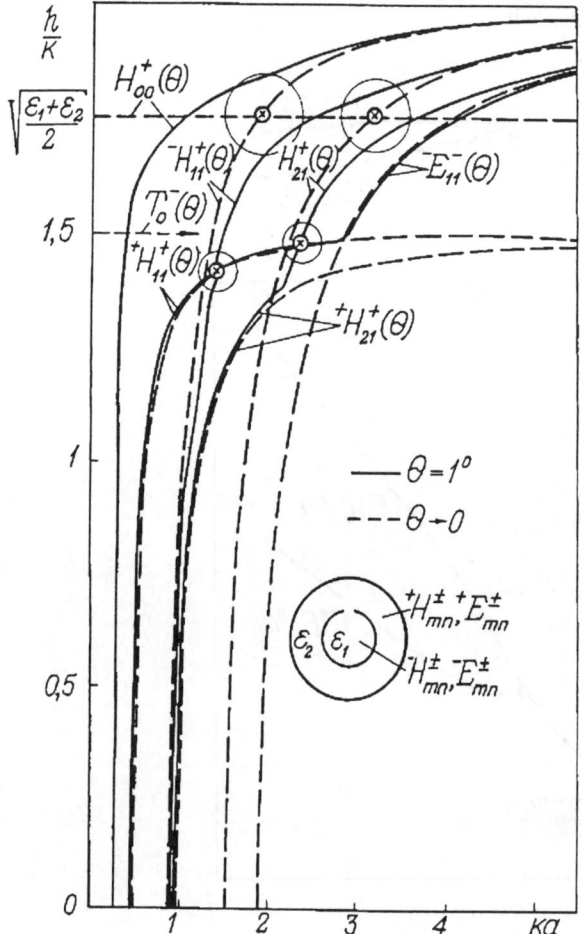

Figure 30.

evolution of isolated critical singular points with respect to the non-spectral parameters together with the study of the eigenvalue evolution.

§ 8. Quasi-statical methods for the study of cylindrical slotted waveguides

The results of the rigorous spectral theory developed for idealized structures allow us to consider the characteristics of open transmission lines. Let us indicate the data which may be obtained as a result of the asymptotical solution to the dispersion relation (2.11) of the CSW with one dielectric layer by the successive iteration method for sufficiently small values of θ. The real roots lie in the interval $k \leq h \leq \sqrt{\varepsilon}$ and their number is finite. There are no real and imaginary roots outside this interval, as well as the complex roots in the domain $\operatorname{Im} g > 0$ ($g = \sqrt{k^2 - h^2}$). When the frequency parameter $\kappa = ka\sqrt{\varepsilon - 1}$ increases and takes a certain characteristic (transfer) value κ_{1i}, one of the complex roots h_i becomes real and approaches $k\sqrt{\varepsilon}$. The principal quasi-H_{00} wave is an exception and $h_{00} \to k\sqrt{(\varepsilon + 1)/2}$ as $\theta \to 0$. For a sufficiently narrow slot (when

$\theta \to 0$) one can determine the critical frequency κ_{2i} for every complex root h_i from the domain $\operatorname{Im} g < 0$. When κ decreases and takes the value κ_{2i}, $|h_i|$ attains its minimum with $\arg\{h_i(x_i)\} \simeq \pi/4$ and $\lim_{\theta \to 0} |h_i(\kappa_i)| = 0$.

Note in this way that every eigenvalue h_i of a closed metallic waveguide filled with a homogeneous dielectric lies either in the interval $0 \le h < k\sqrt{\varepsilon}$ or on the imaginary axis (for the frequency less than the critical value). Only the first four solutions possess the indicated properties and they correspond to the dielectric pivot in the free space [98]. $\operatorname{Im} h_i$ rapidly increases at the frequencies that are less than the transfer (critical) value κ_{2i}.

When the problem of the CSW excitation by concentrated sources is considered, one must deform the integration contour in order to determine the fields (as functions of z) in the form of integrals along the cuts and the sums of the residues calculated in the integrand poles [98]. The residues in the poles situated on the real axis determine the slow surface eigenwaves and the residues in the poles from the domain $\operatorname{Im} g < 0$ correspond to the fast leaky quasi-eigenwaves having the small radiation level for $\kappa > \kappa_{2i}$. The integrals along the cut determine some additional field or the "back-side wave" [11].

If the CSW has a narrow slot ($|\ln^{-1}\theta| \ll 1$) and $ka\sqrt{\varepsilon} < 1$, then we have

$$\kappa = ka\sqrt{\varepsilon - 1} < \kappa_{21} = \sqrt{\frac{\varepsilon - 1}{\varepsilon}}[1.84 + O(\theta^2)], \qquad (2.19)$$

i.e., the line is "over-critical" with respect to the higher order waves. In the empty waveguide [112] the slot wave is one of H waves and its eigenvalue or the "critical frequency"

$$x_0 = \left(-2\ln\sin\frac{\theta}{2}\right)^{-1/2}\left(1 + \frac{i\pi}{16}\ln^{-1}\sin\frac{\theta}{2}\right), \qquad (2.20)$$

tends to zero as $\theta \to 0$.

The zero eigenvalue belongs to the spectrum of the internal Neumann boundary value problem for the Helmholtz equation and the corresponding eigenfunction is constant in the entire waveguide cross-section. It follows from the Maxwell equations that this constant is zero and hence the wave with the zero cut-off frequency of the closed waveguide (the H_{00}-wave) exists conditionally as the wave with the zero amplitude. Therefore, one can consider the H_{11} wave as the principal wave. However, as has been shown above, if there is a longitudinal slot in a cylindrical waveguide, the H_{00} wave may exist as the leaky one [113] with a non-zero amplitude and then it becomes the principal wave. If the waveguide is filled with a dielectric, then the slot wave has a hybrid character with the prevailing H_z component, i.e., it becomes a quasi-H_{00} wave (similar to the quasi-TEM wave in strip transmission lines).

The position of the pole corresponding to the slot wave is defined by the formulas

$$h_0 = h_0' + ih_0'' = k[1 - x_0^2\kappa^{-2}(\varepsilon - 1)]^{1/2}, \qquad (2.21)$$

where x_0^2 may be found from the approximate dispersion relation

$$4\kappa^2 + \kappa^4 - 8\left(-2\ln\sin\frac{\theta}{2}\right)^{-1/2} + x^2(8 + 3\kappa^2 + 2x^2) - x^2(\kappa^2 + 2x^2)\ln\left(-\frac{h}{4k}x^2\right) = 0, \qquad (2.22)$$

where $x = ka\sqrt{1 - (h^2/k^2)}$. The first approximation of (2.22) yields

$$h_0 = k\left[\frac{\varepsilon + 1}{2} + \frac{1}{(ka)^2}\ln^{-1}\sqrt{-2\sin\frac{\theta}{2}}\right]^{1/2} + O(\ln^{-1}\theta). \tag{2.23}$$

It follows from (2.23) that depending on the parameters κ, ε and θ the slot wave may exists both as the fast leaky wave and the non-attenuating surface wave with a low phase velocity. The transfer frequency separating the domains of the existence of these two waves is equal to

$$\kappa_{10} = \left(\frac{2}{\ln^{-1}\sqrt{-2\sin\frac{\theta}{2}}}\right)^{1/2} + O(\ln^{-1}\theta). \tag{2.24}$$

The radiation losses are present at the frequencies that are less than the transfer value

$$h_0'' = \pi(k^2 - h_0'^2)\left(\frac{8h_0'}{\ln^{-1}\sqrt{-2\sin\frac{\theta}{2}}}\right)^{-1} + O(\ln^{-1}\theta), \quad 0 \leq \kappa_{10} - \kappa \ll 1. \tag{2.25}$$

The critical frequency limiting from below the attenuation of the leaky wave is given by the relation

$$\kappa_{20} = \left[\frac{2(\varepsilon - 1)}{(\varepsilon + 1)\ln^{-1}\sqrt{-2\sin\frac{\theta}{2}}}\right]^{1/2} + O(\ln^{-1}\theta), \tag{2.26}$$

and at the critical frequency

$$h(\kappa_{20}) = e^{i\frac{\pi}{4}}k\sqrt{\frac{\pi}{2}}\frac{1}{\ln^{-1}\sqrt{-2\sin\frac{\theta}{2}}} + O(\ln^{-2}\theta). \tag{2.27}$$

These formulas describe the solution close to the transfer frequency κ_{10} but they cannot be applied in the vicinity of the critical frequency κ_{20}.

Let us note that the critical frequencies κ_{10} and κ_{20} tend to zero when $\theta \to 0$ providing thus the wide transmission band of the CSW.

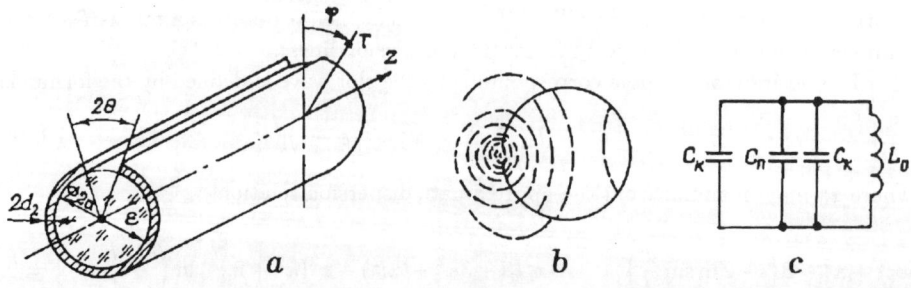

Figure 31.

It has been shown above that principal quasi-H_{00}-wave of the CSW has a distinct quasi-statical character: its (transversal) electric field is concentrated on the slot forming the quasi-concentrated capacitance C_0. The longitudinal component of the magnetic field is practically homogeneous along the metallic screen. The screen plays the role of the inductance coil L_0. C_0 and L_0 form the successive oscillation contour (see Fig.31). Hence, the propagation of the H_{00} wave along the CSW may be approximately described by the transversal resonance method [113], which enables one to determine all the main characteristics of the CSW and consider the lines with finite thickness of the screen.

The longitudinal wavenumber h may be represented as

$$h = \sqrt{k^2 \varepsilon_{\text{eff}} - k_c^2} = \sqrt{k^2 \varepsilon_{\text{eff}} - (\omega^2 L_0 C_0)^{-1}}, \tag{2.28}$$

where ε_{eff} is an effective permittivity and k_c is the wavenumber corresponding the critical wavelength λ_c in the CSW. In the method of the transversal resonance the value C_0 is usually taken to be equal to the capacitance $C_p = d_2/d_1$ of the planar condensator ($d_1 = a \sin \theta$ and $2d_2$ is the thickness of the conducting metallic layer of the CSW). One can take into account the edge capacitance [103]

$$C_k = \frac{1}{\pi} \ln \frac{1}{\sin \theta/2}, \quad C_0 = C_p + 2C_k, \tag{2.29}$$

if to assume that the capacitance of the CSW with the infinitely thin screen is determined only by the edge fields. The inductance L_0 is calculated by the formula for an infinite solenoid. For the CSW with narrow slots ($2\theta < 60°$) relation (2.28) takes the form

$$h = k\sqrt{\varepsilon_{\text{eff}} - \frac{\pi}{2k^2 \eta S_0}}, \tag{2.30}$$

where $\eta = -\ln(\sin \theta/2) + (\pi d_2)/(2a \sin \theta)$ and $S_0 = \pi a^2$ is the area of the CSW transversal cross-section. The effective permittivity $\varepsilon_{\text{eff}} = \sum_{i=1}^{N} \varepsilon_i \nu_i$, where N is the number of layers in the cross-section, ε_i is the permittivity and $\nu_i = C_i/C_0$ is the filling coefficient of the ith layer ($\sum_{i=1}^{N} \nu_i = 1$). For the line presented in Fig.31 $N = 2$ ($\varepsilon_1 = \varepsilon$, $\varepsilon_2 = 1$)

$$\varepsilon_{\text{eff}} = \varepsilon \nu_1 + \nu_2, \quad \nu_2 = 1 - \nu_1, \quad \nu_2 = \left[2 + \frac{\pi d_2}{2a \sin \theta} \left(2\eta - \frac{\pi d_2}{2a \sin \theta} \right) \right]^{-1}.$$

One can see that the coefficient ν_i is close to 0.5 and $\nu_i \to 0.5$ as $d_2 \to 0$. The latter relation is in an agreement with the results from Ref. [114].

Two characteristic wavelengths are obtained from the solution to the dispersion relation: the critical wavelength λ_c such that $h(\lambda_c) = 0$ and the transfer wavelength λ_t for which

$$\lambda_c = 2\pi a \sqrt{2\eta \varepsilon_{\text{eff}}}, \quad \lambda_t = 2\pi a \sqrt{2\eta (\varepsilon_{\text{eff}} - 1)}. \tag{2.31}$$

It follows from (2.30) that $\lambda_c > 20a$ for $\theta \approx 3° \div 10°$. For $\lambda_0 < \lambda_t$ the slow surface wave propagates along the CSW (with the phase velocity $v_\phi < c/\sqrt{\varepsilon_2}$ where ε_2 is the permittivity of the surrounding medium), Here the single-mode band is limited by λ_t (the long-wave boundary) and by the critical wavelength of the first waveguide mode

λ_f. For the CSW presented in Fig.31 $\lambda_f \approx 1.84a/\sqrt{\varepsilon_2}$ (according to the calculations from Ref. [114]). The width of this band is about two octaves. There is no power radiation out of the CSW when the quasi-H_{00} wave propagates in the line. Losses are determined by the attenuation in the metal and dielectric, and the losses in the metal prevail. The losses in the dielectric are given by the formula

$$h_\varepsilon'' = 8.69 \sum_{i=1}^{N} h_{\varepsilon_i}'' = 8.69 \sum_{i=1}^{N} \varepsilon_i \, \text{tg}\, \delta_i \frac{k^2}{2\nu_i h'}, \; [\text{dB/m}] \tag{2.32}$$

where h_{ε_i}'' are the losses in the ith dielectric and tg δ_i is the tangent of the attenuation angle in ith dielectric. For the line presented in Fig.31

$$h_\varepsilon'' = \frac{4.35\varepsilon k^2 \, \text{tg}\, \delta}{\nu_1 h'}, \; [\text{dB/m}]. \tag{2.33}$$

One can determine the wave resistance Z_{CSW} in the quasi-static approximation by the formula $Z = V^2/(2P)$ where V is the distance to the edge of the slot and P is the carried power. For the CSW presented in Fig.31

$$Z_{CSW} = 120\pi^2 \frac{k}{\eta h'}, \; [\text{Ohm}]. \tag{2.34}$$

In the wave range

$$\frac{\Delta\lambda}{\lambda_c} = \frac{\lambda_c - \lambda_t}{\lambda_c} = 1 - \sqrt{1 - \frac{1}{\varepsilon_{\text{eff}}}}$$

the situation is possible when $v_\Phi > c/\sqrt{\varepsilon_2}$. It leads to the Cherenkov-type radiation of the slot wave out of the open slotted structure. The radiation occurs inside the Cherenkov cone with the angle 2α such that

$$\cos\alpha = \frac{c}{v_\Phi \sqrt{\varepsilon_2}} = \frac{h'}{k}. \tag{2.35}$$

The electromagnetic fields that cause the continuous energy flow out of an open guiding structure are usually called the leaky waves. If the eigenvalue problem is solved by the method developed for closed structures [115, 116], then the leaky waves appear as a contribution of poles and the radiation condition for them is not valid. The leaky waves form a part of the continuous spectrum [115, 116]. Note that before to study leaky waves one has to determine their behaviour in the near zone where the field "tears off" the structure.

According to the transversal resonance method, the representation for the real part h' and the imaginary part h'' of the complex propagation constant of the quasi-H_{00} wave have the form obtained in Refs. [117, 118]:

$$h' = k\sqrt{1 - \frac{1}{2\pi(ka)^2}\left(\ln\frac{1}{\sin\theta/2} + \frac{\pi d_2}{2a\sin\theta}\right)}, \tag{2.36}$$

$$h'' = 0,54\frac{\pi}{h'}(h'^2 - k^2)\left(\ln\frac{1}{\sin\theta/2} - \frac{\pi d_2}{2a\sin\theta}\right), \; [\text{dB/m}] \tag{2.37}$$

Formulas (2.36), (2.37) take into account the finite thickness of the walls and (2.37) describes the diffractional losses in CSW.

The critical wavelength λ_c given by (2.31) corresponds to dispersion relation (2.36) and the following representation holds

$$\lambda_c = 2\pi a \sqrt{2\pi \left(\ln \frac{1}{\sin \theta/2} + \frac{\pi d_2}{2a \sin \theta} \right)}. \tag{2.38}$$

It also follows from (2.36) that the quasi-H_{00} wave is the fast wave propagating along the CSW with the phase velocity $v_\Phi = ck/h'$, which is greater than the speed of light c. Hence, the propagation of leaky waves is accompanied by their radiation into the free space. Assuming that the phase values of the beams directed towards the far-field pattern maximum are close to each other, one can determine the corresponding observation angle $\alpha = \arccos(h'/k)$ from formula (2.36): (this expression is similar to the condition of the Cherenkov radiation for a moving particle [110]).

The study of the radiation intensity in the far zone shows that the far-field patterns has the form of the Cherenkov cone with the vertex angle 2α. The form of the main lobe is approximately described by the Kirchhof formula

$$|\psi(\gamma)|^2 = \frac{h''^2}{h''^2 + \xi^2} \left(\cos^2 \frac{L\xi}{2} + \sin^2 \frac{L\xi}{2} \operatorname{ctg}^2 \frac{h''L}{2} \right), \tag{2.39}$$

where $\xi = h' - k \cos \gamma$ and γ is the observation angle in the longitudinal plane.

The main lobe and the positions of the first sidelobes do not change up to the level $z_n = L$. The field radiating at the angle α to the axis Oz is not a leaky wave since this field is not connected with the open structure. This wave propagates with the speed c and does not satisfy dispersion relation (2.36). Hence, angle α does not correspond to the leaky waves region as assumed in Ref. [115] and it does not form the "light–shadow" boundary as stated in Ref. [116]. The experimental studies of the distribution of the E_ϕ-component of the leaky wave show that the field distribution of the waves propagating in the CSW decrease in the transversal direction as fast as the surface wave field. An open guiding structure may support such amplitude distributions for arbitrarily big values of z, while for leaky waves they are observed only close to the source [120]. The field distribution of a leaky wave first gradually broadens (with the observation angle of the radiation field that does not coincide with the angle α from (2.35)) and then a step appears on its boundary which transfers into the lateral field maximum going off the structure. Hence, the complicated transference process occurs close to the CSW, and this phenomenon cannot be described by the Huygens principle.

The radiation field amplitude is determined by the contradictory factors: a decrease proportional to $(gR)^{-2}$ (where R is the coordinate of the Cherenkov cone and $g = \sqrt{k^2 - h'^2}$) due to the spatial divergence and an increase due to the field "pumping" from the structure ($z < L$). Consequently, in the beginning of the radiation process the amplitude grows, then it stabilizes and gradually starts to decrease as $(gR)^{-2}$. The theoretical models developed in Refs. [14, 116, 112, 121] deal with two-dimensional structures for which the Cherenkov cones radiate from the points of the open structure with different coordinates z and transfer to the cylindrical (or plane) waves going off to infinity (note that there are no other waves for two-dimensional structures). Here there is no difference between the propagating and radiating fields. Their superposition forms a leaky wave and is considered as a quasi-eigenfield. Anyway, the leaky wave is a part of a total field and has the physical meaning only in the bounded domain close to the structure. The field intensity of the leaky wave is sufficiently high close to an open waveguide and this field collapses at infinity. On the contrary, the radiation field is much smaller close to the waveguide, but it prevails over the field of a leaky wave as the transversal coordinate increases. Hence, they cannot be united in one field because of different properties. A leaky wave propagates with the speed $v_\phi > c$ and for this wave one can determine $\lambda_g = 2\pi/h'$ close to the waveguide. Far from the structure, where the total field is determined by the radiation field, the value of h' has the meaning of the projection of the wave vector $\mathbf{k}$ on the axis Oz. The radiation field propagates with the speed c and, therefore, it is not possible to measure λ_g far from the structure.

One can indicate two phase fronts of the total field when the leaky waves exist: the front of the leaky waves which is perpendicular to the axis Oz and the front of the radiation field which is inclined at an angle $\pi - \alpha$ to the axis Oz. Here the phase front of the radiation field outruns the leaky wave phase front. There may be an analogy between an electrodynamical structure supporting the leaky wave, a hypothetical elementary particle (tachyon) moving in vacuum and the wave radiating at the Cherenkov angle α. The tachyon radiation field may be compared with the field of a leaky wave in the CSW.

Table 3.

	$\varepsilon_{\text{eff}}, \nu$
1	$e_{\text{eff}} = \nu_1\varepsilon_1 + \nu_2\varepsilon_2,$ $\nu_2 = 1 - \nu_1,$ $\nu_1 = \left[2 + (\pi d_2/d_1)(2\eta - \pi d_2/d_1)^{-1}\right]^{-1}$
2	$e_{\text{eff}} = 0.5(\varepsilon_1 + \varepsilon_2),$ $\nu_1 = \nu_2 = 0.5,$
3	$e_{\text{eff}} = \nu_1\varepsilon_1 + \nu_2\varepsilon_2,$ $\nu_1 = 1 - \nu_2,$ $\nu_2 = \left[2 + (\pi d_2/d_1)(2\eta - \pi d_2/d_1)^{-1}\right]^{-1}$
4	$e_{\text{eff}} = \nu_1\varepsilon_1 + \nu_2\varepsilon_2 + \nu_3\varepsilon_3,$ $\nu_2 = 1 - (\nu_1 + \nu_2),$ $\nu_1 = \nu_3 = \left[2 + (\pi d_2/d_1)(2\eta - \pi d_2/d_1)^{-1}\right]^{-1}$

Some other types of open waveguide structures are also widely applied in the millimeter wavelength range, One of them is the image slot waveguide (ISW) which is a modification of the CSW. A part of its metallic screen serves as a conducting coating and the slot is formed by the edge of the screen and the coating. This open waveguide may contain a circular (semi-circular) dielectric pivot. ISWs may have a semi-elliptical or other form of the transversal cross-section. The ISW with a rectangular cross-section [118] is the simplest from the manufacturing viewpoint. ISWs are characterized by high operation characteristics and a sufficient rigidity. This waveguide may serve as a basic device for various functional circuits. One can apply the planar technology for the ISW-based integrated schemes placed on a single coating and to calculate their parameters applying the algorithms developed for strip lines and dielectric waveguides.

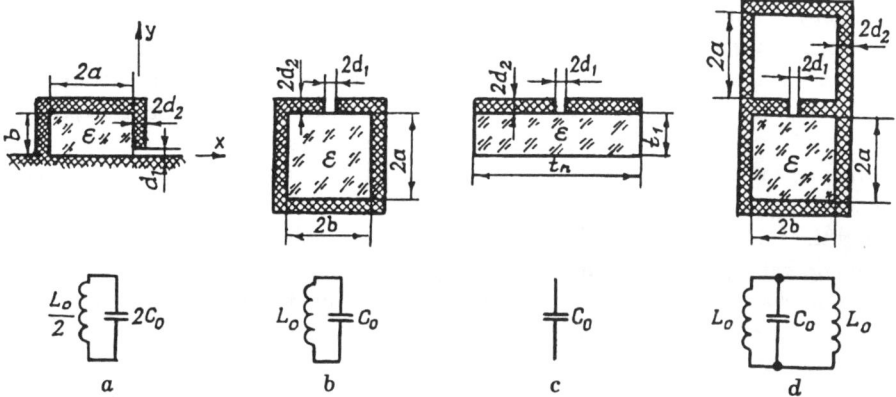

Figure 32.

It is rather difficult to construct the rigorous solution to the problems of wave propagation in the ISW with a rectangular cross-section. However, one can do it within the frame of the transversal resonance method. The transversal cross-sections of various types of ISWs are shown in Table 3 and the equivalent schemes are presented in Fig.32a. One can obtain the dispersion relation for the ISW which is similar to (2.30) if to take

$$\eta = \frac{\pi d_2}{d_1} + 0,693 - \ln\left(1 - \sqrt{1 - \frac{\pi d_1^2}{4ab}}\right) \tag{2.40}$$

and $S_0 = 4ab$. The values of the effective permittivity are presented in Table 3 (corresponding to Fig.32). We find the characteristic wavelengths from relation (2.40):

$$\lambda_c = 4\sqrt{\varepsilon_{\text{eff}}\pi ab\eta}, \quad \lambda_t = 4\sqrt{(\varepsilon_{\text{eff}} - 1)\pi ab\eta}. \tag{2.41}$$

Here the single-mode band is limited by the values

$$\lambda_f = 4b\sqrt{\varepsilon} \quad \text{for} \quad b > a,$$
$$\lambda_f = 4a\sqrt{\varepsilon} \quad \text{for} \quad a > b. \tag{2.42}$$

Expressions (2.32) and (2.37) are also valid for the ISW. The wave resistance of the ISW is determined by the formula

$$Z_{CSW} = 60\pi^2 \frac{k}{\eta h'}, \quad [\text{Ohm}]. \tag{2.43}$$

In order to compare the characteristics of the CSW and ISW, we will write down the dispersion relation and wave resistance for the CSW with the rectangular cross-section (its equivalent scheme and geometrical parameters are presented on Fig.32b):

$$h' = k\sqrt{\varepsilon_{\text{eff}} - \frac{\pi}{4\eta k^2 ab}}, \quad Z_b = 120\pi^2 \frac{k}{\eta h'}, \tag{2.44}$$

where η is determined by expression (2.40).

It follows from these relations that the ISW with the square cross-section ($2a = b$) has the maximal transmission band reaching 2-3 octaves. One can cover the whole millimeter wavelength range with the help of the single ISW (otherwise, one would need about ten standard rectangular waveguides with different cross-sections). The dimensions of the corresponding CSW ($2a = b$) is twice as large as that of the ISW and the CSW transmission band is substantially narrower (the difference attains 20-30%). Note that the transmission band of the CSW with a rectangular cross-section is a little wider than that of the CSW with a circular cross-section.

In order to perform a complete analysis of the advantages of ISWs, we will compare different open slot lines within the framework of a single approach. The planar slot line and the shielded slot line (the fin-line) together with the corresponding equivalent schemes operating in the millimeter wavelength range are presented in Fig.32,c,d.

Planar slot lines are the best for manufacturing printed schemes and circuits. One can obtain the dispersion relation of the planar open slot line depicted in Fig.32c from (2.44) using the formal transformation $a = b \to \infty$, $h'_{sl} = k\sqrt{\varepsilon_{\text{eff}}}$. In the limiting case $2d_2 \to 0$ ($\nu_1 = \nu_2 = 0.5$) equality (2.44) transfers to the relation [125] for planar slot lines

$$h' = k\sqrt{\frac{1}{2}(\varepsilon + 1)}.$$

This expression holds for the slot line with the layer thickness $t_1 \to \infty$. However, the comparison with [126] shows that the relation $h' = k\sqrt{(\varepsilon + 1)/2}$ holds for $t_1 \approx 0.07 \div 0.1\lambda$ (when $2d_2 \ll t_1$). The similar expression may be obtained for the CSW when $\lambda_0 < \lambda_t$, i.e., far from the critical frequencies.

One can determine the wave resistance for the slot line if to calculate the limit in (2.44):

$$Z_{sl} = \frac{120\pi}{\sqrt{\varepsilon_{\text{eff}}}C_0}, \ [\text{Ohm}].$$

Hence, the principal wave propagating in the CSW and ISW in the short-wave part of the wavelength range is analogous to the slot wave of the planar slot line. However, the quasi-H_{00} wave is characterized by additional properties (in particular, this wave may radiate to the free space). The latter broadens the possibilities for creating the element base. The ISW is smaller than the planar slot line, although its transmission band ($\lambda_c \to \infty$) is wider. The width t_n of the strip line with the small thickness t_1 may reach $\lambda_0/2$. The planar slot line [127] applied in the 3-cm wavelength range has the thickness $t_n = 20$ mm for $t_1 = 0.05$ mm and $\varepsilon = 10$, while the size of the ISW operating in the same wavelength range is ≈ 2 mm.

The field concentration and the effective size of the planar slot line are determined in Ref. [125] where the line with $\varepsilon = 16$ is considered at the frequency 36 GHz ($\lambda_0 = 10$ cm). The field intensity at the distance 2.5 cm from the slot center is 28.4 dB less than its average value in the transversal direction. These values correspond to the effective cross-section $0.2\lambda_0^2$ and to the low field concentration. The field concentration in the ISW is denser, the "shielded area" is larger and the noise protection is better.

Let us carry out a more detailed comparison of the ISW with the shielded slot line (fin-line) described in Ref. [128] and depicted in Fig.32d. The dispersion relation and the wave resistance for this line is obtained from (2.28) with the help of its equivalent scheme:

$$h'_{f-l} = k\sqrt{\varepsilon_{\text{eff}} - \frac{\pi}{\eta k^2 ab}}, \quad Z_{f-l} = 120\pi^2 \frac{k}{\eta h}, \; [\text{Ohm}]. \tag{2.45}$$

The comparison of relations (2.30), (2.44) and (2.45) shows that the single-mode band of this line is $\sqrt{2}$ times less than that of the CSW (its size is twice as big as for the CSW). The dimensions of the fin-line are four times greater and the transmission band is 40-50% less than the corresponding ISW parameters. It is proposed in Ref. [129] to cover the 26 GHz–90 GHz band by three fin-lines with the slotwidth $2d_1 = 0.5 \div 0.05$ mm, $\varepsilon = 2.2$ and different cross-sections $2a \times 4b = 7.1 \times 3.56$, 4.8×2.4 and 3.1×1.55 mm^2 ($d_2 = 0$).

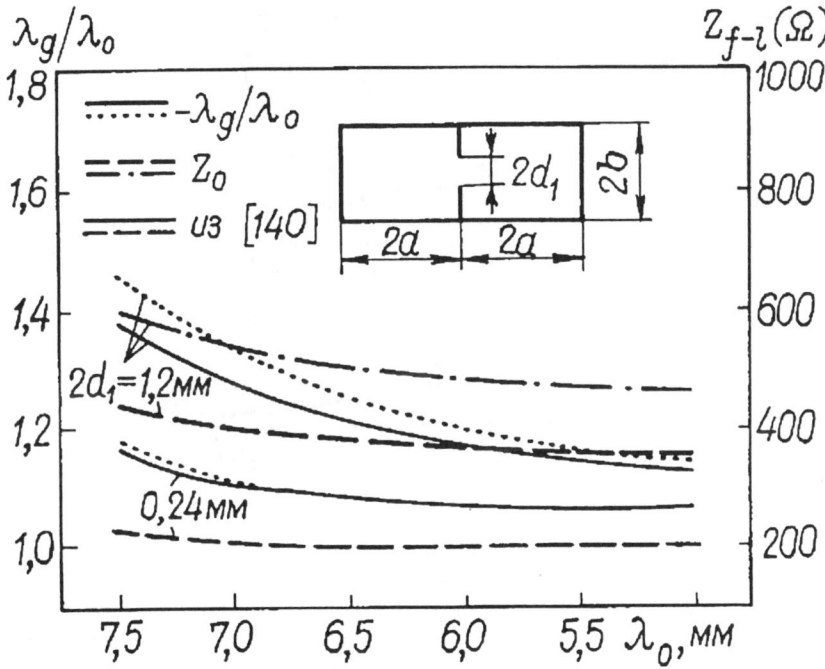

Figure 33.

Let us compare the results of the direct calculations of the dispersion and the wave resistance for the fin-line with the size $2a = 2b = 2.39$ mm, $2d = 0$ and $\varepsilon = 1$ for two values of the slotwidth $2d_1 = 0.24$ mm and 1.2 mm. with the data obtained by formula (2.45) (the bar lines in Fig.33) and the values taken from Ref. [129] (calculated at $\lambda_g = 2\pi/h'$ and shown by the bold lines in Fig.33). For $2d_1 = 0.24$ mm the dispersion curves for the ISW and for the fin-line practically coincide. For $2d_1 = 1.2$ mm the approximate dispersion calculations yield too high results as compared with Ref. [129], but the maximal divergence for $\lambda_0 = 7.5$ mm does not exceed 4%. The data for Z_{f-l} presented in Fig.33 by the bar lines are taken from Ref. [129] and the dot-dashed lines give the curve obtained by (2.45). For $2d_1 = 0.24$ mm both lines coincide (as well as the dispersion curves) and for $2d_1 = 1.2$ mm formula (2.45) gives too high values, for which the divergence reaches 30% and practically does not depend on the frequency.

The wave resistance in the fin-line calculated by (2.45) as a function of the cross-section parameters (with the help of the data from Ref. [130]) is shown in Fig. 33 where it is assumed that $Z_{f-l} = Z_{0\infty}k/h'$ and $Z_{0\infty} = 120\pi^2/\eta$.

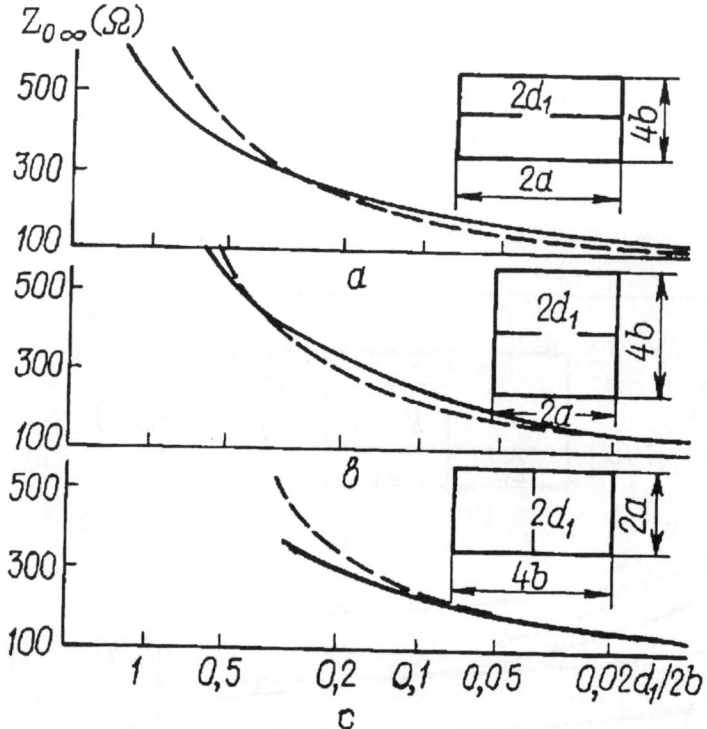

Figure 34.

The data in Fig.34 is evidence of an agreement between the curves corresponding to sufficiently narrow slots ($d_1 < 0.4b$). A divergence is connected not only with the inaccuracy of the transversal resonance method but mainly with the fact that for $2d_1 \sim b$ the slot wave transfers into the H_{01} wave of the rectangular waveguide. In the latter case the line with quasi-concentrated reactive parameters is transformed to the line with distributed parameters. The comparison of the results obtained in Ref. [130] with expression (2.45) shows that (2.45) is valid when $b/(2a)$ is close to 1.5. For $d_1/(2b) < 0.08$ formula (2.45) may be applied in the interval $0.5 < a/(2b) < 5$. When the slot broadens, the admissible interval decreases up to $a/(2b) = 1.5$ for $d_1/(2b) = 0.64$. These properties also hold for the ISW.

The fin-line wave resistance is approximately the same as in the corresponding CSW and twice as big as that in the ISW.

One can study in details the properties of the coupled CSW and ISW by the transversal resonance method. The surface character of the principal wave in these open waveguides allows to direct a part of the energy from one line to another and to construct directional couplers similar to the units where dielectric waveguides are applied [131, 132]. One can make these couplers as integrated schemes constructed on the basis of the ISW. The power transformation coefficient is mainly determined by the distance between the lines and by the length of the interaction region. One has to take into account the quasi-statical character of the CSW as well as the spectrum splitting. In fact, the additional capacitance appears between the external surfaces of the conducting screens of coupled lines, and the influence of this capacitance causes the redistribution of the electric charges on the screens and the deformation of the field distribution.

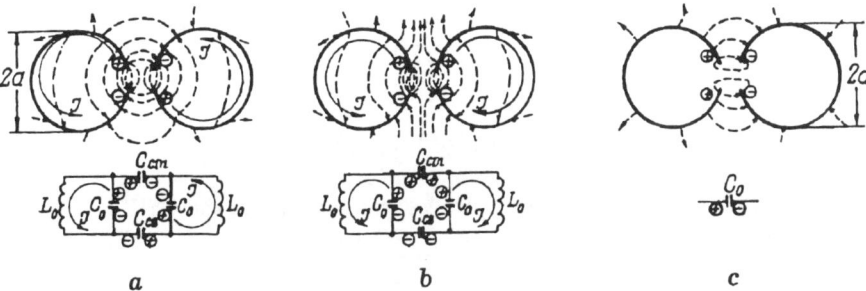

Figure 35.

Let us consider two equal coupled CSW or ISW with the slots facing each other (Fig.35). In this case the degree of communication varies in broad limits. For all other types of orientation of the slotted cylinders the values of the communication coefficient k_{cm} is rather small. For $ka < 1$ (where $k = 2\pi/\lambda_0$ and a is the cylinder radius) three quasi-statical waves exist: two of them is connected with the wave splitting into the syn-phase and counter-phase waves of the coupled CSW caused by the interaction of the principal quasi-H_{00} wave. The third is the TEM wave of a two-wire line. The sign of the edge electric charge, the direction of the transversal current and the qualitative distribution of the electric field (bar lines) are presented on Fig.35 for the counter-phase (Fig.35a), the syn-phase (Fig.35b) and the TEM (Fig.35c) waves. As has been shown above, the transversal oscillations occur in the CSW (and in the ISW) at the critical wavelength λ_c, and the slot lines are similar to two-dimensional quasi-static resonators. The contour with inductance L_0 and capacitance C_0 models an oscillation process in the lines. The analysis of the equivalent schemes presented in Fig.35 enables us to formulate the following conclusions. The slotted cylinders whose slots face each other communicate by means of the transversal electric components concentrated on the slots (the longitudinal magnetic components do not vanish for $\lambda_0 = \lambda_c$ and they are concentrated in the cylinder volume) and this is the capacitance-type communication.

The phase velocity v_Φ of the quasi-TEM wave is equal to c for $\varepsilon_{out} = 1$. The phase velocity of the cylindrical slot wave is greater than c at $\lambda_0 = \lambda_c$. Hence, one can eliminate the quasi-TEM wave when describing the spectrum splitting in the coupled CSW. This property is also valid for the coupled ISW, since the conducting coating causes the short-circuit effect for the waves propagating in the two-wire line. The mutual influence of two resonance volumes (Fig.35) leads to the appearance of two "cut-off" wavelengths λ_{c_1} and λ_{c_2} corresponding to the syn-phase and counter-phase oscillations:

$$k_{c_1} = k_c(1 - k_{cm})^{-1/2}, \quad k_{c_2} = k_c(1 + k_{cm})^{-1/2}, \quad k_c = \frac{2\pi}{\lambda_c}, \quad k_{c_1} = \frac{2\pi}{\lambda_{c_1}}, \quad k_{c_2} = \frac{2\pi}{\lambda_{c_2}},$$

where k_{cm} is the communication coefficient of the resonance volumes. The coupled communication contours shown in Fig.35 correspond to the case of the external capacitance communication with $k_{cm} = C_{cm}(C_0 + C_{cm})^{-1}$. Here C_0 is determined by (2.29), C_{cm} is the communication capacitance of the resonance contours and

$$k_{c_{1,2}} = \left\{ k_c^{-2} \pm \left(k_c^2 + \frac{1}{S_1 C_{cm}} \right)^{-1} \right\}^{-1/2}, \tag{2.46}$$

where $S_1 = S_0$ for the CSW, $S_1 = 2S_0$ for the ISW and S_0 is the cross-section area of the internal cavity. It is convenient to write (2.41) in the form

$$k_{c_{1,2}} = \varepsilon_{\text{eff}}^{-1/2} \left\{ k_{c_0}^{-2} \pm \left(k_{c_0}^2 + \frac{\varepsilon_{\text{eff}}}{S_1 \varepsilon_b C_{cm}} \right)^{-1} \right\}^{-1/2}, \tag{2.47}$$

where $k_{c_0} = k_c$ ($\varepsilon = 1$) and ε_{eff} is the equivalent permittivity of the CSW or ISW.

Two slot waves of the coupled CSW (ISW) correspond to the values k_{c_1} and k_{c_2}. Their propagation constants $h'_{z_{1,2}} = \sqrt{k^2 \varepsilon_{\text{eff}} - k_{ci_{1,2}}^2}$, where $k_{ci_{1,2}}$ is the value of $\{ k_{c_0}^{-2} \pm [k_{c_0}^2 + \varepsilon_{\text{eff}}/(S_1 C_{cm})]^{-1} \}^{-1/2}$. The dispersion relation for the quasi-TEM wave in the coupled CSW without losses may be obtained from the solution to the telegraph equations [103]: $h'_{z\,TEM} = k\sqrt{\varepsilon_{in}}$. As is shown in Ref. [133], the coefficient describing the degree of communication between two open lines may be expressed as $k_s = \Delta h'_z/2$, $\Delta h'_z = h'_{z_1} - h'_{z_2}$. The power flows between the lines have the known form

$$T_1 = e^{-h''_z z} \cos^2 k_s z, \quad T_2 = e^{-h''_z z} \sin^2 k_s z, \tag{2.48}$$

where h''_z determines heat losses in the transmission line. It follows from (2.48) that the wave power may be completely pumped from one line to another in the interval with the length $L_n = \pi k_s/2$ and one can calculate the parameters of the power coupler formed by the coupled CSW and ISW (k_{c_0} and ε_{eff} are determined by the formulas from the present section and C_{cm} is taken as the statical capacitance of two appropriate conducting pivots).

Let us calculate the parameters of two circular coupled CSW. We will use the following notation: the diameter of the internal cylinder is $2a$, the wall thickness is $2d_2$ and the angle slotwidth is 2θ. Relations (2.47) will be applied under the assumption that C_{cm} is the capacitance of two pivots with the external diameter

$$C_{cm} = \frac{1}{2} \ln^{-1} \left\{ \frac{t_3 + 2}{2 + 4t_2} + \sqrt{\left(\frac{t_3 + 2}{2 + 4t_2} \right)^2 - 1} \right\}, \tag{2.49}$$

where $t = d_2/a$, $t_3 = (l - 2a)/a = \delta/a$ and l is the distance between the cylinders. Then one can rewrite (2.47) taking into account that $S_0 = \pi a^2$:

$$k_{c_{1,2}} a = \left\{ (k_{c_0} a)^{-2} + \frac{2\varepsilon_{\text{eff}}}{\pi} \ln \left[\frac{t_3 + 2}{2 + 4t} + \sqrt{\left(\frac{t_3 + 2}{2 + 4t_2} \right)^2 - 1} \right]^{-1} \right\}^{-1/2} \frac{1}{\sqrt{\varepsilon_{\text{eff}}}}. \tag{2.50}$$

Note that for $t_3 = 0$ $\varepsilon_{\text{eff}} = 1$ and (2.50) transfers to the expression obtained by the methods of the rigorous theory [134]. The calculations carried out with the help of

(2.50) show that for $\theta = 5°$ the behaviour of $k_{c_{1,2}}a$ as functions of t_3 is similar to that of the resonant frequencies of lossless coupled oscillation contours. For $k_{cm} = 0$ ($l \to \infty$), $k_{c_{01}} \approx k_{c_{02}} \approx k$ The increase of k_{cm} (or the decrease of l) causes the spectrum splitting. The corresponding divergence of the spectral curves becomes noticeable for $t_3 \approx 6$. When $k_{cm} \to 1$ ($l_{cm} \to 2a$), $k_{c_{01}} \to \infty$ and $k_{c_{02}} \to k_{c_0}/\sqrt{2}$. For $t_3 \to 0$ the critical wavelengths of two coupled CSW do not transfer to the critical wavelength λ_c of the fin-line. Namely, in the first case λ_{c_1} is $\sqrt{2}$ times less than λ_c of the CSW and λ_{c_2} is $\sqrt{2}$ times greater than this value. In addition to this, $\lambda_{c_1} \to 0$ as $t_3 \to 0$. The data given by (2.50) practically coincide with the results obtained by the analogous formula taken from the rigorous theory for $t_3 < 7$. This coincidence may be explained by the specific properties of h'_g in the case of two slotted cylinders: for the counter-phase oscillations $h''_g \to 0$, while for the syn-phase oscillations h''_g do not vanish and the losses practically do not influence k_{c_2} [136].

We have indicated for $t_3 > 5$ the periodical oscillations (up to the zero level) of $(k_{c_1} - k_{c_2})$ considered as a function of t_3. The period of these oscillations is equal to one half of the resonant wavelength and they are caused by the interference of electromagnetic fields reflecting from the cylinders. Expression (2.50) obtained by the transversal resonance method may be applied in the quasi-statical case, and the value of $t_3 \approx 4.5$ corresponds to $\delta = \lambda_c/4$ for $\theta = 5°$. In other words, one can separate two regions differing by the way of the slotted cylinder interaction: the quasi-statical region $t_3 < 5$ of the communication of quasi-statical fields corresponding to capacitance C_{cm} and the interference region $t_3 > 5$ where the communication is caused by the interference fields. The problem under study cannot be solved in the latter region by the transversal resonance method and the periodical change of $(k_{c_1} - k_{c_2})$ for $t_3 > 5$ cannot be described by formula (2.50).

Let us consider the behaviour of $(k_{c_{1,2}}a)$ with respect to θ for the single CSW and the coupled CSW. One can calculate $(k_c a)$ by the asymptotic formula $(k_c a) \approx \left[2\ln\left(\sin(\theta/2)\right)^{-1}\right]^{-1/2}$. However, the quasi-statical relation

$$(k_{c_0}a) = \left\{2\left[\left(1 - \frac{\theta}{\pi}\right)^2 + \frac{\theta}{\pi}\right]\ln \operatorname{ctg}\frac{\theta}{4}\right\}^{-1/2} \tag{2.51}$$

(for $2d_2 = 0$) is more precise. This expression coincides with the asymptotic formula when $\theta < 1°$. The results calculated by (2.51) repeat for $1° < \theta < 180°$ the data obtained by the methods of the rigorous theory [7] as well as the experimental data (the error reaches 12% for $\theta < 35°$). The increase of θ leads to the decrease of $C_0 = [2\ln \operatorname{ctg}(\theta/4)]/\pi$ and yields the relative increase of k_{cm} (when C_{cm}/C_0 grows).

Now we may conclude that formulas (2.46) through (2.50) describe the characteristics of the CSW in the region $\theta < 35° \div 40°$ and $\delta < 1/2\lambda_c$.

Let us consider the interaction of two slotted cylinders with the rectangular cross-section $(2a \times 2b)$, the wall thickness $2d_2$, the slotwidth $2d_1$ and the permittivity ε. The distance between the front edges of the CSW is denoted by δ. Let us introduce the parameter $t = d_1/a$. Substituting the values $S_0 = 4ab$ and C_{cm} obtained in Ref. [137] into (2.47) and taking into account that

$$C_{cm} = \frac{\frac{4b}{a} + bt_2}{t_3} + \frac{2}{\pi}\left(\frac{2 + 4t_2}{t_3}\right)\left[1 + \ln\left(1 + \frac{t_3}{1 + 4t_2}\right)\right], \tag{2.52}$$

we finally get

$$k_{c_0,1,2} = \left\{ k_{c_0}^{-1} \pm \left(k_{c_0}^2 + \frac{\varepsilon_{\text{eff}}}{4abC_{cm}} \right)^{-1} \right\}^{-1/2} \frac{1}{\sqrt{\varepsilon_{\text{eff}}}}. \tag{2.53}$$

Expressions for ε_{eff} known for rectangular CSWc in the quasi-statical approximation have the form

$$k_c\sqrt{ab} = \frac{1}{2}\left\{ \pi \left[2\varepsilon_{\text{eff}}\left(\frac{\pi d_2}{2d_1} \right) + \ln\left(4\sqrt{\frac{ab}{\pi}}d_1 \right) \right]^{-1} \right\}^{1/2}. \tag{2.54}$$

The calculations for the square CSW ($a = b$) at $\varepsilon = 1$ and $\varepsilon = 3.8$ when t_3 varies show that this slot line keeps the properties of the coupled circular CSW: the splitting of the spectral line in (2.53) is observed for $t_3 < 4$, i.e., for the values of t_3 which are less than the corresponding threshold values of the circular CSW. It is connected with "pulling" the electric field into the dielectric and with corresponding increase of C_0 by the factor of $\sqrt{\varepsilon_{\text{eff}}}$. The level of pulling increases with growing ε. When t_1 increases, the splitting begins at large values of t_3. The following conclusion is valid for the considered types of lines: the splitting of a spectral curve becomes noticeable when $t_3 \geq \lambda_c/4$. This wavelength region is of principal importance for the construction of directional couplers.

§ 9. Experimental study of cylindrical and image slot lines

The experimental study of CSWs is carried out by means of both the traditional approaches well-known in the radio engineering and the original methods. Among them are the method applied for the determination of dispersion based on the independent wavelength measurement by the standing wave picture. The substitution method for the determination of losses consists in placing the CSW section and a calibrated attenuator in the gap between the generator and detector. It is not easy to apply this method because the design of a reliable junction between the CSW and a rectangular waveguide is an independent complicated problem. An open character of the CSW allows to measure both the longitudinal field distribution and the attenuation (deceleration) coefficients. In this section we will consider the application of CSWs and diffractional gratings in some new methods of the experimental study of the field structure and the dispersion characteristics in open slot lines. Note that, in addition to the slot waves, there may be other types of waves in the CSW, in particular, the Sommerfeld waves [103]. On the other hand, a dielectric waveguide, which is a part of CSW, has the zero critical oscillations for two principal modes with orthogonal polarization [131, 138]. Hence, the corresponding waves may propagate along the CSW, i.e., in a circular partially coated dielectric pivot.

The analysis of the CSW mode composition performed in Refs. [8, 139, 140] is based on the account of the different phase velocities of the waves propagating in open waveguides. One can use a dispersion structure, for example, a diffractional grating to transform these differences into the spatial wave generation.

It is known [141-143] that as a result of the inhomogeneous plane wave diffraction by the grating in a dielectric waveguide the infinite discrete spectrum of spatial harmonics appear containing the slow (surface) waves, in particular, the fast (volume) radiating waves. Their angles of maximal radiation ψ_n are connected with the grating period l and the surface wave deceleration coefficient $\lambda/\lambda_g = c/v_\Phi$ by the relation

$$\cos \psi_n = \frac{\lambda}{\lambda_g} + \frac{n}{\kappa}, \quad (n = -1, -2, -3, \ldots) \tag{2.55}$$

where λ is the free space wavelength, λ_g is the wavelength in the dielectric waveguide, $\kappa = l/\lambda$ is the frequency parameter and ψ_n is the angle between the wave vector of the nth radiating spatial harmonic and the wave vector of the surface wave in the dielectric waveguide.

If different types of waves have different phase velocities, the number of the main lobes in the far-field patterns of the system "surface-wave waveguide–diffractional grating" corresponds to the number of propagating waves. It is assumed that only the minus first harmonic ($n = -1$) radiates. The quantity λ_g, i.e., the relative phase velocity, is determined by the measured values of l, λ and ψ_n. The deceleration coefficient of the slot wave must coincide with the value that is determined by the period of the standing waves picture.

From the viewpoint of the antenna theory the above radiation system must be characterized by the aperture amplitude distribution and the width of the far-field pattern. For the homogeneous amplitude distribution (when a weak communication of the grating and the dielectric waveguide takes place) the width W_F of the far-field pattern is about $51°\lambda/L_p$, where L_p is the length of the grating. The value of W_F determines the scale accuracy of the diffractional grating (two types of waves with the deceleration coefficients $(c/v_\Phi)_1 = 1.1$ and $(c/v_\Phi)_2 = 1.07$ are distinctly separated for $L_p = 100$ mm and $\lambda = 4$ mm). The maximal radiation angles ψ_1 and ψ_2 differ by the value of W_F ($\psi_1 - \psi_2 = \Delta\psi = 2°$). The reciprocity theorem yields the equality of the radiation far-field pattern and that of the receiving antenna. Hence, the scale accuracy of the grating remains the same when it transfers from the output of the diffractional energy to the diffractional input.

Note that the scale accuracy of the grating does not depend on its period l. An increase of the dispersion (i.e., the degree of the dependence of the radiation angle on the wavelength in the line) is accompanied (with decreasing l) by an equal decrease of the radiating aperture. Consequently, the far-field pattern broadens. The radiating aperture decreases because the spatial radiating harmonics are transformed to the propagating harmonics. In the normal radiation sector $90° < \psi_{-1} < 130°$ the width of the far-field pattern is minimal and there exists only one spatial radiating harmonic ($n = -1$). Note that here the Bragg's second-order diffraction conditions ($l \neq \lambda g$) do not hold

The distortion of the longitudinal field distribution for running waves is an additional evidence of the existence of several waves in the CSW. For example, in the case of two waves the spatial beating appears with the period $L = 2\pi/|k_2 - k_1|$, where $k = \pi/\lambda_{g_{1,2}}$. If the number of waves is greater than two, the beatings become nonperiodical. Let us note that by adding the waves with the wavenumbers determined by the diffractional methods, one can reconstruct the longitudinal field distribution measured by a sensor.

It is easy to determine the wavelength of one type of the wave by the beating period. This technique is much more accurate than the diffractional method. If, for example, the deviation of the amplitude of the unknown wave is -15 dB with respect to the wave that can be effectively excited, a weak signal will be lower than the sidelobes of the strong signal far-field pattern. Existence of the accompanying wave is clearly seen from the picture of longitudinal beatings even in the case when its relative amplitude is $-20 \div 30$ dB as compared with the main signal. The beating amplitudes are 0.83 dB and 0.26 dB, respectively. Hence, the developed methods for determining the losses, the mode composition and the deceleration coefficient of the slot wave are based on two independent measurements of the longitudinal field distribution for small and large values of the standing wave ratio (SWR) and the far-field patterns of the system "CSW–diffractional grating". The diffractional input of energy into the CSW may be applied in the measurement scheme for the separate excitation of one wave. The experimental plants have been designed on the basis of these methods.

The study of the CSW and ISW requires the construction of wide-band devices for the measurements of the attenuation, dispersion and field distribution. In the centimeter wavelength range the waveguides, coaxial lines and various automatized measurement units are applied for the SWR determination. The waveguide measurement lines that are used in the millimeter wavelength range do not fit the requirements of the CSW and ISW measurements since in the latter case a high-quality waveguide-slot junction must be provided. Otherwise, the measurement errors are inevitable. In fact, one should use for such measurements the CSW and ISW themselves. These lines will be both the objects under study and the measuring devices. In particular, the measuring ISW section may be applied as the universal unit for investigating open waveguides and transmission lines.

The properties of the CSW and ISW as open structures provide a free access to the field of the slot wave. Therefore, it is not necessary to use a waveguide-slot section as an element of the proposed measuring line. One can place the sensor in every point of the CSW and move it along the whole scheme. The sensor head should be constructed in the form of a wide-band ISW fixed on the two-coordinate manipulator moving parallel to the scheme plane. The sensor should admit the possibility to be lifted over the ISW and shifted from the slot to provide the appropriate SWR value in the sufficiently wide wavelength range. Such a two-coordinate measuring line described in Ref. [114] gives additional possibilities for creating the required three-dimensional field distribution in an open waveguide, measuring the communication and gain factors, the characteristics of directional couplers and other parameters. One can follow the signal along a multi-channel integrated scheme with an arbitrary number of channels by the same detector. The two-coordinate line allows to perform the complex quality control of the integrated schemes based on the use of the ISW and to find the local inhomogeneities that indicate the SWR increase.

An experimental plant for the study of the slot wave must provide the effective excitation in the given wavelength range, and a small sensor applied for measurements should admit an easy shift along the CSW in the direction perpendicular to its axis. The scheme of the plant satisfying these demands is presented in Fig.36. The clystron generator 1 feeds CSW 11 through measuring attenuator 3, flexible dielectric waveguide 7 and the excitation device 9. The G4-104 generator with the removable clystron blocks covers the wavelength range $\lambda = 3.8 \div 5.7$ mm.

The directional coupler 2, the wave-meter and the measuring device with detector 6 are used for the wavelength measurement and for the stability control of the clystron generation zone. The generator's regime is chosen in such a way that the oscillation

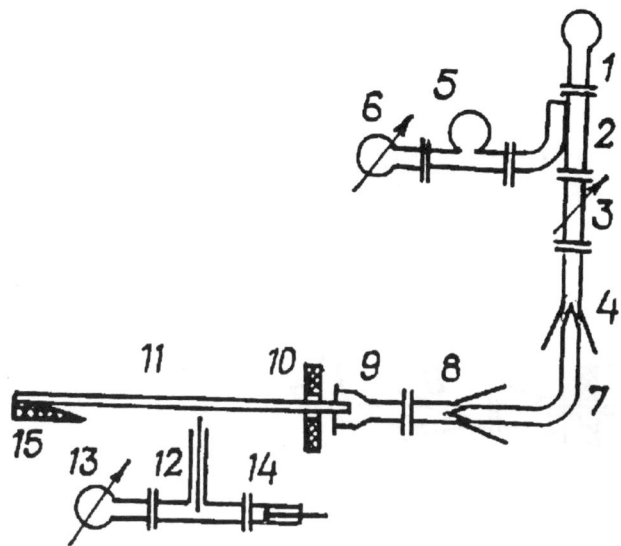

Figure 36.

power does not change in the process of the adjustment of measuring attenuator 3 .

The excitation device 9 has the form of a smooth junction between the rectangular and circular waveguides. The metal diaphragm with a hole is placed at the end of the circular waveguide. Washer 10 made of an absorbing material protects sensor 12 from the radiation field appearing in the place of excitation, and wedge 15 provides the running wave propagation.

The dielectric waveguide and rectangular metal waveguide 8 are coordinated with the excitation device. Hence, one can control the depth of the CSW penetration inside the excitation device and to choose the conditions of the effective slot wave excitation.

The electric field longitudinal distribution is measured by coaxial sensor 12 ending with the quarter-wave dipole. The sensor is adjusted by the short-circuit piston 14. The micrometer head controls the contact of the sensor and CSW. The maximal distance between them is $0.02 \div 0.03$ mm in the interval $L = 100$ mm.

A signal from detector 13 comes through the amplifier to the recorder. The longitudinal field distributions are calibrated by measuring attenuator 3.

The scheme of the experimental plant applied for the study of the mode composition in the CSW by the diffractional method is depicted in Fig.37. The bar diffractional grating 12 is placed close to CSW 11. Antenna 13 and the detector section 14 is fixed on a rotating rod to register the far-field patterns. The error of the angle determination does not exceed $\pm 0.25°$. The far-zone condition is fulfilled for all measurements.

The plant described above allows us to study the slot wave and the higher order waves successively by two methods and with the equal accuracy. The error of the wavelength determination is proportional to the grating length and to the length of the measured line section. The excitation unit 9 is small enough to provide the efficient utilization of the sample length in the measurements.

Several factors influence the proper choice of the CSW characteristic dimensions that are appropriate for experimental investigations of the slot wave: the study of the

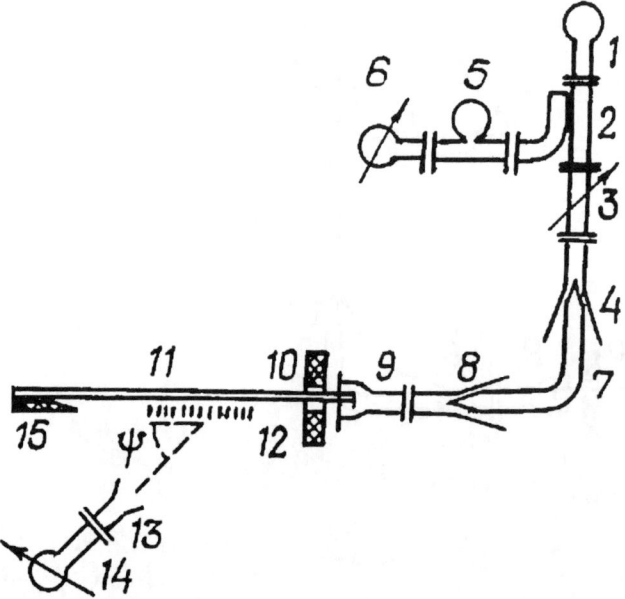

Figure 37.

dispersion requires determination of the transference frequency from the fast wave to the slow wave, together with the simultaneous search for higher order waves and the optimal (minimal) CSW wall thickness. Here one has to take into account that the walls cannot be thinner than the metal skin-layer, and the slot must be sufficiently narrow in order to provide the possibility to apply the small parameter technique. At the same the time, the wall thickness is often comparable with the slotwidth.

To eliminate manufacturing difficulties, we use the polyethylene with the small permittivity $\varepsilon = 2.25$ and the low losses ($\mathrm{tg}\,\delta = 0.5 \cdot 10^{-3}$) as a dielectric substrate. The pivot diameter is 1.1 mm and, hence, the transference wavelength $\lambda_b = 5.6$ mm belongs for $\theta = 28°$ to the operation range of the clystron blocks (shown in Fig.36).

The proper choice of the wall thickness and the type of the metal was more complicated. The modern vacuum precipitation methods allow to obtain the films with the thickness equal to several microns and a higher resistance because here the film thickness is comparable with the skin layer depth. The metal thickness may reach tens of microns when electro-chemical methods are applied, but in this case the slot cutting becomes rather complicated. In our experiments the slot was milled in the hard German silver tubes and the soft copper strips were used that repeated the form of a slotted cylinder. The choice of this low-conductivity metal is caused by several reasons. To rise the measurement accuracy one has to increase either the sample length or the attenuation. The latter is more convenient and here one can use the analytical expressions obtained in section 8 to calculate the attenuation in the metal and in the dielectric substrate and then to determine the losses in the German silver coating with the help of the relations from Ref. [116].

The sample dimensions that are optimal for the study of slot waves have the following values: the diameter of the internal dielectric pivot is 1.1 mm, the angle slotwidth $\theta = 28°$ and the German silver tube with the wall thickness 0.2 mm may be used as a metallic coating.

The electric field of the slot wave are mainly concentrated close to the slot. The metallic cylinder is soldered on the plate lying in the plane of symmetry. First, the 0.265 mm slot is milled in the screen and then the inner cavity is filled with a dielectric. Other samples have the 0.05 mm copper walls.

The experimental study of the slot waves is carried out with the help of the plant presented in Fig.37. The transversal distribution of the dominant E_ϕ-component at the CSW open end determined for different values of the distance Δ between the sensor and the plane of the open end shows a gradual transformation of the field of the slow slot wave to the radiation field when Δ increases. The amplitude distribution is registered by a coaxial sensor with the quarter-wave dipole placed perpendicular to the slot and directed from one metallic edge to another. For the small distances ($\Delta = 0.1$ mm) the electric field maximum is concentrated in the slot. The calculated curve (see section 7) has the form similar to the amplitude distribution in the plane $\phi = 0$. The slot wave energy is carried along the small part of the cross-section in the region of the slot. Note that this behaviour has been prescribed theoretically in the previous sections.

The radiation from the CSW is weakly directed. In fact, the transversal field maximum falls when the distance from the line reaches several wavelengths ($\Delta = 4$ mm). The equivalent radiator may have the form of either the electric dipole placed perpendicular to the slot or the end of a two-wire line. In both cases the equivalent aperture ($\approx 0.07\lambda$) is comparable with the slotwidth. Therefore, such radiators are practically equivalent to the point sources having wide far-field patterns and they may be applied in the antennas formed by short-focus mirrors or in the lens antennas.

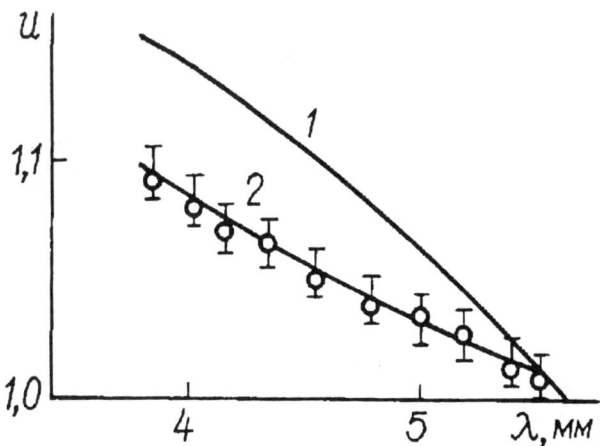

Figure 38.

The wavelength measurements in the CSW are carried out by the graphical analysis of the longitudinal field distributions obtained at high values of the SWR. The corresponding dispersion curves and the deceleration coefficient against the wavelength are presented in Fig.38. The experimental points for the deceleration coefficient is lower than its calculated values. It is explained by the inevitable differences between the experimental sample and the theoretical model.

The finite thickness of the metal walls does not cause the shift of the transference (boundary) frequency. The best agreement of the experimental and calculated data is observed in the regions of small deceleration. The physical explanation of these result is as follows: in the case of low deceleration the surface field slowly falls in the external domain, and the currents have a uniform distribution on the external surface of the metal cylinder. Finally, the equivalent total capacitance of the slot increases and becomes greater than the capacitance of the additional condensator formed by the opposite edges of the thick metallic coating. The relative contribution of this additional capacity decreases and does not influence the slot wave deceleration.

The experimental investigation of the cylindrical slot lines with a thin metallic coating (0.05 mm) show the good coincidence with the measured and calculated values of the deceleration coefficient. For the lines with the parameters we have obtained the following results:

$$2a=1.50\,\text{mm},\ \theta=55°,\ \varepsilon=2.25 \text{ for } \lambda=4.1\,\text{mm},\ \left(\frac{c}{v_\Phi}\right)_{\text{cal}}=1.166,\ \left(\frac{c}{v_\Phi}\right)_{\text{exp}}=1.158\pm0.015,$$

$$2a=1.35\,\text{mm},\ \theta=33°,\ \varepsilon=2.25 \text{ for } \lambda=4.1\,\text{mm},\ \left(\frac{c}{v_\Phi}\right)_{\text{cal}}=1.77,\ \left(\frac{c}{v_\Phi}\right)_{\text{exp}}=1.173\pm0.005.$$

These experimental data demonstrate the decrease of the deceleration coefficient due to the finite thickness of the CSW metallic walls. The boundary frequency has no shift and, therefore, the SWR value does not depend on the coating thickness. The theoretical and experimental results have small deviations when the sample characteristics are close to that applied in the theoretical model.

The experimental determination of the slot wave losses is performed with the help of the graphs of the longitudinal field distribution measured for the small SWR values. The average losses reach 19.5 dB/m.

The use of the German silver for manufacturing the prototype CSW could be hardly justified because of its high losses. The attenuation in this metal can be calculated by the data obtained for the better conducting metals, for example, the silver or the copper.

According to the results of Ref. [131], the homogeneous plane wave attenuation at the wavelength $\lambda = 4$ mm is 5.1 dB/m in the medium with the parameters $\varepsilon = 2.25$ and $\text{tg}\,\delta = 0.5 \cdot 10^{-3}$ (polyethylene). Assuming that one half of the power is carried in the pivot and the other half in the external domain, we find that the attenuation caused by the dielectric losses is equal to 2.5 dB/m. Hence, the losses in the walls are $19.5 - 2.5 = 17$ dB/m. The conductivity of the German silver is $16 \div 20$ times less than that of the copper ($5.7 \cdot 10^5$ cm/m) and $17.5 \div 22$ times less than that of the silver ($6.28 \div 10^7$ cm/m). Since the attenuation is inversely proportional to the square root of the metal conductivity, the losses in the metal for the same waveguide having the copper or the silver-plated walls are $4.25 \div 3.8$ dB/m and $4.05 \div 3.6$ dB/m, respectively. The total losses in the copper waveguide with the polyethylene filling are ≈ 6.5 dB/m.

The loss measurements are also performed for the samples with thin (0.05 mm) copper walls by registering the SWR at the beginning and at the end of the line in the case when the slot wave is excited by the diffractional method. In the CSW with the parameters $2a = 1.35$ mm and $\theta = 33°$ the losses are 5.0 ± 0.5 dB/m and the calculated value is 6.98 dB/m ($\lambda = 4.1$ mm). In the CSW with the parameters $2a = 1.5$ mm and $\theta = 74°$ the losses are 2.8 ± 1.5 dB/m and the calculated value is 7.18 dB/m ($\lambda = 4.1$ mm).

The experimental studies show that the losses in the CSW decrease as the slotwidth increases. It is caused by the more uniform current distribution on the metal surface. An especially large divergence is observed for the cylinder with the parameters $2a = 1.5$ mm, $\varepsilon = 2.25$ and $\theta = 74°$. Note that theoretical methods developed in the previous sections cannot be applied for wide slots and cannot predict this decrease.

It is interesting to compare the CSW parameters with the characteristics of the analogous microstrip, slot and coupled microstrip transmission lines. We will analyze the degree of coincidence of the experimental and theoretical data obtained for the CSW and the planar slot line [116]. The permittivity of the substrate is $\varepsilon = 13 \div 20$ and the line is formed by metallic strips made of the 0.02 mm aluminium foil pasted on the dielectric surface or drifted on the copper layer of the same thickness. In the first and in the second cases the differences between the experimental and theoretical values of the deceleration coefficient are $(+3\%) \div (+6\%)$ $(+3\%) \div (-2\%)$, respectively. A better coincidence is observed for the CSW. For thin copper covers (0.05 mm) the difference is less than 1%. A good agreement of the theoretical and experimental results enables one to apply the analytical calculation formulas for modeling the cylindrical slot lines.

The properties of the slot lines with thick conductors are considered in Ref. [145] and, in particular, the influence of the thickness of metallic strips on the dispersion and wave resistance. In the slot line with the substrate permittivity $\varepsilon = 20$, the coating thickness 1 mm and the slotwidth 0.5 mm the deceleration coefficient decreases up to 1-4% in the frequency range $3 \div 10$ GHz when the strip thickness increases from 0 to 0.05mm. The wave resistance falls up to the level $4 \div 8\%$. A similar behaviour of the deceleration coefficient is characteristic for the CSW with a thick metallic cover, but the influence of such a cover on the value of the deceleration coefficient is less pronounced (see Fig.38). The deceleration coefficient of the CSW with the parameters $2a = 1.1$ mm, $\varepsilon = 2.25$ and $\theta = 28°$ has a $1 \div 5\%$ decrease as compared with the calculated cover thickness (0.2 mm) and the slotwidth (0.265 mm). Hence, the size of the CSW is less sensible to the parameter stability than that of planar slot lines. The CSW manufacturing accuracy can be substantially improved without a noticeable worsening of their properties.

The comparison of the losses in CSWs and in other types of transmission lines shows that CSWs yield to such classical lines as empty and dielectric waveguides, but excel planar slot lines and strip transmission lines. A standard empty copper 1.8×3.6 mm rectangular waveguide has the attenuation 1.5 dB/m at the wavelength $\lambda = 4$ mm. A polyethylene dielectric waveguide ($\varepsilon = 2.25$, $\text{tg}\,\delta = 0.5 \cdot 10^{-3}$ and $\lambda = 4$ mm) has the attenuation $1 \div 4$ dB/m that depends on the deceleration factor [138]. The attenuation in the CSW exceeds these values and is equal to $5 \div 6$ dB/m for the CSW with a copper cover and the polyethylene dielectric pivot. These data are in a good agreement with the theoretical assumptions about the nature of the CSW losses. The losses in the metal cover is higher than the attenuation in an empty metallic waveguide because of the higher surface current density, especially close to the edges.

The dielectric waveguides applied in integrated dielectric schemes [138] have much in common with CSWs. The losses of the image line (in the form of a dielectric pivot placed on a metallic coating) are comparable with the CSW losses. In the case of the copper coating and the polyethylene strip with the 2×2 mm cross-section, the attenuation values in the dielectric and in the metal are 4.4 dB/m and 2 dB/m, respectively (at $\lambda = 4$ mm). The total losses are 6.4 dB/m.

The losses experimentally obtained in different types of slot lines are presented in Ref. [146]. The typical slot line parameters are as follows: the conductor width is 12.3 mm, their thickness is 0.01 mm, the slotwidth is 0.76 mm and the permittivity of the coating is $\varepsilon = 16$. The losses are $6 \div 12$ dB/m in the frequency range $4 \div 8$ GHz, i.e., they are even greater than the CSW losses at the frequency 75 GHz ($\lambda = 4$ mm) due to the greater permittivity and smaller conductor thickness. The experimental data demonstrate a sharp increase of the losses when the slotwidth decreases up to 0.12 mm.

The detailed theoretical study of the losses with respect to the slotwidth is carried out in Ref. [147], where the system of two coupled microstrip lines is considered that differs from the planar slot line by the presence of that additional metallic conductor on the lower plane of the coating. The field structure of the even wave in coupled microstrip lines is similar to that in the slot line. Therefore, they may have close spectral characteristics. Consider, for example, the line with the width of the conductor 0.595 mm and the thickness of the dielectric layer 0.53 mm ($\varepsilon = 9.7$, tg $\delta = 0.5 \cdot 10^{-3}$). When the slotwidth of this line decreases from 0.7 mm to 0.2 mm the losses increase from 5.5 dB/m to 7.5 dB/m (at the frequency 8 GHz).

The loss calculations in microstrip and coplanar transmission lines at higher frequencies are carried out in Ref. [148]. The lines under study are situated on the 0.635 mm dielectric layer ($\varepsilon = 10$) and have the wall thickness is 6.35 μm. The slotwidth in the coplanar line is 0.27 mm. The losses in these microstrip and coplanar lines at the frequency 60 Ghz are 24D b/m and 43 dB/m, respectively.

We can conclude that the CSW losses are higher than in empty metallic and dielectric waveguides and are comparable with the losses in the transmission lines used in dielectric integrated schemes. The CSWs considered are characterized by smaller losses and better agreement between the experimentally determined and theoretically calculated values of the deceleration coefficient if to compare with a majority of the slot, microstrip and coplanar transmission lines. Hence, CSWs may be applied as a transmission line with relatively low losses and the high energy concentration in the transversal cross-section.

Let us study the mode composition of the CSW assuming the existence of several propagating waves together with the principal slot H_{00} mode. With this purpose we will follow the transformation of the eigenmode of a metal cylinder with a narrow longitudinal slot to the waves of a dielectric pivot (situated inside the cylinder) with a narrow metallic strip on its surface. The latter open waveguide is, in fact, a strip transmission line in the form of a narrow metallic cylindrical strip.

The field intensity of the single-mode CSW falls in the longitudinal direction because of attenuation in the metal and dielectric. The presence of one more wave perturbs the longitudinal field distribution, and the interference of two waves leads to the energy redistribution in the transversal cross-section. The field intensity and the energy flow density are maximal in the domains of constructive interference described in section 8. In the regions of destructive interference the field intensity vanishes or sharply falls, depending on the interfering wave amplitudes. The energy flow density in this domain also vanishes or becomes very small. If the phase velocities of two waves are not equal, the phase relations between the interfering field components are different in the different parts of the line cross-section. As a result, the domains of constructive and destructive interference exchange their positions and beatings of the longitudinal field distribution are observed (this effect is very distinct in the system of two identical coupled waveguides or two coupled strip lines where the even and odd waves of a symmetrical system interfere).

A similar picture of the longitudinal beatings of two different waves in one transmission line has been discovered in the CSW. One can clearly follow the beating corresponding to the interaction of the slot wave with the wave which is still unknown. The distortion of the longitudinal field distribution in the CSW may be caused by its inhomogeneous filling and by the slotwidth and pivot discontinuities. Special measurements of the slotwidth along the longitudinal coordinate (Fig.38) show that the maximal slot deviations from the average position do not correspond to the maxima and minima coordinates in the longitudinal field distribution. In addition to this, the frequency shift is accompanied by the change of the beating period.

Let us consider the picture of beatings in a more detailed way. When the free-space wavelength is $\lambda = 4.02$ mm and the line wavelength is $\lambda_{g_1} = 2.71$ mm, the slot wave deceleration coefficient is $c/v_\Phi = 1.085$. The longitudinal beating period is $\mathcal{L} = 50$ mm, and this value corresponds to the lengths $\lambda'_{g_2} = 4.00$ mm or $\lambda''_{g_2} = 3.45$ mm of the unknown wave indicated above. Its deceleration coefficient is equal to $(c/v_\Phi)_1 = 1.005$ or $(c/v_\Phi)_2 = 1.165$. The indeterminate (two-valued) wavelength is connected with the indeterminate expression for the period of the spatial beatings of two waves $\mathcal{L} = \pm \left(1/\lambda_{g_1} - 1/\lambda_{g_2} \right)^{-1}$. It is not possible to find this unknown wavelength without the use of additional information.

The diffractional analysis of the mode composition allows us to remove this indeterminate situation. The measurements of the deceleration coefficients of the slot waves and the accompanying waves are carried out with the help of the plant presented in Fig.37. The gratings made of rectangular rods is used as a diffractional array. Its thickness provides the maximal transference [6, 24] and the period is $l = 4.25$ mm. One of the far-field patterns is presented in Fig.39. Two sidelobes are related to two different kinds of waves. The maximal radiation angles $(\psi_{-1})_1$ and $(\psi_{-1})_2$ correspond to the deceleration coefficients $(c/v_\Phi)_1 = 1.085$ and $(c/v_\Phi)_2 = 1.033$ defining thus the deceleration coefficient $(c/v_\Phi)_2$. The value of $(c/v_\Phi)_2 = 1.005$ determined by the picture of longitudinal beatings is more precise. The accuracy of the diffractional analysis depends first of all on the wavelength and the effective radiating aperture. The largest error is caused by the far-field pattern broadening due to the slotwidth variation. This broadening reaches 0.5° (Fig.38). In some cases the direct influence of a diffractional grating on the phase velocity of a surface wave is more essential [6, 141], and the displacement of the direction of the maximal radiation may reach several degrees. The sequence of experiments of the phase velocity determination in a rectangular dielectric waveguide with $\varepsilon = 2.45$ was carried out to improve the research technique, namely, to optimize the choice of the target distance, the grating's length and the communication of the grating with the CSW [149, 150]. When the waveguide width increases, the deceleration coefficients asymptotically approaches the deceleration coefficient of

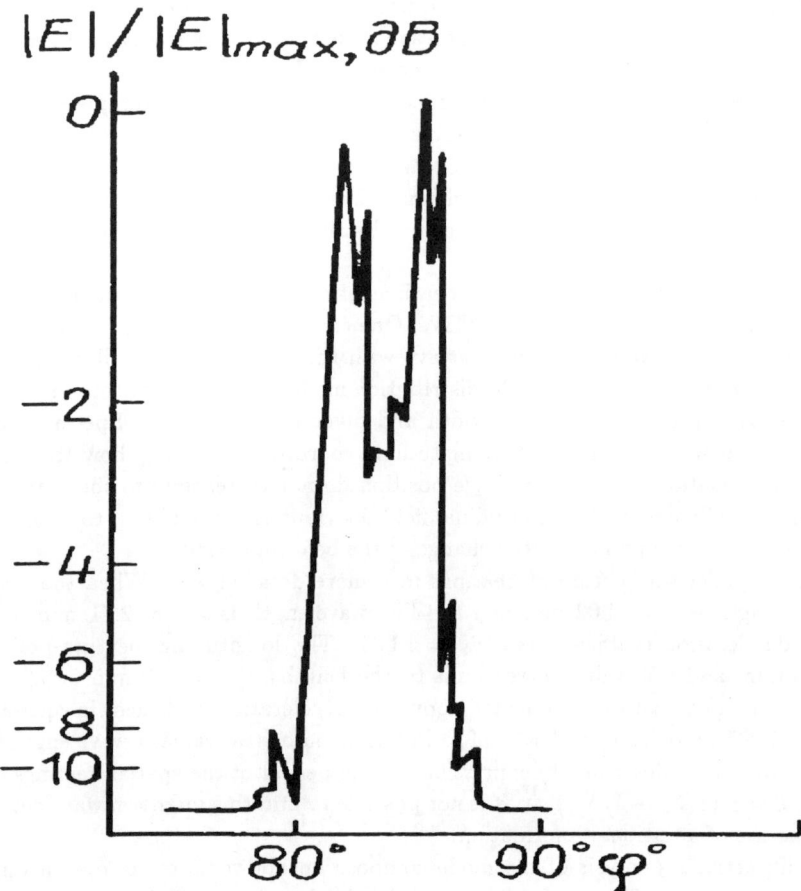

Figure 39.

the principal H_1 (or E_1) wave of a planar dielectric waveguide. As a result of the parameter optimization, the accuracy of measuring the deceleration coefficients by the diffractional method becomes comparable with the accuracy obtained from the picture of standing waves and beatings. Thus, together with the slot wave, we have proved the existence of the additional wave in the CSW.

The results of our studies show that this wave has very low deceleration. The assumption was made that it is similar to the Sommerfeld wave propagating along a circular metallic conductor [103]. The deceleration of the Sommerfeld wave is caused by the finite conductivity of the metal. This wave has the non-zero azimuthal, longitudinal and radial electric components. The longitudinal current has only one longitudinal component on a metal surface. The similar field structure is characteristic for the field in the metal cylinder with a narrow longitudinal slot filled with the dielectric. The edges of the coating with the large surface curvature as well as the high pivot permittivity strongly influence the deceleration of the Sommerfeld wave in the CSW. The study of this influence is of great importance not only for the CSW itself but also for modeling the slot and coplanar transmission lines formed by the conductors having similar cross-sections in the vicinity of the slot.

Consider the CSW waveguide properties when the angle slotwidth varies in the whole interval $(0°, 360°)$. The limiting values correspond to either the circular dielectric or the circular metallic waveguides filled with the dielectric. The latter waveguide does not carry the energy in the long-wavelength region ($\lambda > 3.41\sqrt{\varepsilon}a$). However, the line itself may support the Sommerfeld wave in the external domain. Narrow slots lead to the appearance of the slot H_{00} wave and to the distortion of the Sommerfeld wave (its phase velocity and the field structure are destroyed). The H_{11} wave [134] is the principal wave of the circular dielectric waveguide. Narrow metallic strips destroy the polarization degeneration of the HE_{11} wave. Two waves denoted in what follows by $H^+E_{11}^-$ and $H^-E_{11}^+$ have the prevailing orthogonal components of the vector **E**.

Experimental study of the CSW eigenwave transformation to the modes of a circular dielectric waveguide caused by the gradual increase of the slotwidth allows one to describe in details the wave structure and to formulate recommendations concerning the practical applications of such open lines. This problem is studied theoretically in the previous sections.

The surface waves with arbitrary phase velocities and polarizations can be separately excited by the input of the diffractional energy into the waveguide [149]. In the case of inhomogeneous communication between the waveguide and the diffractional grating (when the waveguide supports surface waves) the coordination of the incident volume wave with a receiving system is close to 100%. The proper choice of the incident wave polarization and the diffractional grating rises the probability of the excitation given wave.

The experimental plant for the study of the CSW (depicted in Fig.40) is used for the separate excitation of all possible types of waves in the CSW 8 and for the investigation of the longitudinal field distribution for the arbitrary SWR values. Generator 1 excites a surface wave in sample 8 through antenna 4 and diffractional grating 7. The field distribution is registered by sensors 9 or 12 adjusted by the short-circuit piston 14 with respect to the maximum signal in detector 13 . The signal from detector 10 (13) comes to the recorder through the amplifier. The amplitude is calibrated by measuring attenuator 3. The standing wave is supported by the short-circuit metallic plate 11. The wavelength is controlled by the resonance wave-meter 5 placed between the directional coupler 2 and detector 6.

The excitation and reception of the considered wave is provided by the choice of the diffractional grating, the excitation angle ψ, the antenna polarization and the type of the sensor–receiver. The transversal electric field components are measured by coaxial sensor 12 with the quarter-wave dipole placed at its end. The diameter of the internal conductor is 0.15 mm. The longitudinal electric field components are registered by the narrow end of waveguide 9 with the 0.1 mm communication slot. The working wavelength is $\lambda = 4.1$ mm and the sensor accuracy is sufficiently high so that it can operate at the large SWR values.

The diffractional grating aperture is $100 \div 120$ mm and the width ψ of the effective excitation sector for the considered wave is equal to the width of the far-field pattern $1.7° \div 2.0°$. The CSW slot faces the grating. Two types of diffractional gratings with the narrow and wide channels are applied. The channel depths (that are chosen to be close to the resonance value) are approximately equal to a quarter of the principal

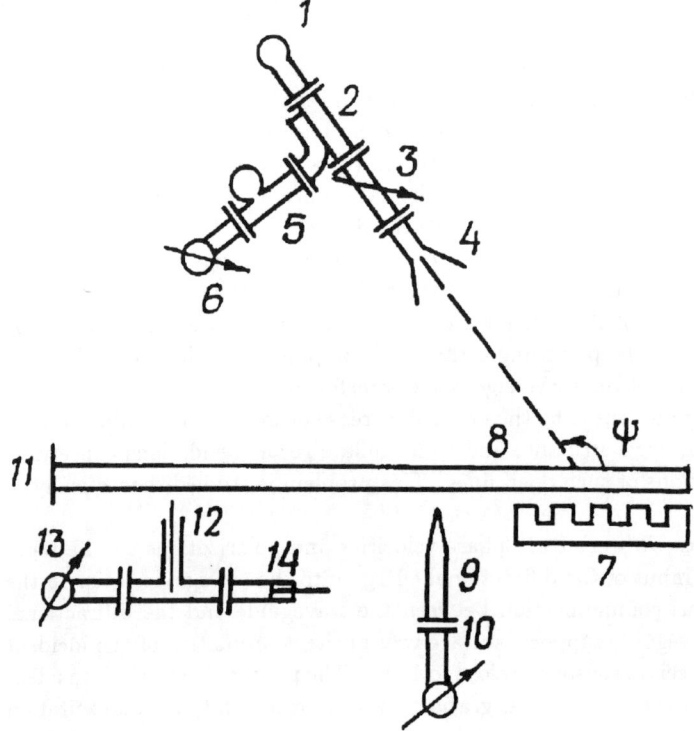

Figure 40.

wavelength in the line formed by two parallel metallic planes (by the opposite walls of the channel). The grating with narrow slots ($l = 3.0$ mm) effectively excites the E-waves, while a grating with wide slots ($l = 3.2$ mm) is applied in the case of the E-polarized incident wave, in particular, for the slot wave. Antenna 4 provides the excitation of grating 7 not only by H- and the E-polarized waves but also by the linearly polarized wave incident at an angle of 45° to the grating elements. Equal parts of the power are carried by the H- and E-polarized waves. Here H and E waves are simultaneously excited in CSW 8. Such a way of excitation is necessary to obtain a more distinct picture of the beatings of two waves having orthogonal polarizations. A similar effect is observed when the sensitive elements of sensors 9 and 12 are fixed at an angle of 45° to the line axis. All the CSW samples are made of circular polyethylene pivots covered by the copper foil with the thickness 0.05 mm.

   The deceleration coefficients of the partially shielded dielectric pivot with the diameter $2a = 1.5$ mm and $ka = 1.15$ as functions of the slotwidth are presented in Fig.41. Curve 1 corresponds to the wave with the prevailing longitudinal currents (on the metallic strip). This wave is identical to the Sommerfeld wave of the shielded pivot ($\theta \to 0$), which has very small deceleration due to the finite conductivity of the cover. An increase of the slotwidth yields the greater Sommerfeld wave deceleration due to a more substantial influence of the dielectric The deceleration coefficient is maximal for the minimal stripwidth when the "shielding" factor is minimal and the effect of the finite metal conductivity increases, as well as in the case of a circular conductor with

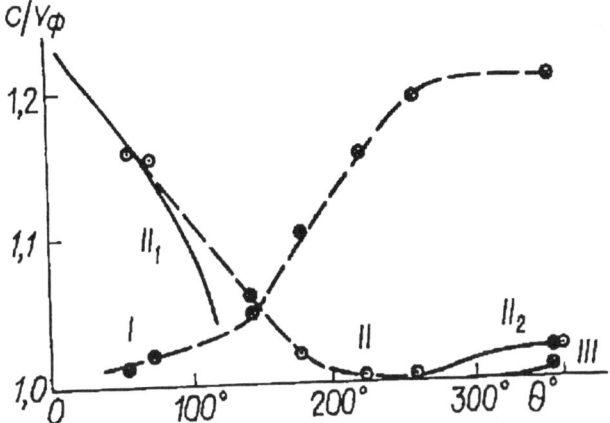

Figure 41.

a small radius.

The curve II in the domain of small slotwidths corresponds to the slot wave with the prevailing transversal components of the vector **E**. The slot wave deceleration coefficient obtained by formula (2.23) is shown by bold line II_1. There is a good agreement of the theoretical and experimental data in the interval $\theta < 80°$, although the condition $\theta < 1°$ does not hold. Hence, one can use (2.23) for $ka \approx 1$ and broaden thus the limits of its application.

The slot wave deceleration coefficient tends to $\sqrt{(\varepsilon + 1)/2}$ $(\theta \to 0)$, i.e., to the deceleration coefficient of the slot wave in the planar slot line (for low frequencies) because the strong shielding of a dielectric pivot leads to the higher field concentration near the slot. As a result, the parts of the metal cover situated far from the slot practically do not take part in the field formation and do not influence the slot wave attenuation. Note that there is a divergence between the theoretical and experimental data for $\theta > 80°$.

The slot wave deceleration coefficient monotonically decreases with respect to the increasing slotwidth up to the value $\theta = 200° \div 220°$. Then the H_{00} wave continuously transfers to the $H^+E_{11}^-$ wave and the value of its deceleration increases up to the deceleration of the HE_{11} wave in the circular dielectric waveguide. The $H^+E_{11}^-$ wave deceleration coefficient obtained by formula (2.23) is depicted by bold line II_2. Here an agreement of the theoretical and experimental results is observed.

The curve III in Fig.41 corresponds to the $H^-E_{11}^+$ wave which is excited like the Sommerfeld wave having the nonzero longitudinal component of the vector **E**. The bold line shows the calculated curve for the deceleration coefficient against the slotwidth obtained by formula (2.23). The experimental points are presented for $\theta = 360°$ and $\theta = 356°$. One can see a sufficiently good coincidence of the theoretical and experimental data.

The curve II in Fig.42 corresponds to rather small deceleration coefficients $c/v_\phi \approx 1.005$ for $\theta = 220° \div 240°$. The transference of the slow surface wave to the fast leaky wave may occur in the interval separating the experimental points. In order to verify the identity of the H_{00} wave and the $H^+E_{11}^-$ wave, we have performed several experiments for the large partially covered dielectric pivot with the parameters $2a =$

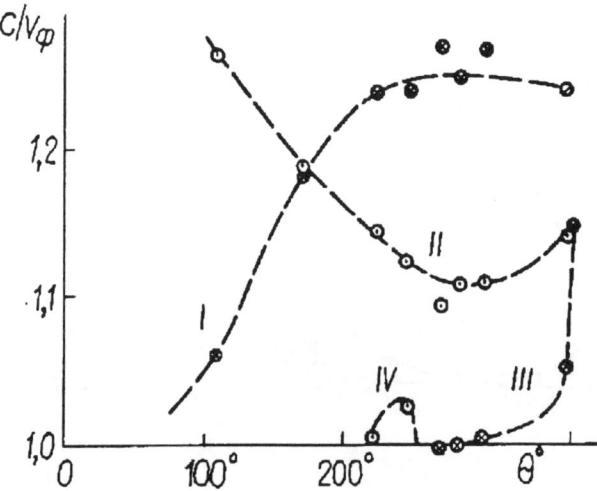

Figure 42.

2.3 mm and $ka = 1.76$.

Large values of ka correspond to the increased values of the surface wave deceleration coefficients. It is confirmed by the experimental data presented in Fig.42 that are analogous to the results shown in Fig.41. As follows from the graphs (curve II in Fig.42), the slot wave continuously transfers to the $H^+E_{11}^-$ wave for $\theta = 240° \div 300°$. A greater divergence of the experimental points as compared with the data obtained for the pivot having the diameter $2a = 1.5$ mm ($ka = 1.15$) is explained by the inaccuracy of the sample manufacturing. The deceleration coefficients calculated by formula (2.23) coincide with the experimental data within the 5% accuracy.

In addition to the waves considered, one of the higher order waves appears (curve IV) in the interval $\theta = 220° \div 250°$, which has the same polarization as the $H^+E_{11}^-$ wave and is excited in a similar manner. Its existence is proved by the separate excitation with the help of a diffractional grating and is confirmed by the direct wavelength measurement using the standing wave picture and observation of beatings registered together with the slot wave.

The level of losses of the principal and the higher order waves is an important characteristic of transmission lines. The losses are measured through the determination of the standing wave coefficient. The Sommerfeld wave losses are measured in two characteristic cases. In the first case the line parameters are: $2a = 1.35$ mm, $ka = 1.03$ and $\theta = 33°$. The measured losses are 26 dB/m and the working section has the length $L = 250$ mm. Such a large value is caused only by the wave radiation and not by the ohmic or dielectric losses. In fact, the dependence of the standing wave coefficient has several relatively horizontal parts and several splashes corresponding to the Sommerfeld wave diffraction by the sample inhomogeneities. The losses in the dielectric are assumed to be very small because of a strong shielding factor. It is natural to assume that the losses in a metal are comparable with the Sommerfeld wave losses in a metallic cylinder. The calculated value of these losses is 0.198 dB/m for the copper cylinder with the parameters $2a = 1.35$ mm and $ka = 1.03$. The Sommerfeld wave

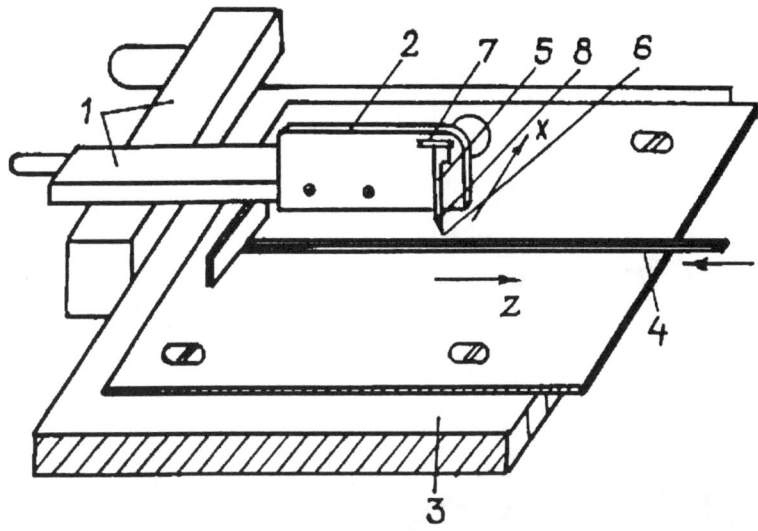

Figure 43.

electric field components using the head rotation. The λ_g-wavelength in the CSW is measured by determining the distance between two minima of the standing wave field in the waveguide-slot CSW junction containing the short-circuit reflector pasted on a conducting foam plastic coating with the slot on the top and the polyethylene dielectric filling ($\varepsilon = 2.25$, $\mathrm{tg}\,\delta = 3 \cdot 10^{-4}$). The measurement data is presented in Table 3 from Ref. [154], where λ_{gc} denotes the value of λ_g taken from Ref. [155], λ_{gk} is calculated by formula (2.25) and λ_{gexp} is the wavelength determined experimentally. It follows from the data presented in this Table that the agreement between the values of λ_{gk} and the experimental data is better than expected (divergence is not more than $5\% \div 7\%$).

To construct the set of basic units (the element base) created on the basis of the CSW one has to know the transversal amplitude field distribution. The two-coordinate measuring line described above enables us to obtain these data and to register the field distribution in the external domain. The information thus obtained concerning the rate of the field decrease out of the slot is of great importance for creating the couplers and optimizing the line dimensions.

The amplitude field distributions in the CSW with the parameters $2a = 0.8$ mm, $2d_2 = 0.1$ mm, $\theta = 2.9°$ and $2d_1 = 0.05$mm, different values of the permittivities ($\varepsilon = 1$, 2.25, 3.8) and the working wavelengths ($\lambda_0 = 7.7$ mm, 3.9 mm) are presented in Fig.44. The registered E_ϕ field component defines the CSW effective size. The measurement results are depicted by the light and dark circles, and different orientations of the triangles correspond to two successive measurements. The transference from the slow surface wave ($h' > k'$) to the fast leaky wave ($h' < k'$) takes place in the CSW at the 7.7-mm wavelength when ε varies from 1 to 3.8. The slow surface wave is measured for $\varepsilon = 3.8$. As can be seen from Fig.44, the field rapidly decreases in the direction off the slot and the rate of this decrease has the order of $(a/r)^4$. The calculated amplitude distribution taken from Ref. [155] is shown in Fig.44 by the bold line (curve 1). For small values of y (small r) a good agreement with the experiment is observed due to

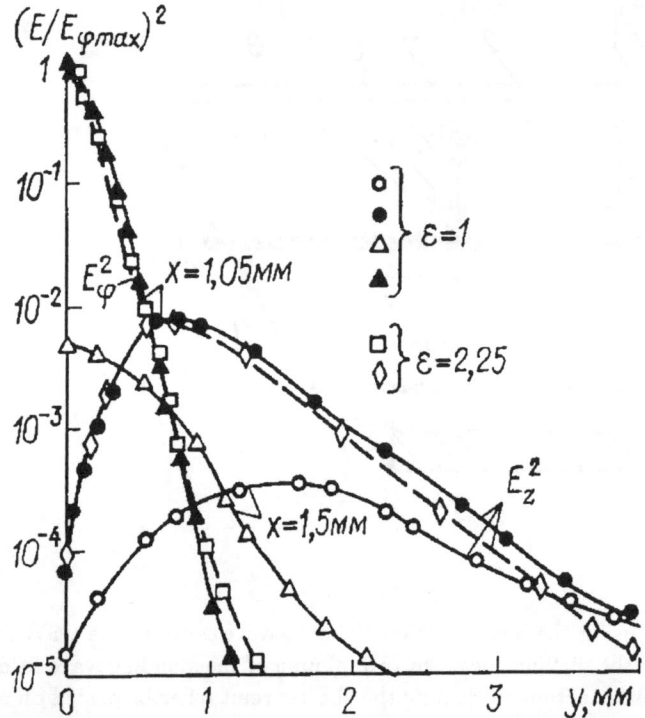

Figure 44.

the asymptotical character of the measured field. The rate of the E_ϕ decrease weakly depends on the frequency and permittivity far from the critical points. For $r \gg a$ the field decreases exponentially, and this behaviour is characteristic for surface-wave lines. Hence, the field in the CSW has a distinct quasi-statical character in the near zone and is similar to the fields in statical condensators. According to the Meixner condition [41], E_ϕ diverges in the vicinity of the edges ($r = a$, $\phi = \pm\theta°$).

Since the electric field is concentrated on the slot, its level is substantial at the distances of several slotwidths from the slot for $\phi \approx 0°$, so that for $\phi = 180°$ the field intensity is very small already for $r = a$. Thus, by analogy with Ref. [137], one can write the semi-empirical formula for the effective CSW diameter

$$t_{\text{eff}} \approx 2a + 4d_2 + 4a\sin\theta. \tag{2.58}$$

Since $2a\sin\theta \ll a$ (usually $\theta \approx (2\div10)°$), t_{eff} practically coincides with the geometrical diameter ($2a + 4d_2$).

When the permittivity decreases, the amplitude field distribution sharply changes in the domain of fast leaky waves ($h' < k'$). The measuring sensor registers the radiation field of leaky waves. The curves of the field distribution in Fig.44 correspond to this case when $\varepsilon = 2.25$ and $\varepsilon = 1$. The rate of decrease corresponding to theoretical curve 2 has the order of $1/(gr)$ (where $g = \sqrt{k^2 - h^2}$), and this value is characteristic for an infinite cylindrical line. A finite slotted cylindrical radiator (with the length $L = 80$ mm) radiates not the cylindrical wave but the wave of the Cherenkov cone with the angle α. For $\varepsilon = 1$ this radiator works close to the critical point (2α), that is, when $h' \to 0$ ($\alpha \to 90°$), the field radiates in the direction perpendicular to the longitudinal cylinder axis. The rate of the field intensity decrease is proportional to $(gr)^{-2}$. When $\varepsilon = 2.25$, $h' > 0$, $\alpha < 90°$ and the Cherenkov cone approaches the cylinder surface, the rate of the decrease remains the same in the direction of the cone radiation and increases perpendicular to the cylinder axis (see Fig.44).

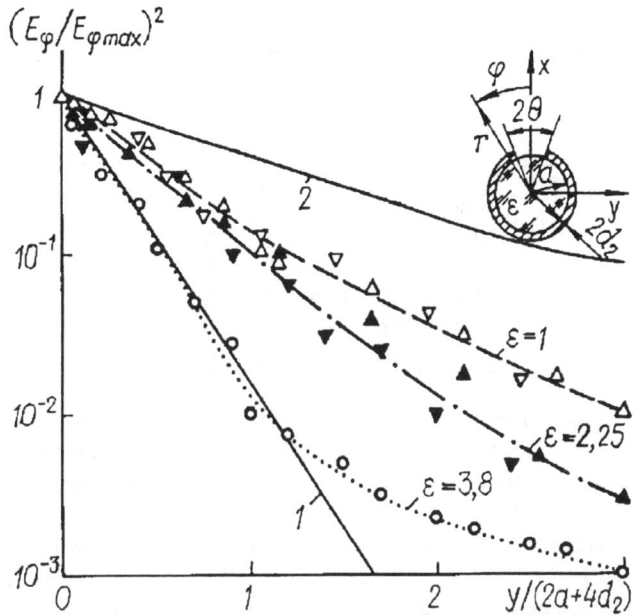

Figure 45.

The amplitude field distribution in the CSW having the same size and registered at the wavelengths that are far from the critical values ($\lambda_0 \ll \lambda_{cr}$) weakly depends on ε (if to compare with the quasi-statical case shown in Fig.45). The following results illustrate the degree of the field concentration on the slot: a 1-mm shift of the sensor yields a five-order fall of the field intensity. We have also established that the rate of decrease becomes smaller when the distance between the sensor and the slot increases: the field intensity falls up to three orders of magnitude within the interval of the distances 1 mm –2 mm.

The experiments show that the difference between the E_z^2 and E_ϕ^2 components reach two orders of magnitude. Hence, the quasi-H_{00} classification for the principal wave in a slotted cylinder is justified, and the CSW is similar to a planar slot line (where the transversal electric field components exceed the longitudinal component in ten times). When the distance to the slot increases, the rate of the field decrease becomes smaller and the specific weight of the E_z component grows: at the distance 1 mm from the slot E_z^2 is only one order of magnitude less than E_ϕ^2. The study of the reliefs of the E_ϕ and E_z transversal distributions allows to estimate the field concentration caused by the slot.

Table 4.

	$2a$, mm	θ	$2d_1$, mm	$2d_2$, mm	λ_0, mm	α_c	α_k	$\alpha_{\exp}$
1	1.3	3.5°	0.08	0.2	4.1	23.9°	18°	18.5°
2	1.3	3.5°	0.08	0.2	5.7	31.9°	25°	26.5°
3	1.3	3.5°	0.08	0.2	8.2	49.9°	37.5°	38.0°
4	0.8	6°	0.08	0.2	4.1	41°	31.5°	35.0°
5	0.8	6°	0.08	0.2	5.7	66°	47.1°	50.0°
6	0.8	6°	0.08	0.2	7.8	–	80.7°	80.0°
7	2.9	10°	0.5	0.2	7.8	22.8°	20.2°	21°
8	1.3	44°	1.0	0.1	4.1	45°	41.7°	35°
9	0.8	2.8°	0.04	0.1	4.2	39.8°	27.8°	30°

Table 5.

	$2a$, mm	θ	$2d_1$, mm	$2d_2$, mm	λ_0, mm	ε	λ_{gc}	λ_{gk}	$\lambda_{g\exp}$
1	0.8	2.8°	0.04	0.1	4.2	2.25	3.8	3.98	4.1
2	0.8	2.8°	0.04	0.1	3.8	2.25	2.33	3.54	3.7
3	0.8	2.8°	0.04	0.1	4.2	1	5.45	4.74	4.8
4	0.8	2.8°	0.04	0.1	3.8	1	4.65	4.18	4.4
5	0.9	26.4°	0.4	0.1	4.2	2.25	4.48	4.3	4.5
6	0.9	26.4°	0.4	0.1	3.8	2.25	3.77	3.77	3.8

To measure the CSW dispersion in the millimeter wavelength range one can register the electromagnetic power radiation from the line supporting the leaky wave. The formula $\cos\alpha = h'/k = c/(v_\Phi\sqrt{\varepsilon})$ enables one to conclude that the Cherenkov radiation angle α is determined only by the wave dispersion in a slotted cylinder. We will examine cylindrical slotted radiators having the form of empty slotted cylinders covered by the German silver and excited by the CSW filled with the melted quartz ($\varepsilon = 3.8$, $\mathrm{tg}\,\delta = 10^{-4}$). The connection between the line and the radiator is provided by decreasing the cylinder diameter $2a$, the wall thickness $2d_2$ or the permittivity ε of the dielectric filling, or by increasing the slotwidth 2θ. In the present experiments we varied the permittivity. The scheme of the experimental set-up is as follows: The spectral waveguide-slot junctions connect samples with the clystron generator and the waveguide measuring section. The measurement data for the angle α (defining the height of the main lobe of the far-field pattern measured from the longitudinal axis of the cylindrical slotted radiator) is presented in Table 4. (the sample length is $60 \div 80$ mm). Other geometrical parameter of the slotted generator and the working wavelengths are presented in Table 5 where α_k and $\alpha_{\exp}$ denote the calculated and experimental angle values, respectively. The experimental error does not exceed 1° and for point 9 in Table 4 it is not greater than 0.3°. The divergence between $\alpha_{\exp}$ and α_k is much less than that between $\alpha_{\exp}$ and α_c due to the influence of the screen finite thickness on the CSW capacitance. The break at point 6 for α_c shows that this value corresponds to the over-critical case for an infinitely thin cylinder. On the whole, the divergence between $\alpha_{\exp}$ and α_k lies within the several percent interval, except for point 8, at which the divergence reaches 17%. It is probably connected with the large slotwidth ($\theta = 44°$). For $\alpha > 25° \div 30°$ the expression $C_k = [\ln^{-1}\sin(\theta/2)]/\pi$ describing the capacitance of the edge field gives a greater error [155]. For $\theta = 44°$ this error reaches 15%.

The wave propagation process in the ISW are similar to that in the CSW. Consider

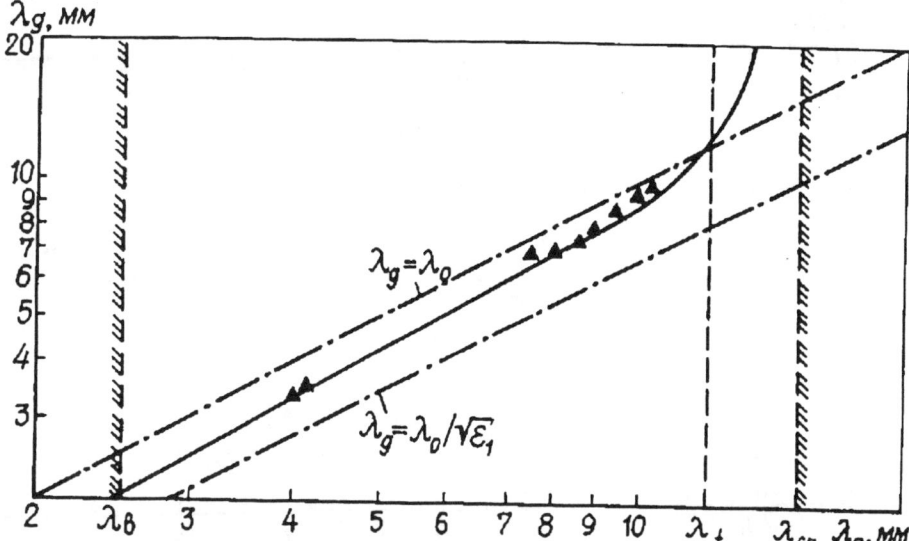

Figure 46.

some experimental findings that confirm this assertion. Two special samples are made to study the dispersion in the ISW. They consist of a smooth or the Chebyshev step junction connecting the waveguide and the ISW (with the tephlon dielectric core, $\varepsilon = 2.0$, $\mathrm{tg}\,\delta = 10^{-4}$ and these values correspond to point 3 in Table 5) or the ISW section and a short-circuit reflector. The wavelength λ_g as a function of λ_0 is presented in Fig.46 for the single-mode ISW (with the dimensions $2a = 0.8$ mm, $b = 0.85$ mm, $2d_2 = 0.35$ mm and $d_1 = 0.1$ mm). The theoretical and experimental results are shown by the bold line and dark triangles, respectively. The divergence does not exceed 8% in the long-wavelength range, which is evidence of the validity of the obtained dispersion relations.

The measured ISW attenuation and the SWR values are presented in Fig.47 (by dark circles). One can see that the losses are $3 \div 4$ dB/m in the 8-mm range and 10 dB/m in the 4-mm range. This level may be decreased by improving the quality of the metal surfaces.

One may conclude that losses in the ISW are somewhat greater than in a standard waveguide and much less than in a coaxial line (up to one order of magnitude). Hence, they are less than in a strip line: the losses in the ISW in the 4-mm range have the same order as in strip lines in the centimeter wavelength range. We conclude that ISWs may be applied in the integrated schemes of the millimeter wavelength range.

Let us determine the amplitude field distribution in the ISW. The electric field in the ISW is transversal and concentrated in the small vicinity of the slot. The field has a quasi-statical character in the near zone and the slot capacitance is found using the formula similar to that obtained for the CSW. The expression for the field distribution in the rectangular ISW may be obtained by the conformal mapping of the cylindrical coordinate system to the Cartesian system.

The results of the experimental study of the amplitude field distribution in the

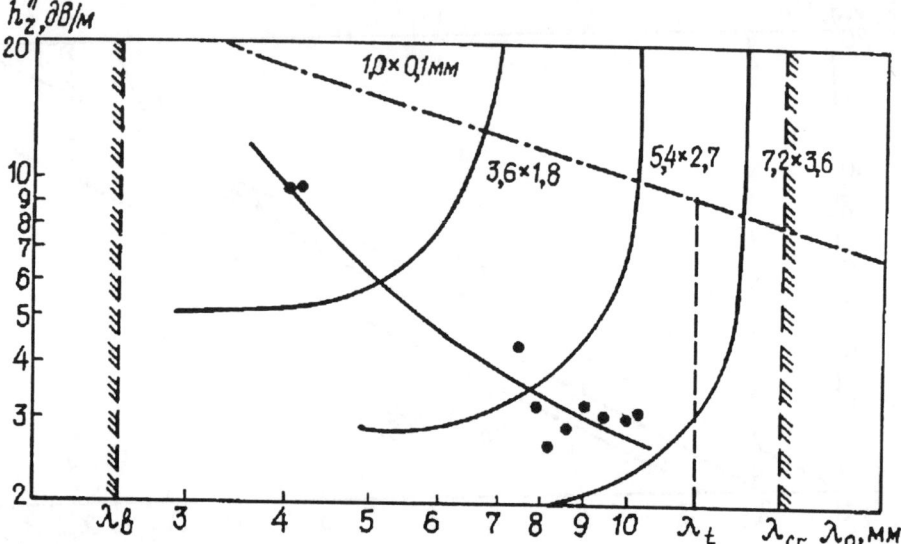

Figure 47.

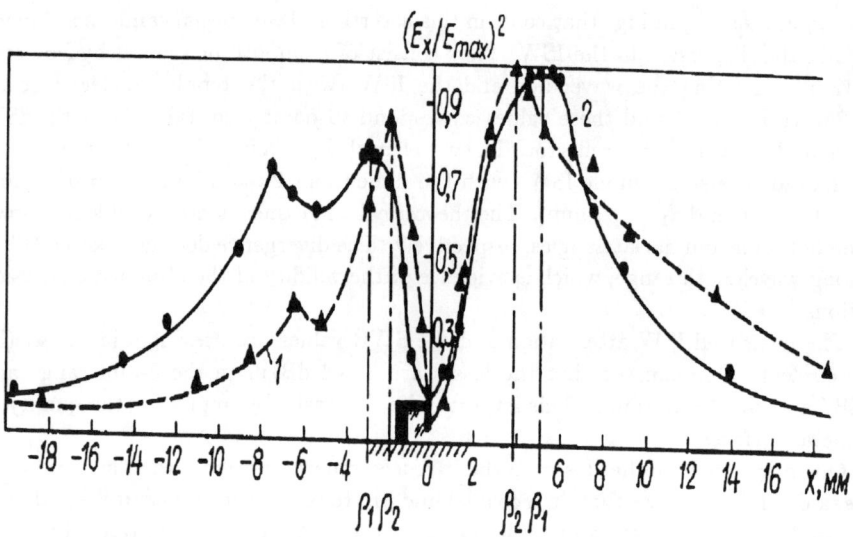

Figure 48.

ISW with the parameters $2a = 0.8$ mm, $b = 0.85$ mm, $2d_2 = 0.1$ mm, $d_1 = 0.35$ mm and $\varepsilon = 2.1$ are presented in Fig.48. Selective properties of the sensor applied in the two-coordinate measuring line enable us to register only the E_x field component. Two characteristic curves for two wavelengths of the exciting field, $\lambda_0 = 7.6$ mm and $\lambda_0 = 10.25$ mm, are presented in Fig.48. The function of x has two maxima. When the frequency increases, the field concentration in the line volume slowly increases and becomes noticeable close to the transfer wavelength λ_t. Unfortunately, it is not possible to determine the effective line radius from the measurements of the E_x distribution since $E_x \ll E_y$.

Leaky waves in the CSW may tear off the structure and go to the free space in the form of volume waves. Usually, when leaky waves are studied experimentally, the main attention is paid to their antenna characteristics (see [116, 117]) and the fields are measured in the far zone. But these waves are formed near the waveguide and, hence, the most valuable information about the properties of leaky waves should be obtained in the near zone.

We will present the results of the experimental study of leaky waves of the copper-foil CSW in the 4-mm and 8-mm wavelength ranges (the foil thickness is $2d_2 = 0.1$ mm, the angle slotwidth is $2\theta = 2.8°$ and $\varepsilon = 1$). Formulas (2.25) and (2.38) may be applied for calculating the propagation constant of the H_{00} wave where the value of h'' defines the diffractional losses. For the chosen values of θ they considerably exceed the ohmic losses and the latter are neglected. A two-coordinate measuring line is used for registering the near field. The line operates in the $2.5 \div 12$-mm wavelength range providing the dynamical range of 50 dB and the SWR is not greater than 1.15. These properties of the measuring line are caused by the use of a small sensor and high sensibility of the CSW-based detector. The area covered by the line manipulator is 80×25 mm^2 and the coordinate scale accuracy is 0.01 mm.

The critical wavelength λ_{cr} in the CSW is determined from (2.40) and for the considered sample $\lambda_{cr} = 9.14$ mm. The H_{00} wave is the leaky wave with the phase velocity v and this assumption is confirmed by the measured wavelength $\lambda_g = 2\pi/h'$.

Propagation of leaky waves in the CSW is accompanied by their radiation to the free space [116, 117] at an angle $\alpha = \arccos(h'/k)$ coinciding with the Cherenkov propagation angle. The far-field patterns presented in Fig.49 have the form of the Cherenkov cones. The distance between the CSW and the receiving antenna z_p is 510 mm and the length L of the CSW is 80 mm. The cone forms the lobes observed for $\gamma = \alpha = 30°$ (the theoretical value is $\alpha = 27.3°$ where γ is the observation angle in the longitudinal plane). The measured forms of the far-field patterns and especially of their main lobes are in a satisfactory agreement with the data obtained by (2.41).

The E_ϕ component of the leaky (slot) wave prevails and only this component does not vanish and forms the radiation far-field pattern.

The E_ϕ maximum at $\phi = 0$ corresponds to the field of the leaky wave concentrated on the slot. The amplitude distribution for $x = 1.05$ mm and $z = 3$ mm (the field on the slot close to the source) rapidly falls with respect to the coordinate y. E_ϕ has the same distribution for $x = 1.5$ mm and $z = 3$ mm, but the absolute value of the maximum and the rate of decrease is less than in the previous case because the field of the slot H_{00} wave is concentrated on the slot. If to fill a slotted cylinder with a dielectric (in the experiment we have taken $\varepsilon = 3.8$), then the fast leaky wave will transform to the slow surface wave. The experiments show that the radiation to the free space is stopped and λ_g becomes less than λ_0. However, the E_ϕ amplitude distribution and the rate of the field decrease virtually do not change. This phenomenon, which is interesting from the theoretical viewpoint, has been already pointed out above: the CSW supports the leaky wave, and the field decreases in the transversal direction, as well as in the case of the surface wave propagation. The only difference is that the surface wave exists for all values of z and the leaky waves are observed for sufficiently small z.

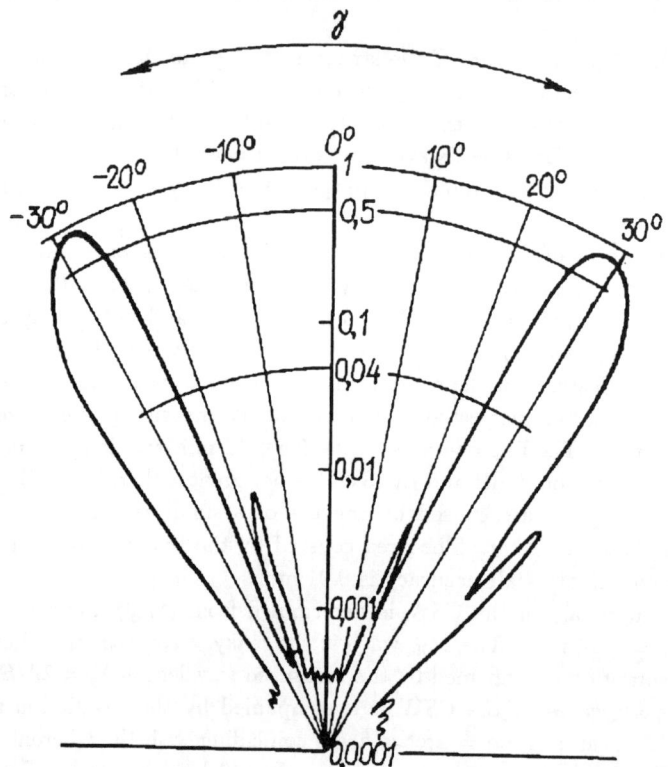

Figure 49.

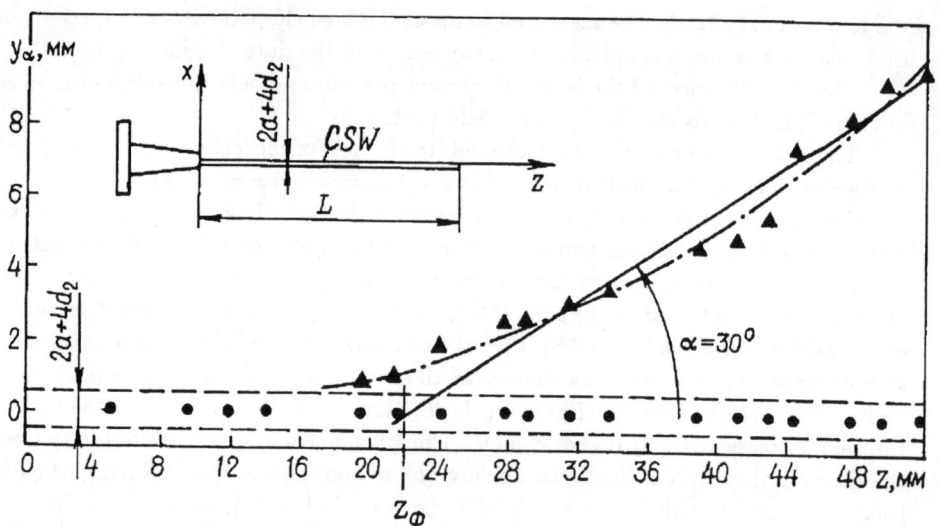

Figure 50.

The E_ϕ amplitude distribution of the quasi-H_{00} slot wave along the axis z for $x = 1.5$ mm is measured by a specially constructed sample in the form of a slotted cylinder. The slot wave is excited through the waveguide-slot junction . The device is fixed on the foam plastic coating placed on a heavy table together with the two-coordinate measuring line. The measurement data allow us to conclude that the field maximum at $y = 0$ corresponds to the slot wave concentrated on the slot ($x = 1.5$ mm). The field exponentially falls along the line with the 44 dB/m attenuation, which is three orders of magnitude greater than the ohmic losses. Approximate formula (2.39) gives the diffractional losses that exceed experimental values. Two lateral maxima in Fig.49 correspond to the Cherenkov cone (with the cross-section $x = 1.5$ mm). This boundary field is formed by the slot wave and then tears off at the Cherenkov angle. In the beginning of the process the angle of the radiation field maximum γ_n is not equal to the angle α. The coordinate of the field maximum as a function of z is presented in Fig.50. One can see that for $z \approx 20$ mm $\approx 5\lambda_0$, $\gamma_n < \alpha$ and for $z \approx 30$ mm $\approx 7\lambda_0$, $\gamma_n > \alpha$ (for $\lambda_0 = 3.9$ mm). These data are evidence of the complicated wave transformation process observed close to the open waveguide ($y > 2\lambda_0$). In order to find the phase origin z_ϕ of the far-field pattern, we turn the receiving antenna around the points with different z-coordinates ($x = y = 0$) and determine the Cherenkov observation angle for two values $z = z_p$. The obtained value $z_\phi \approx 22$ mm corresponds to the field picture registered by the two-coordinate measuring line. The value of z_ϕ depends on λ_0, i.e., on the angle α. With increasing α the phase origin approaches the source and with decreasing α z_ϕ increases.

The leaky wave amplitude distribution depends on z: for $z < z_\phi$ the Cherenkov cone is not generated. The analysis of (2.41) reveals a weak dependence of the far-field pattern on the sample length. For comparatively short lines ($L < 5\lambda_0$) the axis radiation is observed with the far-field pattern maximum at $\alpha = 0$. For long samples the far-field patterns look like the Cherenkov cone. Hence, the field of the leaky wave propagating along the line covers the whole interval from $L = 0$ to $L = \infty$.

We note that the obtained pictures of the leaky wave field distributions are not caused by the finite size of the line sections. Theoretical assumptions concerning the qualitative properties of the amplitude distribution are valid for $L \to \infty$ and also in the far zone (when the line is excited by a concentrated source). Relation (2.41) for $L \to \infty$ takes the form

$$|\psi(\gamma)|^2 = \frac{h''^2}{h''^2 + \xi^2}.$$

This expression describes the main lobe of the far-field pattern with the width that does not vanish as $L \to \infty$ (because $h'' > 0$ and the field decays exponentially along the line). Here there are no terms responsible for the sidelobe formation. Note that sidelobes appear in the near field and to register them one has to use much more sensible lines.

The experiments show that the leaky wave may be considered as the wave that is excited separately. Such a wave in the "empty" CSW with $\varepsilon = 1$ is always leaky because covering the slot destroys it. When the slot is cut, the higher order (waveguide) waves that can exist in the empty cylinder, may radiate to the free space. However, the picture of their field distribution considerably differs from the one obtained for the slot H_{00} wave.

The CSW parameters are strongly influenced by the surrounding medium since the conductivity and other properties of a metal cover depend on the mechanical and climate factors. Hence, the CSW covered with a dielectric, i.e., the three-layered CSW, may be a reliable line to be applied in the millimeter wavelength range. One can determine from (2.11) its relative deceleration as

$$\frac{1}{2}(ka)^2\left(\frac{\varepsilon_1 + 1}{2h_0^2} - 1\right)(\varepsilon_2 - 1)\ln\frac{b}{a},$$

where b is the CSW external radius and ε_1 and ε_2 are the permittivities of the internal and external media. Hence, the deceleration increases together with the thickness and the cover permittivity. It is evident that the losses must also increase and the additional contribution may be estimated in the following way: in the CSW without the dielectric cover this contribution consists of the losses in the metal and in the dielectric with the permittivity ε_1, $\eta_0 = \eta_{\varepsilon_1}^0 + \eta_\sigma$, and in the covered CSW the additional losses are caused by the dielectric with the permittivity ε_2, $\eta_0 = \eta_{\varepsilon_1}^0 + \eta_\sigma + \eta_{\varepsilon_2}$. Since the deceleration slowly varies, $\eta_{\varepsilon_1}^0 \approx \eta_{\varepsilon_1}$ and $\eta_\sigma^0 \approx \eta_\sigma$. Therefore, if to take into account that η_ε is proportional to $\varepsilon\,\mathrm{tg}\,\delta$ and the power are divided in two equal parts between the internal and external domains for $\theta \ll 1$, one can find the maximum value of these additional losses

$$\frac{\Delta\eta}{\eta_0} \simeq \frac{\varepsilon_2\,\mathrm{tg}\,\delta_2}{\varepsilon_1\,\mathrm{tg}\,\delta_1}\frac{1}{1 + \frac{\eta_\sigma}{\eta_\varepsilon}}, \tag{2.59}$$

where $\eta_\sigma/\eta_{\varepsilon_1}$ is calculated by the formulas obtained above.

We have experimentally analyzed the influence of the cover thickness (made of an amorphous carbon film) on the deceleration and the specific attenuation in the CSW at the frequencies $25 \div 30$ GHz. The CSW samples are taken in the form of cylindrical pivots with the length of about 100 mm and the diameter 3 mm, made of the melted quartz ($\varepsilon = 3.77$, $\mathrm{tg}\,\delta = 7.7 \cdot 10^{-4}$). Their surfaces are covered with the burnt silver paste and the slot with the width 200 μm is cut by the focused laser beam. Then the protecting cover in the form of an amorphous diamond-like carbon film is drifted on the surface.

The specific attenuation and deceleration are measured in the unprotected and protected CSW. The SWR distribution along the line is registered by the waveguide sensor moving in the surface wave region. The phase constant and the specific attenuation are determined by measuring the period of spatial beatings and the SWR variation. The analysis of the interference of the incident and reflected fields enables one to show that under the assumption of small losses the following relation holds for the SWR and the specific attenuation at the wavelength distance from the line:

$$\eta\left(\frac{dB}{cm}\right) = \frac{4.343}{z_n - z_0}\ln\frac{1 - k_n}{1 + k_n}\cdot\frac{1 + k_0}{1 - k_0},$$

where z_0 and z_n are the longitudinal coordinates of the initial and the nth maximuma and k_0 and k_n are the values of the initial and the nth coefficients. Averaging of the series of measurements gives a sufficiently good accuracy of the determination of propagation characteristics against the cover thickness. Note that when the averaging is applied, it is not necessary to fulfil the strict demands concerning the accuracy of the sensor position providing its permanent communication with the line.

In our experiments the CSW is excited by the waveguide-slot junction and the specific attenuation is measured for 10 samples. The results show that the use of the protecting cover yields the increase of the specific attenuation. Hence, the deceleration grows proportionally the thickness at least for thin films. The nature of these losses mạy be explained by the influence of the film dissipation. This conclusion is confirmed by the behaviour of the regression of the deceleration increments in the protected CSW: the regression curve is close to a linear function and the calculated value of the correlation coefficient is $R_\eta = 0.9$.

Such a low magnitude of the relative additional contribution $\Delta u/u \lesssim 2 \div 3\%$ is evidence of the weaker influence of the protecting cover on the phase constant (due to a small difference between the film and the CSW permittivities).

The measurements show that the deceleration increases together with the cover thickness. The measurement data exceed the theoretical values because the following inequalities hold

$$ka\left[\varepsilon_1 - \left(\frac{v_\Phi}{c}\right)^2\right]^{1/2} > 1, \quad ka\left[\varepsilon_2 - \left(\frac{v_\Phi}{c}\right)^2\right]^{1/2} > 1,$$

for the CSW considered in our experiments. Therefore, in order to calculate the deceleration factor with the given accuracy, one has to use the exact dispersion relations obtained within the framework of the rigorous spectral theory.

We conclude that the attenuation factor in the protected CSW varies in the interval $0.25 \div 0.3$ dB/cm, which is acceptable for the majority of practical applications in the 25–30-Ghz frequency range. The protecting cover causes the increase of the specific attenuation, while the deceleration coefficient is not affected. The increase of the dissipation losses in the millimeter wavelength range may be substantial for certain types of the dielectric cover. For example, the cover in the form of an amorphous carbon film provides a considerable increase of the total attenuation by a factor of $1.5 \div 3$ if to compare with the CSW without a dielectric protection.

Chapter 3

Theoretical and experimental studies of diffractional gratings

Foundations of theory of the wave diffraction on gratings were developed in the beginning of the present century. This theory, together with the modern simulation technique, have been applied in the last two decades for the studies of the resonance wave scattering. Limiting ourselves to the countries of the former USSR, we indicate the results obtained in Refs. [5-12, 156-158] and, in particular, in Ref. [24].

Every new problem solved within the framework of the theory of gratings produces new questions that cannot be answered if to use only classical notions of the theory. In fact, it is not always possible to construct a sufficiently complete mathematical model, which adequately describes physical processes. The approach developed in this book is based on the analysis of the singularities of analytical continuations of the traditional diffractional problems to the complex (non-physical) domain. This chapter is devoted to elaborating such methods and, in particular, to the development of rigorous analytical and numerical-analytical methods for solving boundary value problems of the electrodynamical theory of gratings and for practical applications in the resonant wave scattering by periodical one-dimensional structures. The main attention is paid to the study of gratings considered as open periodical resonators (waveguides) and to the problems of excitation by mono-chromatic, periodical and compact sources.

We also consider several important problems of the mathematical diffraction theory connected with separating the physical solutions of the stationary boundary value problems in the "non-canonical domains" and working out algorithms for the solution to the problems of the plane wave diffraction. Intertype oscillations are studied using the theory of the Morse critical points.

The transformation of surface waves to volume waves are of a considerable interest in the millimeter and submillimeter technology. In this chapter the developed mathematical approach will be applied to the investigation of the volume-surface and surface-surface wave transformation. Similar situation takes place in the experimental studies: the data describing these transformations are applied in the analysis of the volume wave diffraction. The mathematical model of the wave scattering by idealized gratings in supplemented by experimental results taking into account the real parameters of scatterers (the finite size, conductivity, manufacturing features, etc.) and incident fields (for example, deviation of the wave front from that of the plane wave and the influence of repeating reflections). The choice of the millimeter and submillimeter waves for the experimental studies has of principle significance: only in these wavelength ranges one can take into account all the necessary conditions for organizing the laboratory experiment with real gratings and electromagnetic fields directly on the table.

Theoretical and experimental studies presented in this chapter are carried out by the author and his co-workers Yu.K. Sirenko, A.E. Poyedinchuk, Yu.A. Tuchkin, V.V. Yatsik, A.A. Kirilenko, S.A. Masalov, C.D. Andrenko, A.P. Evdokimov, V.V. Kryzhanovskij, A.S. Provalov, and Yu.B. Sidorenko.

§ 10. Some results of the classical theory of gratings and the research perspectives

All the processes in this chapter are considered in the coordinate system $\{x, y, z, t\}$, where the length of the grating period is equal to 2π. The time dependence is $e^{-i\omega t}$, where $\omega = \kappa c$ is the circular frequency, κ is the frequency parameter characterizing the ratio of the period l to the excitation wavelength λ and $c = 1/\sqrt{\varepsilon_0 \mu_0}$ is the absolute value of the speed of light in the medium containing the grating ($\operatorname{Im} c = 0$). The magnitudes of x', y', z', t' and ω' are given by the formulas $x' = xl/(2\pi)$, $t' = tl/(2\pi)$ and $\omega' = \omega 2\pi/l$. The periodicity direction coincides with the axis Oy (Fig.51) and the structures and sources are homogeneous and infinite along the direction Ox. A point on the plane $\{x, y\}$ is denoted by g. The expression $\operatorname{int} S = \emptyset$ means that the domain bounded by the contour S is an empty set. The function $\varepsilon(y, z)$ such that $\varepsilon(y + 2\pi, z) = \varepsilon(y, z)$ and $\varepsilon(y, |z| > 2\pi\delta) \equiv 1$ describes the relative permittivity of the medium. The condition $\operatorname{Re} \varepsilon > 0$, $\operatorname{Im} \varepsilon \operatorname{Re} \omega \geq 0$ enables us to consider the absorbing structures. In this chapter only the planar problems are considered. The sources that are assumed to be infinite along the axis Ox are called the point (compact) sources if they intersect the plane $x = \text{const}$ at one point (or if the points of intersection form a compact set).

Two direct methods are used in the theory of gratings. The first is based on the potential theory and on the reduction of the boundary value problems to the Fredholm integral equations [159-161, 24] and to infinite systems of ordinary differential equations [160]. Both these methods are universal and rigorous approaches (one can prove the existence and the uniqueness theorems and estimate the accuracy of the calculation results), but they are not sufficiently effective for practical computations. In fact, computer simulations require big amounts of time, and already for $\kappa > 1$ it is difficult to obtain the numerical data describing diffractional phenomena even with the help of modern powerful computers.

The method of partial domains is also one of such universal approaches. The problem is reduced to the operator equations of the first or of the second kind with respect to the infinite vectors of the unknown amplitudes of the scattered field. This method is used when one can separate the "regular" partial domains, where the field representations are known. It is not possible to prove the solvability of the corresponding operator equations and to justify the numerical procedures obtained within the framework of this approach. Correct methods of the approximate solution (that is, the rules for the truncation of infinite systems) are described with the help of numerical experiments and by means of the semi-inversion technique [162, 163, 24].

The ill-posed functional and matrix equations of the method of partial domains

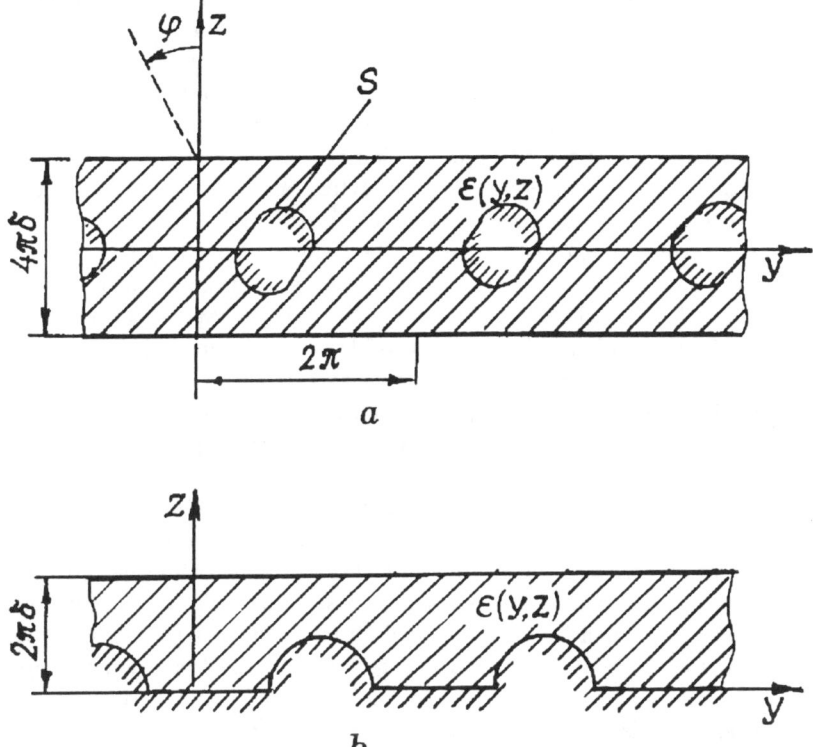

Figure 51.

(or the method of "sewing") are applied in such semi-analytical approaches as the semi-inversion method [5, 156, 162, 164, 165] and the modified residue method [166, 167]. The creation of these semi-analytical techniques for the effective numerical treatment of the resonant wave scattering by periodical gratings is initiated in Ref. [168], where the theory is developed on the basis of the classical functional analysis. The diffraction problems are reduced to the canonical Fredholm equations by means of the inversion (or the left regularization) of the statical operator. This procedure formally completes the solution because for the Fredholm equations the uniqueness yields the existence and the operators compactness provides the convergence of numerical algorithms based on the truncation of infinite systems of linear algebraic equations. The idea of semi-inversion turned out to be fruitful for the regularization of the convolution-type matrix equations in the method of partial domains. The rigorous solutions to the problems of the plane wave diffraction by the following (basic) types of gratings have been obtained in Refs. [156, 162, 169, 24]: perfectly conducting and absorbing echelettes, knife, shutter and plane strip gratings, gratings formed by unclosed thin screens, by circular, rectangular, dielectric and metallic bars, etc. The semi-inversion methods have been constructed that enable us to take into account finite dimensions of gratings and the spatial structure of exciting waves [170, 171]. We note that the semi-inversion method, which is based on the solution to the conjugation problems of the theory of analytical functions, is completely justified. This method guarantees the efficiency of numerical algorithms and does not require big computational resources. Other semi-inversion methods (like the semi-inversion of convolution-type matrix operators, the modified residue method and the modified Wiener-Hopf-Fock method [166, 172]) are less powerful. All these approaches are comparable in efficiency (and similar operator equations are considered) but the first method is more flexible and allows one to take into account the structure parameters. Semi-analytical methods are always oriented to the particular features of scatterers. Their efficiency is mainly caused by the "completeness" of the (explicitly extracted) part of the solution defined by the free terms in the final equation of the second kind. In other words, it is important to estimate the possibility to substitute the structure under study (with respect to a reasonable physical metric) by the one, for which it is possible to obtain an explicit solution of the corresponding diffraction problem (sometimes, this structure is purely hypothetical). For example, the semi-inversion of the convolution-type matrix operators in Refs. [156, 162] applied for the electrodynamical analysis of knife-type gratings leads to the very effective algorithms, when the regular waveguide channels are sufficiently large. But when the height of the channels decreases, the numerical error for the equal-order systems increases, and the numerical procedures do not converge when the regular part is removed (in the shutter-type gratings). The situation becomes clear if to recall that in the analytical solution to the problem one separates the component, which describes the wave diffraction by the periodical system of planes [173, 174, 24] and does not contain the information about the interaction of the lower and the upper apertures. This interaction grows with decreasing height and finally becomes prevailing. As has been established, this interaction causes the major contribution to the diffraction field. Hence, the "physical distance" between the gratings situated on the finite and on the infinite planes increases. The same situation occurs in the modified residue method, where the solution is separated, which has an asymptotically correct behaviour of the complex amplitudes of the higher order attenuating harmonics. This procedure provides the definite singularity of the field intensity close to the edge points. If the prevailing contribution to the scattered field is caused by the interaction of the separate elements at the lower order harmonics (all propagating waves and a part of attenuating waves), then the efficiency of numerical algorithms substantially falls. Hence, in this case the necessity of the correct substitution also must be taken into account. All these factors limit possible applications of the semi-analytical methods in such a way that only a finite number of gratings with simple shapes may be effectively treated. The results obtained with the help of preliminary analytical processing of initial boundary value problems provide the reliable basis for testing the algorithms of direct methods. Mathematical models for the semi-analytical methods have a number of advantages when applied in the global study of the physics of various phenomena and the qualitative picture of the wave diffraction by gratings, as well as for solving inverse problems depending on a small number of parameters. A powerful tool to raise the universal character of these methods is the application of generalized scattering matrices [5, 156, 162, 175, 24]. By means of these matrices one can analyze inhomogeneities in the regular interaction domains. The interaction is taken into account by solving the additional problem described by the canonical Fredholm system of linear algebraic equations with exponentially decreasing elements. In the semi-inversion methods the boundary value problems are explicitly solved for some specific values of the parameters, and the corresponding results (obtained, for example, in Refs. [176, 177]) form the basis for the creation of new semi-analytical methods.

The properties of gratings depend on the ratio between the period and the excitation wavelength. We consider three domains of the variation of the frequency parameter κ: the long-wave ($\kappa \ll 1$), the short-wave ($\kappa \gg 1$) and the resonant ($\kappa \sim 1$). The value of κ determines the choice of the most effective methods for analyzing the diffractional phenomena. The approaches described above are mainly applied in the resonant frequency range. In addition to this, they enable us to obtain exact numerical results for small values of κ. In one of these approaches the approximated boundary conditions are used [157] (and also in the case of anisotropic media).

In the theory of frequent-periodical gratings the shape and dimensions of conductors and the presence of dielectric inclusions are taken into account. This theory allows to carry out a correct consideration of the limiting case when the conductors are infinitely close. The reflection and transmission coefficients are determined for $\kappa \ll 1$ by means of the statical problem method. This approach has been used in the problems of the long-wave diffraction on periodical gratings [5, 156, 162, 158, 178].

The analysis of the acoustic and electromagnetic wave diffraction on periodical obstacles for $\kappa \gg 1$ is one of the most complicated problems. The main question here is to justify and estimate the lower boundary for possible applications of the physical optics (when the diffractional phenomena on local parts of a scatterer surface are not taken into account) and find the solutions, which may be easily constructed analytically and numerically [179]. Two methods are applied for the solution to this problem. The first is connected with the parallel computer simulations based on the rigorous semi-inversion methods and heuristic models [162]. In the semi-inversion method [5, 162] the limitations are essentially caused by the type of a computer, because the amount of computations providing the required accuracy rapidly increases together with κ. The second is in creating reliable mathematical methods for the analysis of the short wave diffraction. Such methods have been applied for the first time in Ref. [180] for the plane strip grating. The Kirchhoff approximation was mathematically justified and the estimates for the integrals of the scattered field amplitude expansion coefficients were obtained. The short wave case is analyzed in Ref. [156] for the plane strip grating and in Ref. [181] for the knife-type grating by the separation of the single grating element. In these works the problem is reduced to the integral equation of the second kind with the operator having the small norm (for sufficiently large values of κ), the solution is represented in the form of the converging Neumann series and the asymptotical behaviour of the scattered field for $\kappa \to \infty$ is determined. The analysis of the short-wave asymptotic is closely connected with the study of the stationary Green function of the grating in the complex domain κ and the behaviour of the solutions to the non-stationary problems for large values of the time t [182]. Such an approach requires the construction of a special theoretical background outlined in the next section.

The physical results of the theory of the stationary wave diffraction on gratings are presented in Refs. [158, 24].

In the diffractional electronics [6] several new problems are set connected with the radiation properties of reflecting structures excited by the inhomogeneous plane wave. The non-resonant case (when the phase velocity of the incident wave is less than one half of the speed of light) is considered in Refs. [6, 183], where the comparative analysis of the diffractional radiation efficiency is carried out for different types of gratings (the cylinder grating, the echelette and the comb). In the resonant case (for the model problems of the relativistic diffractional electronics, where the phase velocity of the inhomogeneous plane wave is close to the speed of light) and in the studies of the transformation of surface waves in planar dielectric waveguides to volume waves the results of the same quality are not available.

Monograph [158] contains the analytical, graphical and numerical data about the frequency, amplitude, polarization and other properties of the wave scattering by diffractional gratings of various profiles and types. It completes together with Refs. [5] and [156] the definite stage of the electrodynamical theory of gratings. Let us note, however, that in Refs. [5, 156, 158] the questions "what happens" and "what may happen" are answered, and there is only a slight touch to the mechanism of the resonance scattering In fact, the data obtained for real values of κ (the "diffraction" domain) cannot give the complete information about the properties of the solutions to boundary value problems. These fundamental properties are obtained when analytical continuations of the resolvents of these problems to the complex domain κ are considered.

The construction of such analytical continuations is the principal problem of the spectral theory of gratings. Non-trivial solutions of the homogeneous (spectral) problems have a distinct physical sense. The homogeneous problems with respect to the spectral parameter κ for the fixed parameter Φ describe the free field oscillations in open periodical resonators (formed by an inhomogeneity placed in the Floquet channel). All the results obtained within the frames of the L_2-theory allow us to extract a sufficiently complete information about the surface (slow) eigenwaves [103, 184, 185]. This classical approach based on the solution to self-adjoint problems gives us the possibility to analyze other types of waves, namely, the leaky and piston waves. We construct the continuation of solutions "through" the continuous spectrum (in the form of integration with respect to the spectral parameter).

We create the theory of open periodical guiding structures applying the solutions to non-self-adjoint spectral problems. Their complete statements (on the Riemann surface) with the appropriate radiation conditions eliminate subjective elements that inevitably appear when these problems are considered in the physical domain of the spectral parameter Φ.

The study of the free field oscillations in gratings was not as evident as the necessity to consider eigenwaves. There are very few works, where these spectral problems are investigated, and the majority of the studies are devoted to physical interpretation of the results [156, 160, 158, 186, 187, 189]. The development in this direction was limited both by serious mathematical difficulties and the absence of the justified physical approach. Spectral problems of the theory of gratings have been considered for the first time in Ref. [190]. A number of qualitative results characterizing the point spectrum of the periodical (in one dimension) metal-dielectric gratings is obtained with the help of the generalized radiation conditions after the continuation to the Riemann surface of the complex frequency spectral parameter κ. Further development of this approach is connected with the study of the two-parameter spectral problems with respect to the frequency (κ) and the quasi-periodicity parameter (Φ, $\Phi = \kappa \sin \phi$. For example, if $\phi = \pi/2$, the function $e^{i\kappa y}$ is the solution to the simplest spectral problem (the case of the H-polarization for the grating coinciding with the perfectly conducting plane $z = 0$) for all values of κ. This result contradicts the theorem from Ref. [190], where it is stated that the fundamental frequencies of metallic gratings form a countable set of points.

The recent theoretical studies of the spectral problems arising in electrodynamics [15, 20] considerably intensifies a full-scale attraction of the modern achievements of the theory of non-self-adjoint operators and operator-valued functions [28, 47, 191, 192]. The recent results in this field form the basis of the spectral theory of gratings obtained on a qualitatively new level.

Gratings are mainly studied as periodical structures excited by quasi-periodical monochromatic sources. The local and non-stationary excitation problems were practically not considered.

The theory of the plane wave diffraction left unsolvable a number of basic problems concerning the nature of the resonance and anomalous phenomena, the short-wave asymptotic, etc. The mathematically correct statements of the stationary problems in unbounded domains filled with lossless media are still not given, and, in particular, because they require the proof of the limiting absorption and the limiting amplitude principles. The lack of the corresponding theoretical results influenced the efficiency of the practical use of gratings as microwave devices in the antenna technology, electronics, acoustics, optics, etc. The treatment of these problems is based on the analysis of the singularities of the analytical continuation of the solutions to traditional diffraction problems.

In this chapter the rigorous analytical and numerical-analytical methods are developed for the analysis of the boundary value problems arising in the dynamical theory of gratings, as well as for the study of the resonant wave scattering by periodical structures. The singularities of analytical continuations (from the real axis of the frequency parameter κ to the complex domain) of the solutions to the problems of the quasi-periodical monochromatic wave diffraction are studied. The main attention is paid not to the regular points, where the corresponding operators are bounded and invertible, but to the complementary set Ω, which is defined as the spectrum (spectral set), in particular, to the spectrum distribution on the complex plane. The natural boundaries for the domain $\mathcal{H}$ of the variation of the complex frequency spectral parameter κ are determined. It turns out that the complete mathematical statement of the problems must include the consideration of the infinite-sheet Riemann surface $\mathcal{H}$ with the real algebraic second-order branch points. A one-to-one correspondence is established between the set Ω and the point spectra of the corresponding eigenvalue (homogeneous) problems. The local uniqueness theorems are proved by means of the known energy relations. It opens the way to describe the domains of the spectrum location for different types of gratings (planar and volume, ideal and non-ideal) and to apply the spectral theory of finite-meromorphic operator-valued functions for investigating the spectrum qualitative characteristics. It is shown that Ω consists of not more than a countable set of points (without finite accumulation points). It means that the Green function $G(g, g_0, \kappa)$ of a grating placed in the field of the point quasi-periodical source (and all its diffractional characteristics that linearly depend on $G(g, g_0, \kappa)$) are meromorphic functions with respect to κ in every finite part of the surface $\mathcal{H}$. The residues $\operatorname{Res} G(g, g_0, \kappa)$ in the poles $\bar{\kappa} \in \Omega$ determine the non-trivial solutions of the eigenvalue (spectral) problems corresponding to eigenvalues $\bar{\kappa}$. Although in the general case $\operatorname{Im} \bar{\kappa} \neq 0$, one can consider physical processes only for real values of the frequency parameter that define free oscillations (or quasi-stationary states) of the electromagnetic fields in periodical gratings. The analysis of the operator singularities enables us to prove that the spectrum Ω of periodical structures with sufficiently smooth boundaries S of the cross-sections placed in a dielectric layer is not empty. All these results are obtained without substantial restrictions on the geometrical and material parameters. A resonant nature of various phenomena is shown. The local expansion theorem is proved for the functions characterizing the diffractional properties of gratings. The essence of these results is in separating the "main" (resonant) and the "background" parts of diffractional characteristics. In many situations one can explicitly determine the former with the help of the methods of spectral theory and technique of scattering matrices. These approaches give a simple analytical description of the diffractional properties of gratings and are applied, in particular, for the computer-aided design of microwave devices containing periodical structures. We develop the new spectral method for the analysis of the complete transformation of the plane wave packages by open periodical resonators, which is based on investigating real fundamental frequencies on non-physical sheets of the complex domain. At these frequencies the grating "pump" the energy from one package to another, when both packages contain different spatial harmonics of the diffractional spectrum.

Determination of the complex spectra of non-self-adjoint boundary value problems forms the basis of the spectral theory of gratings. Two classical ways of separating physical solutions to the problems of the quasi-periodical wave diffraction are analyzed by the methods of spectral theory: the radiation principle and the limiting absorption principle.

The interaction of free oscillations in gratings are also studied. The oscillations with super-high q-factors are discovered in the gratings with open radiation channels. This phenomenon is caused by the interaction of free oscillations that belong to one class but different families. The intertype interaction has been observed in closed resonators with non-ideal boundaries and in open resonators with compact boundaries. Here we analyze the intertype oscillations in open resonators with non-compact boundaries, i.e., in periodical gratings.

We also consider several model problems of the mathematical theory of the plane wave diffraction in the resonant frequency range. Special attention is paid to the anomalous resonant scattering: the Wood anomalies, the total reflection and transmission in semi-transparent gratings, the non-mirror wave reflection and the anomalous increase of the radiation response for reflecting gratings. All these effects are of great interest for theoreticians, experimentalists and engineers. In this chapter we analyze all the anomalous phenomena from the single viewpoint of the spectral theory of gratings.

§ 11. Spectral theory of gratings

This section is devoted to the construction of the non-self-adjoint spectral theory of periodical gratings in terms of the spectral theory of operator-valued functions developed in Refs. [28, 47, 191, 192]. This approach is based on the reduction of initial problems to the Fredholm operator equations and allows us to obtain the spectrum qualitative characteristics, to analyze numerically the variation of complex fundamental frequencies and to study and classify free oscillations in open periodical resonators. The analysis of singularities of the resolvent analytical continuation to the complex domain forms the basis for the study of the resonant grating excitation by non-stationary and compact sources.

Let ω be the real value. We will give the complete formulation of the problem of excitation of an arbitrary grating (periodical in one dimension) by a monochromatic and quasi-periodical (with respect to y) source. We introduce the communication domain $Q = \{g \in R : |z| \le 2\pi\delta\}$ situated in the strip $R = \{g = \{y, z\} : 0 \le y \le 2\pi, |z| \le \infty\} \setminus S$. One has to determine the solution $U(y, z)$ of the two-dimensional equation

$$\left[\frac{\partial^2}{\partial y^2} + \frac{\partial^2}{\partial z^2} + \kappa^2\varepsilon(y,z)\right]U(y,z) = f(y,z), \quad g \in R \tag{3.1}$$

satisfying the generalized boundary conditions

$$E_{\text{tg}} = 0 \quad \text{on} \quad S.$$

E_{tg} and H_{tg} are continuous on the gap lines of $\varepsilon(g)$,

$$U\left\{\frac{\partial U}{\partial y}\right\}(2\pi, z) = e^{i2\pi\Phi}U\left\{\frac{\partial U}{\partial y}\right\}(0, z), \tag{3.2}$$

the solutions satisfy the edge condition

$$\int_{C\setminus B}(|U|^2 + |\operatorname{grad}U|^2)dv < \infty,$$

where $C \in \mathbb{R}^2$ is an arbitrary compactum, and the radiation condition

$$U(y,z) = \sum_{n=-\infty}^{\infty}\begin{Bmatrix}a_n\\b_n\end{Bmatrix}e^{i[\Phi_n y \pm \Gamma_n(z\mp2\pi\delta)]}, \tag{3.3}$$

$$z \gtrless \pm2\pi\delta, \ \operatorname{Im}\Gamma_n \ge 0, \ \operatorname{Re}\kappa\operatorname{Re}\Gamma_n \ge 0, \ n = 0, \pm1, \pm2, \ldots.$$

Here $U(y,z) = E_x$ for the E-polarization and $U(y,z) = H_z$ for the H-polarization, E_x, H_x, E_{tg} and H_{tg} are the electric and magnetic fields components, δ and Φ are the real parameters of the problem, $\Phi_n = n + \Phi$, $\Gamma_n = (\kappa^2 - \Phi_n^2)^{1/2}$ and S are the piece-wise smooth contours of the transversal cross-sections of the perfectly conducting parts. The the piece-wise smooth (piece-wise constant for the H-polarization) complex-valued function $\varepsilon(y, z)$ such that $\operatorname{Re}\varepsilon > 0$, $\operatorname{Re}\kappa\operatorname{Im}\varepsilon \ge 0$, $\varepsilon(y + 2\pi, z) = \varepsilon(y, z)$ and $\varepsilon(y, |z| > 2\pi\delta) \equiv 1$ is the relative permittivity of the grating. The source function $f(y, z)$ satisfying the condition $f(y + 2\pi, z) = e^{i2\pi\Phi}f(y, z)$ has a compact support $B \in Q$, that is, $f(y, |z| > 2\pi\delta) \equiv 0$. Expression (3.3) is a mathematical formulation of the radiation conditions and it guarantees the absence of waves coming from the domains that do not contain sources.

In order to correctly construct the continuation of spectral problem (3.1)-(3.3) to the complex domain κ, it is necessary to determine natural boundaries of the domain $\mathcal{H}$ of the variation of the complex frequency parameter κ and to formulate radiation conditions (3.3) in this domain (we assume that for absorbing gratings $\operatorname{Im}\varepsilon\operatorname{Re}\kappa \ge 0$).

Consider a scattering object with the definite cross-section S and permittivity $\varepsilon(y, z)$ placed in the strip $0 \le y \le 2\pi$, $|z| < \infty$, where the problem is considered. Perturbations caused by the scatterer radically change the electrodynamical characteristics of initial (canonical) problem (3.1)-(3.3). One can follow these changes if the structure of the domain $\mathcal{H}$ does not depend on the geometrical and material parameters of the scatterer. This approach allows us to connect the choice of $\mathcal{H}$ with the determination of the natural boundaries of the analytical continuation of the Green function

$$G(g, g_0, \kappa, \Phi) = \frac{i}{4\pi} \sum_{n=-\infty}^{\infty} e^{i[\Phi_n(y-y_0)+\Gamma_n|z-z_0|]} \Gamma_n^{-1}. \tag{3.4}$$

This Green function corresponds to unperturbed problem (3.1)-(3.3) (we will call it the canonical Green function). Parameter κ enters (3.4) as an argument of an infinite number of the double-valued functions $\Gamma_n(\kappa)$. Hence, in order to reach the natural boundaries of the analytical continuation of G_0, one has to consider an infinite-sheet Riemann surface with the second order branch points κ_n:

$$\Gamma_n(\kappa_n) = 0, \quad n = 0, \pm 1, \pm 2, \ldots.$$

The first sheet of the surface $\mathcal{H}$ (the pair κ, $\{\Gamma_n(\kappa)\}_{n=-\infty}^{\infty}$) is completely determined by the conditions $\operatorname{Im}\Gamma_n \geq 0$, $\operatorname{Re}\Gamma_n \operatorname{Re}\kappa \geq 0$ and by the cuts

$$(\operatorname{Re}\kappa)^2 - (\operatorname{Im}\kappa)^2 - \Phi_n^2 = 0, \quad \operatorname{Im}\kappa \leq 0; \ n = 0, \pm 1, \ldots. \tag{3.5}$$

Other sheets differ from the first by the fact that the signs of $\Gamma_n(\kappa)$ are changed to the opposite ones for a finite number of indices n. It completes the definition of the domain of the variation of κ (the surface $\mathcal{H}$) for the diffraction problem with radiation conditions (3.3) considered at complex frequencies.

Let us define the resolvent set P of (3.1)–(3.3) as the points $\kappa \in \mathcal{H}$, where there exists the unique solution to this problem. Consequently, the complementary set Ω is called the spectral set (the spectrum) of problem (3.1)–(3.3).

The radiation conditions considered on the real axis of the spectral parameter is the only criterion defining the continuation of the diffraction problem to the complex domain κ and they determine the structure of the Riemann surface. Indeed, the choice of cuts is not possible, if to consider only the first sheet and to apply only this criterion. In order to overcome this difficulty, it is necessary to introduce an artificial notion of the continuous spectrum connected with the integration along cuts. Relation (3.5) together with the conditions $\operatorname{Im}\Gamma_n \geq 0$ and $\operatorname{Re}\Gamma_n \operatorname{Re}\kappa \geq 0$ determine only the third and the fourth quarters of the complex planes κ, where there must be the cuts (with arbitrary configuration on the Riemann surface and given by (3.5)).

Consider the first sheet of the surface $\mathcal{H}$ (Fig.52). For $0 < \arg\kappa < \pi$ we have $\operatorname{Im}\Gamma_n > 0$ and $\operatorname{Re}\Gamma_n \geq 0$ in the domain $0 < \arg\kappa \leq \pi/2$ and $\operatorname{Re}\Gamma_n \leq 0$ in the domain $\pi/2 < \arg\kappa < \pi$ for all $n = 0, \pm 1, \ldots$. The values of a finite number of functions $\Gamma_n(\kappa)$ (for which $d_n = (\operatorname{Re}\kappa)^2 - (\operatorname{Im}\kappa)^2 - \Phi_n^2 > 0$) are determined by the conditions $\operatorname{Im}\Gamma_n < 0$, $\operatorname{Re}\Gamma_n > 0$, while other functions $\Gamma_n(\kappa)$ satisfy the condition $\operatorname{Im}\Gamma_n > 0$, $\operatorname{Re}\Gamma_n \leq 0$. In the domain $\pi < \arg\kappa \leq 3\pi/2$ the situation is the same, if to change the signs of $\operatorname{Re}\Gamma_n$. Radiation field (3.3) is formed both on the first and on the other sheets of surface $\mathcal{H}$ by the groups of terms (waves) satisfying the following conditions:

$\operatorname{Re}\Gamma_n \operatorname{Re}\kappa \geq 0$, $\operatorname{Im}\Gamma_n > 0$, $\operatorname{Im}\kappa > 0$, the waves do not come to the grating from infinity, attenuate in space and increase in time;

$\operatorname{Re}\Gamma_n \operatorname{Re}\kappa \leq 0$, $\operatorname{Im}\Gamma_n < 0$, $\operatorname{Im}\kappa > 0$, the waves do not go off the grating and increase in space and time;

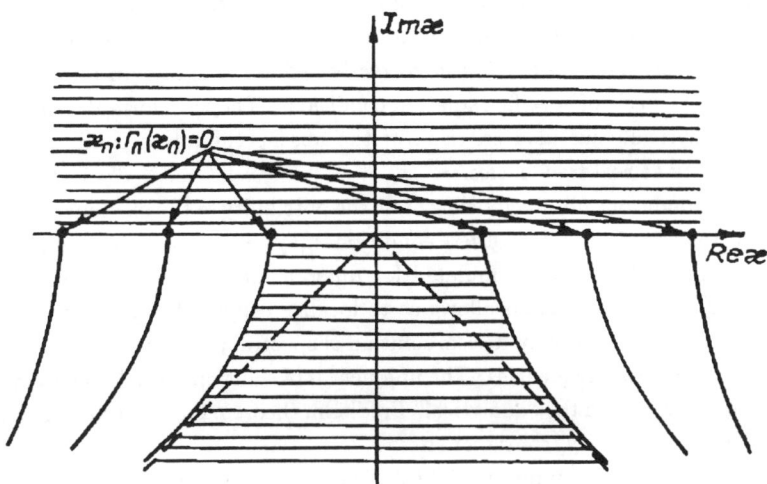

Figure 52.

$\operatorname{Re}\Gamma_n \operatorname{Re}\kappa \geq 0$, $\operatorname{Im}\Gamma_n < 0$, $\operatorname{Im}\kappa < 0$, the waves do not come to the grating, increase in space and attenuate in time;

$\operatorname{Re}\Gamma_n \operatorname{Re}\kappa \leq 0$, $\operatorname{Im}\Gamma_n > 0$, $\operatorname{Im}\kappa < 0$, the waves do not go off the grating and attenuate in space and time.

These definitions of the wave sets enable one to clear up sense of the radiation conditions applied in the formulations of diffraction problems for real and complex frequencies.

The simple poles κ_n: $\Gamma_n(\kappa_n) = 0$ $(n = 0, \pm 1, \dots)$ are the only singularities of $G(g, g_0, \kappa, \Phi)$ (considered as a function of the local coordinates) on the surface $\mathcal{H}$. We prove that $G(g, g_0, \kappa, \Phi)$ satisfies all the properties of the Green function at the points $\kappa \in \mathcal{H}$, where G is holomorphic with respect to the spectral parameter. Therefore, G may be used as a kernel of the generalized (Green) potentials in the complex frequency domain.

Let us define the point spectrum of (3.1)-(3.3) as the points κ of the surface $\mathcal{H}$ where the problem

$$\left[\frac{\partial^2}{\partial y^2} + \frac{\partial^2}{\partial z^2} + \kappa^2 \varepsilon(y, z)\right] v(y, z) = 0, \quad g \in R, \tag{3.6}$$

has a non-trivial solution ($v(y, z)$ satisfies generalized boundary conditions (3.2) and radiation conditions (3.3)). The point spectrum is a subset of the spectrum Ω of (3.1)-(3.3). It will be shown below that, in fact, the point spectrum coincides with the whole spectral set. Therefore, we will denote the point spectrum by Ω and call (3.6) the spectral problem. One can consider every periodical structure as an open resonator with infinite boundaries. Hence, it is natural to call the non-trivial solutions $v(y, z)$ of (3.6) eigenoscillations and the corresponding values of κ fundamental frequencies. Note that the field oscillations are called the free oscillations, because their existence is not caused by any external sources. The term "quasi-stationary" reflects the fact that the amplitude of corresponding oscillations is time-dependent, while and the spatial fields configuration is stationary.

The spectral characteristics of open resonators are governed by the radiation conditions determining a diffractional communication between the resonator's volume and the surrounding space. Therefore, one can correctly define the spectrum and free oscillations only on the basis of specific radiation conditions. They are formulated above and reflect the physically justified criterion of the absence of waves coming to the grating from infinity at real frequencies.

A number of conditions obtained below limit the possibility of the existence of non-trivial solutions to (3.6). In principle, these conditions may be treated as the uniqueness theorems for problem (3.1)-(3.3) formulated for different parts of the surface $\mathcal{H}$. A preliminary study of the uniqueness allows us to localize roughly the point spectrum and to substantially improve the numerical solution of spectral problems by facilitating the search for fundamental frequencies. The uniqueness of solutions enables us to apply the Fredholm alternative and the "analytical" Fredholm theorem for the diffraction and spectral problems under study. characteristics.

Let us point out that every non-trivial solution of (3.6) must satisfy the complex power theorem.

We will divide the solutions of (3.6) into three groups according to the conditions $\operatorname{Re}\Gamma_n \operatorname{Re}\omega > 0$, $\operatorname{Re}\Gamma_n \operatorname{Re}\omega \geq 0$ and $\operatorname{Re}\Gamma_n \operatorname{Re}\omega \leq 0$ ($n = 0, \pm 1, \pm 2, \dots$).

If to consider only the first sheet of the surface $\mathcal{H}$, one can prove that the spectral problem has only trivial solutions in the domains where $0 < \arg\kappa < \pi$ and $5\pi/4 < \arg\kappa < 7\pi/4$. In other words, these domains do not contain the point spectrum. Moreover, in order to guarantee the uniqueness, it is not necessary to introduce the absorption of the grating material or the absorption in the medium surrounding the grating.

The conditions for the existence of non-trivial solutions of the spectral problem are formulated in terms of the signs of $\operatorname{Re}\omega$, $\operatorname{Im}\omega$, $\operatorname{Re}\Gamma_n$, $\operatorname{Im}\Gamma_n$, i.e., in terms of the position of κ on the surface $\mathcal{H}$. It allows us to consider the results obtained as uniqueness theorems for problem (3.1)-(3.3).

The uniqueness is proved for semi-transparent gratings (where the oscillation fields are non-zero both in the reflection, $z > 2\pi\delta$, and in the transmission, $z < -2\pi\delta$, domains) and for reflecting gratings (where the oscillation fields exists only in the reflection domain).

Before to construct the foundations of the spectral theory, we note that analytical properties of the operator-valued functions $A(\kappa)$ are uniquely determined by the analytical properties of the functional $(A(\kappa)f, \psi)$, where f and ψ are arbitrary elements of appropriate functional spaces [47]. This assertion may be easily verified for integral operators. When a complex parameter varies in the Riemann surface, one may introduce the local variable [193], and the properties of $A(\kappa)$ will be formulated with respect to this local variable (when $\kappa \in \mathcal{H}$).

The spectrum qualitative characteristics for periodical structures will be obtained for the H-polarization, because the E-polarization is treated in a similar manner.

Assume that functions $\varepsilon(y, z)$ and $f(y, z)$ are once continuously differentiable in Q and $U(y, z)$ is the solution (for $\kappa \in \mathcal{H}$). to the diffraction problem (3.1)-(3.3) with int $S = \emptyset$, i.e., for a grating without metallic inclusions. Following Ref. [194], we will construct the sequence of functions such that

$$(\Delta + \kappa^2)U_0 = f, \quad (\Delta + \kappa^2)U_1 = (1 - \varepsilon)\kappa^2 U_0 + f, \quad \ldots,$$
$$(\Delta + \kappa^2)U_n = (1 - \varepsilon)\kappa^2 U_{n-1} + f, \quad \ldots,$$
$$\Delta = \frac{\partial^2}{\partial y^2} + \frac{\partial^2}{\partial z^2}. \tag{3.7}$$

Using the properties of the canonical Green function $G = G_0$, we rewrite (3.7) in the form

$$U_0 = -\int_Q G_0 f \, dg_0, \quad U_1 = -\int_Q G_0[(1 - \varepsilon)\kappa^2 U_0 + f] dg_0, \quad \ldots,$$
$$U_n = -\int_Q G_0[(1 - \varepsilon)\kappa^2 U_{n-1} + f] dg_0, \quad \ldots. \tag{3.8}$$

Then one can prove that there exists a domain $\mathcal{K} \subset \mathcal{H}$ where the sequence (3.8) converges to the function $U(g, x)$ uniformly with respect to κ and g in $\mathcal{K}$ and Q, respectively. Function $U(g, \kappa)$ is determined from the equation

$$U(g, \kappa) = -\int_Q G_0(g, g_0, \kappa)[\kappa^2(1 - \varepsilon(g_0))U(g_0, \kappa) + f(g_0)] dg_0, \quad g \in Q, \tag{3.9}$$

$U(g, \kappa)$ may be continued from $\mathcal{K}$ onto the surface $\mathcal{H}$ as a meromorphic function of κ, and this "continued" function will keep all the properties of a solution to (3.1)-(3.3).

The smoothness conditions for $\varepsilon(y, z)$ may be weakened if to assume that they hold on subsets of Q. Then equalities (3.8) and (3.9) remain valid due to the continuity of $U(g, \kappa)$. Finally, one can formulate the following assertion.

The spectrum Ω of problem (3.1)-(3.3) for the dielectric gratings with piece-wise continuous complex permittivity consists of not more than a countable set of isolated points κ in every finite domain of surface $\mathcal{H}$. The resolvent of the problem has the finite order poles at these points.

One can write equality (3.8) in the form $U_n(g, \kappa) = -\int_Q G_n(g, g_0, \kappa) f(g_0) dg_0$, where

$$G_n(g, g_0) = G_0(g, g_0) - \kappa^2 \int_Q G_0(g, g_1) G_{n-1}(g_1, g_0)[1 - \varepsilon(g_1)] dg_1. \tag{3.10}$$

We prove in a similar way that there exists the domain $\mathcal{K} \subset \mathcal{H}$, where sequence (3.10) uniformly converges to the function

$$G(g, g_0) = G_0(g, g_0) - \kappa^2 \int_Q G_0(g, g_1) G(g_1, g_0)[1 - \varepsilon(g_1)] dg_1. \tag{3.11}$$

for all $\kappa \in \mathcal{K}$, $g, g_0 \in Q$ and

$$U(g, \kappa) = -\int_Q G(g, g_0) f(g_0) dg_0, \tag{3.12}$$

The latter relations allow us to define $G_0(g, g_0, \kappa)$ in $\mathcal{H}$ as a meromorphic function of the complex parameter κ for all points $g \in \mathbb{R}^2$ and $g_0 \in Q$.

Let us analyze the spectrum of (3.1)-(3.3) for arbitrary gratings. Application of the potential theory requires certain restrictions concerning the smoothness of contours S. One can weaken these restrictions by means of conditions (3.2). We will assume in what follows that the integral equations of the second kind obtained on the basis of the potential theory are equivalent to the initial problem with the accuracy up to the resonances of the internal domain [41].

The spectrum Ω of the grating made of metallic bars placed inside a dielectric layer with piece-wise continuous $\varepsilon(g)$ forms not more than a countable set of isolated points in every finite domain of the surface $\mathcal{H}$. The resolvent of the problem (the Green function) has the finite order poles at these points.

The Green function residues calculated at the (finite order) poles of the resolvent's analytical continuation coincide with the required non-trivial solutions. On the other side, assuming the existence of non-trivial solutions to (3.6) at the points κ, which are not the poles of the Green functions of (3.1)-(3.3), we come to contradiction (if to use representations similar to (3.12)). Hence, the complex spectrum Ω of the considered diffraction problem may be determined with the help of the solution to (3.6). Here, we have excluded the gratings with edges due to the limitations caused by the requirements of the potential theory [195]. The spectrum qualitative characteristics of such gratings may be studied by means of operator equations obtained, for example, by the semi-inversion method. This approach will be used below in the spectrum computations.

One can prove the existence of singularities of the resolvent's analytical continuation for the simplest electrodynamical objects. However, every such example may give the background for the proof of the "global" existence theorem. Note that there are several types of gratings, where the existence of the free field oscillation is obvious.

Let us consider the diffraction problem (3.1)–(3.3) for the grating formed by metallic bars, $\varepsilon(y, z) \equiv 1$, denoting by θ the characteristic size of metallic inclusions in the transversal cross-section $S(\theta)$. Let κ_n be one of the branch points of $\mathcal{H}$ (that is, the spectrum element of the canonical problem) We call (3.1)–(3.3) with $\theta \neq 0$ the perturbed problem (relative to the canonical problem). One can prove that if γ is a closed curve in $\mathcal{H}$ which surrounds κ_n and does not contain other spectrum elements, then there exists (for sufficiently small θ) at least one fundamental frequency of the perturbed problem lying in the double-sheet circle bounded by γ.

This result yields the possibility to apply explicit solutions of the canonical problem (3.6) for the proof of the existence of free oscillations in "weakly perturbed" periodical structures.

In real situations one has to consider complicated gratings but often they may be described by the continuous "modification" of a simple geometry or by the continuously varying parameters of dielectric inclusions. Let us choose the following scheme of such a continuous modification. We transfer from the canonical problem to a "weakly perturbed" one (assuming that perfectly conducting boundaries have smooth transversal contours $S_1 \neq \emptyset$ and $\varepsilon(y, z) \equiv 1$). Namely, we reach the required value of S_2 by continuously varying S. Then we will transfer from the structure described by the parameters S_2 and $\varepsilon(y, z) \equiv 1$ to the grating with S_2 and $\varepsilon(y, z)$, where $\varepsilon(y, z)$ is the given piece-wise continuous permittivity of the layer $|z| \leq 2\pi\delta$.

One can deduce from the previous results that there exists such a contour $S = S_1$ that the spectrum of (3.1)-(3.3) with S_1 and $\varepsilon \equiv 1$ contains at least one point. The transference from the structure with $S = S_2$ and $\varepsilon(y, z) \equiv 1$ to the grating with the parameters $S = S_2$ and $\varepsilon_2(y, z)$ is mathematically justified by the possibility of a continuous transformation (with respect to the operator norm) of the operator $A(\kappa)[U]$ in (3.9), where the Green function of the problem (3.1)–(3.3) with $S = S_2$ and $\varepsilon(y, z) \equiv 1$ is taken in place of $G_0(g, g_0, \kappa)$. We prove (using the Schwarz inequality) that the difference of two operators $A(\kappa, \varepsilon)[U]$ corresponding to two different Green functions satisfies the continuity requirement, and the operator family $A(\kappa, \varepsilon)$ is continuous with respect to the parameter $\|\varepsilon(y, z)\|$.

Hence, the grating spectrum Ω cannot "disappear" in a finite domain of the surface $\mathcal{H}$ when the parameters vary in a "regular" region. On the other hand, all the spectrum components may tend to infinity for certain values of the parameters. As an example, one can take a perfectly conducting layer $|z| \leq 2\pi\delta$, where there are no free oscillations for any finite value of $\kappa \in \mathcal{H}$ (in the case of the E-polarization).

The Green function G of the diffraction problem (3.1)–(3.3) depends on κ^2 and on the sequence $\{\Gamma_n(\kappa)\}_{n=-\infty}^{\infty}$. We use this property of G to prove the existence of the points $\kappa \in \Omega$ such that $\operatorname{Im}\kappa > 0$ and the following assertion holds: Assume that there exists the point $\kappa \in \Omega$ with the negative imaginary part $\operatorname{Im}\kappa < \theta$ (corresponding to the resonance attenuating in time). Then one can find such a sheet $\mathcal{H}^-$ of the surface $\mathcal{H}$ that the point $(-\kappa) \in \mathcal{H}^-$ belongs to the spectrum (and this point corresponds to the resonance that increases in time). Hence, every free oscillation attenuating in time has a counterpart, that is, the free oscillation increasing in time. We arrive at a contradiction: the energy of the free oscillation (the source function $f(y, z) \equiv 0$) grows in time. But the analysis of the radiation conditions

$$v(y, z) = \sum_{n=-\infty}^{\infty} \begin{Bmatrix} a_n \\ b_n \end{Bmatrix} e^{i[\Phi_n y \pm \Gamma_n(z \mp 2\pi\delta)]}, \quad z \gtrless \pm 2\pi\delta, \ \kappa \in \mathcal{H} \tag{3.13}$$

for (3.6) allows us to resolve this problem. In fact, the real and the imaginary parts of $\Gamma_n(\kappa)$ are determined by the position of κ on the surface $\mathcal{H}$. In turn, the signs of $\operatorname{Re}\Gamma_n$ and $\operatorname{Im}\Gamma_n$ determine the "dynamics" of the corresponding plane wave with the index n (i.e., the nth partial component of free oscillations). For $\operatorname{Re}\Gamma_n \operatorname{Re}\kappa > 0$ ($\operatorname{Re}\Gamma_n \operatorname{Re}\kappa < 0$) the waves propagate off the grating (to the grating) and attenuate (increase) for increasing values of $|z|$ when $\operatorname{Im}\Gamma_n > 0$ ($\operatorname{Im}\Gamma_n < 0$). The free oscillation emerges at the frequency κ with $\operatorname{Im}\kappa \neq 0$, when a part of waves in (3.13) comes to the grating and another part tears off the grating. Hence, the free oscillation has an "internal" source in the form of the plane waves with $n \in N_1$ such that $\operatorname{Re}\Gamma_n \operatorname{Re}\kappa < 0$. The radiation condition (3.13) and the wave separation described above allows us to conclude that a complete transformation of the wave package coming to the grating ($n \in N_1$) to the wave package going off the grating ($n \in N_2 = \{n : \operatorname{Re}\Gamma_n \operatorname{Re}\kappa \geq 0\}$) occurs at the frequency κ.

Three classical principles, the radiation (A. Sommerfeld), the limiting absorption (I.V. Ignatovskij) and the limiting amplitude (A.N. Tikhonov and A.A. Samarskij) are applied to separate a unique "physical" solution of boundary value problems in infinite domains. The justification of these principles and possibilities of their applications are complicated mathematical problems, and the solutions have been obtained only for a limited number of specific domains [59, 60]. Very few results are known for the problems with infinite (non-compact) boundaries. A sufficiently complete analysis has been carried out only when explicit solutions can be obtained [59, 60, 196]. We will

study below two ways of the separation of a unique solution to (3.1)–(3.3) in the case
of the excitation by the E-polarized field of a quasi-periodical monochromatic source
(or by the E-polarized plane wave). For the sake of simplicity, we will assume that
contours S and functions $\varepsilon(g)$ and $f(g)$ are sufficiently smooth.

Note that the radiation conditions (3.3) have been used from the times of Lord
Rayleigh, but the corresponding principle has been rigorously justified only for an
absorbing medium in Refs. [156, 164].

Let us prove the existence and uniqueness of the solution to (3.1)–(3.3). Assume
that $\kappa > 0$ (this condition indicates the domain of the diffraction theory). Denote by
$G(g, g_0, \kappa, \Phi)$ the Green function of the grating in the field of a quasi-periodical point
source (the Green function of the problem (3.1)-(3.3)):

$$U(g, \kappa) = -\int G(g, g_0, \kappa, \Phi) f(g_0) dg_0. \tag{3.14}$$

Considering G as a function of the parameter κ, we construct its analytical continuation
to the surface $\mathcal{H}$ in such a way that relation (3.14) remains valid. As a result, we obtain
a meromorphic function on $\mathcal{H}$ with the poles $\bar{\kappa} \in \Omega$. There exists the unique solution
to (3.1)–(3.3) for $\kappa \in \mathcal{H} \setminus \Omega$, which coincides with $\operatorname{Res} G(g, g_0, \kappa)$ for $\kappa = \bar{\kappa}$ (when
$f(g) \equiv 0$).

Now we assume that the poles $G(\kappa)$ are simple (this condition is verified for real
poles) and formulate the radiation principle. Let $\operatorname{Im} \varepsilon(g) > 0$ on an arbitrary non-
empty subset of Q. Then the radiation condition (3.3) determines the unique solution
to the problem (3.1)–(3.4) for all $\kappa > 0$. If $\operatorname{Im} \varepsilon(g) \equiv 0$, there exists the unique solution
of (3.1)–(3.3) for all values of $\kappa > 0$ except for not more than a countable set $\{\bar{\kappa}_n\}$
without finite accumulation points. The solution of the problem (3.1)–(3.3) at the
point $\kappa \in \{\bar{\kappa}_n\}$ may be obtained only in the case when

$$\int \operatorname{Res} G(g, g_0, \kappa, \Phi) f(g_0) dg_0 \equiv 0. \tag{3.15}$$

Assume that $0 < \arg \kappa < \pi/2$. The solution of (3.1)–(3.3) is defined as the limit

$$U(g, \operatorname{Re} \kappa) = \lim_{\operatorname{Im} \kappa \to 0} U(g, \kappa) \tag{3.16}$$

(with respect to the norm in $L_2(\bar{Q})$, where Q is an arbitrary finite domain of $\mathbb{R}^2$). Here
the function $U(g, \kappa)$ is the solution to the equation

$$P[U] \equiv \left[\frac{\partial^2}{\partial y^2} + \frac{\partial^2}{\partial z^2} + \kappa^2 \right] U(g, \kappa) = f(g) + \kappa^2 [1 - \varepsilon(g)] U(g, \kappa), \quad 0 < \arg \kappa < \frac{\pi}{2}, \tag{3.17}$$

where the operator $P[U]$ is defined on the set of functions from $L_2(\mathbb{R}^2)$ satisfying
conditions (3.2). Then one can assert that there exists a unique solution of (3.17),

$$U(g, \kappa) = -\int_Q G(g, g_0, \kappa, \Phi) f(g_0) dg_0, \quad 0 < \arg \kappa < \frac{\pi}{2}, \tag{3.18}$$

which coincides with the solution of (3.1), (3.3) and satisfies the radiation principle
(the radiation condition (3.3)) for corresponding κ. Now we can formulate the limiting

absorption principle. If $\mathrm{Re}\,\kappa \notin \{\bar{\kappa}_n\} \subset \Omega$, the limit in (3.16) exists and (3.16) defines the unique solution of the problem (3.1), (3.2) satisfying the radiation condition (3.3). The limiting adsorption principle is equivalent to the radiation principle, because they both separate the same solution to the problem (3.1), (3.2) for $\kappa > 0$.

The restrictions concerning possible applications of these principles may be formulated in terms of eigenvalues and eigenfunctions of the problem (3.17), as well as with the help of the classical approach [60] developed within the frames of the L_2-theory (for self-adjoint problems). In fact, according to the uniqueness theorems, non-trivial solutions to the problem (3.1), (3.2) belong to the set $L_2(\mathbb{R}^2)$ for the real values $\{\kappa_n \in \Omega : \Gamma(\kappa_n) \neq 0$.

The results of the present section allow us to consider infinite periodical structures as open resonators, because we have proved that there exists discrete spectrum of free oscillations, which determines the diffractional properties (because the spectral points are the poles of the resolvent operators for stationary boundary value problems). The spectrum on the Riemann surface $\mathcal{H}$ has been analyzed, as well as the eigenvalues situated in the "physical" domains of variation of the complex frequency parameter κ (on the first sheet of $\mathcal{H}$). Now we can study the physical properties of the resonant wave scattering by periodical gratings. In particular, we will investigate the formation of resonant responses, which can be taken into account if to consider all the singularities of the problems.

§ 12. Free oscillations in waveguide-type gratings

If only general qualitative characteristics of spectral sets are known, it is not possible to obtain a concrete information about particular features of complex fundamental frequencies as functions of different parameters and to correctly classify eigenoscillations. A computer simulation (numerical experiment) enables one to solve some of these problems.

The integral equations obtained in the previous section may be used for determination of the spectra of periodical structures. The numerical-analytical methods allow us to collect a sufficiently complete amount of numerical information and qualitatively and quantitatively characterize the phenomena considered. The rigorous algorithms for calculating the spectra and free field oscillations are constructed on the basis of the semi-inversion method for the waveguide-type gratings (Fig.53,a,b), the strip gratings (Fig.53c) and the strip gratings in a dielectric layer (Fig.53,d,e).

In Refs. [5, 156, 162, 164, 165, 169, 170] the semi-inversion method for the plane wave diffraction problems is justified. The results connected with the analysis of operator equations of the second kind may be applied for spectral problems, if to treat corresponding operators as operator-valued functions of the spectral parameter $\kappa \in \mathcal{H}$. The structure of $\mathcal{H}$ is mainly determined by the multi-valued function $\Gamma_n(\kappa)$ entering the expression (3.4) for $G_0(g, g_0, \kappa)$. If we formally apply the results of Refs. [5, 156, 162, 164, 169] for spectral problems, the question arises about the role of other quantities that are analogous to $\Gamma_n(\kappa)$ and coincide with the propagation constants of waves in the regular parts connecting the transmission $(z < -2\pi\delta)$ and the reflection $(z > 2\pi\delta)$ zones.

Let us represent the solution of the problem (3.6) for the waveguide-type grating with a complicated period (Fig.53a) in the form satisfying the homogeneous equation

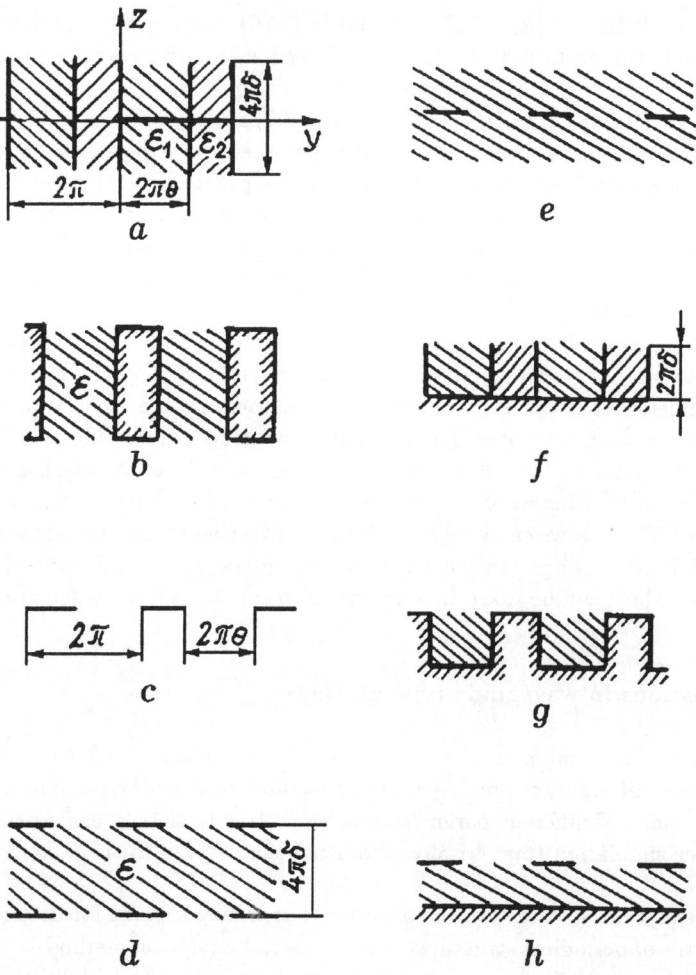

Figure 53.

(3.6) and other conditions. The direct matching of E_{tg}, H_{tg} in the planes $|z| = 2\pi\delta$ leads to an infinite system of linear algebraic equations of the first kind, whose regularization by the semi-inversion method [197, 198] is not possible. Matching by means of the additional systems of functions $\left\{\sin(my/2)\right\}_{m=1}^{\infty}$, which is complete on the period, yield three infinite systems (two of them are coupled) with respect to new unknowns belonging to the same space of sequences. The equivalence of the spectral problems (3.6) is evident. The operator matrix determines the compact finite-meromorphic operator-valued function for $\kappa \in \mathcal{H}$.

The resolvent set of the inhomogeneous problem (that is, of the excitation problem for a waveguide-type grating with a complicated period) is non-empty in $\mathcal{H}$ (due to the uniqueness theorem). Therefore, in accordance with the Fredholm theorem for finite-meromorphic operator-valued functions, one can assert that the set Ω of the fundamental frequencies of waveguide-type grating (i.e., the points κ where the corresponding homogeneous problem has non-trivial solutions) forms not more than a countable subset of $\mathcal{H}$ with the only possible accumulation point at infinity.

The last statement is not a simple particular case of the results from section 11, because the methods of potential theory cannot be applied for the gratings with sharp edges. This remark applies equally to other results obtained below.

Let us describe the algorithm for the calculation of Ω and estimate its efficiency. The corresponding numerical procedures may be constructed on the basis of a reduced ("truncated") problem for the operator equation under study. The results of Ref. [20] yield the existence of the sequence $\{\kappa_N\}$ such that $\kappa_N \in \Omega_N$ and $|\kappa_N - \kappa| \to 0$ for $N \to 0$ if $\kappa \in \Omega$. At the same time, $\kappa \in \Omega$ if $\kappa_N \in \Omega_N$ and $|\kappa_N - \kappa| \to 0$ for $N \to \infty$.

The last statement determines the structure of the algorithm, which is based on the possibility of approximating the true spectrum by the sequences of characteristic numbers of finite-dimensional operator-valued functions. These approximate characteristic numbers (from Ω_N) may be obtained by the numerical solution of the equation $\det\{I - R_N(\kappa_N)\} = 0$.

As has been shown in the previous section, basic properties of operator-valued functions do not depend on geometrical and material parameters. Hence, we can apply all the above results to the case $\mathrm{Im}\,\varepsilon_2 \gg 1$, i.e., for the grating formed by metallic bars with waveguide channels filled with a dielectric (these channels depicted in Fig.53b are considered as communication domains of the transmission and reflection zones) Such gratings are widely used in the millimeter wave technology. Free field oscillations in such gratings may be analyzed within the frame of the algorithm, where the metal losses are taken into account. Corresponding mathematical model is obtained for sufficiently large but finite values of $|\mathrm{Im}\,\varepsilon_2|$, when a non-ideal metal is considered as a non-ideal dielectric.

Let us construct the algorithm for calculating the spectra Ω of planar gratings made of perfectly conducting infinitely thin strips (Fig.53c). Some qualitative characteristics of Ω are applied for the analysis of eigenoscillations in more complicated structures (shown in Fig.53,d,e).

The problem (3.6) of determining the spectrum of the planar grating is reduced (in the case of the H-polarization) by the method of the Riemann-Hilbert problem [5, 168] to the equivalent operator equation

$$\{I - A(\kappa)\}b = 0, \quad b = \{b_n\}_{n=-\infty}^{\infty} \in \tilde{l}_2. \tag{3.19}$$

Here I is an identity operator and $A(\kappa)$ is an infinite matrix with the elements

$$a_{mn}(\kappa) = [|\Phi_n| + i\Gamma_n(\kappa) + 2\delta(n,j)\Phi_n]\Phi_m^{-1}C_{mn}, \quad m,n = 0,\pm 1,\pm 2,\dots,$$

$$C_{mn} = V_m^n - R_m R_\sigma^{-1} V_\sigma^n, \quad \delta(n,j) = \begin{cases} 1, & 0 \le n \le j \\ 0, & n < 0, \ n > j, \end{cases} \tag{3.20}$$

$\Phi > 0$ and $\Phi - j \neq 0$, the values V_m^n, V_σ^n, R_m and R_σ given in Ref. [5] depend on the geometrical parameter θ. Similar to chapter 1, one can show that the matrix $A(\kappa)$ defines a kernel holomorphic operator-valued function $A(\kappa) : l_2 \to l_2$, $\kappa \mathcal{H}$. The uniqueness theorem (proved in section 11) and the Fredholm theorem for analytical operator-valued functions yield the discreteness of the spectral set of $I - A(\kappa)$ (the spectral points are the finite order poles of the operator-valued function $\{I - A(\kappa)\}^{-1}$).

Non-trivial solutions $b = \{b_n\}$ of the spectral problem (3.19) belong to $\tilde{l}_2$ ($|b_n| = O(|n|^{-3/2})$).

The algorithm for calculating the spectrum is constructed on the basis of equation (3.19). Since $A(\kappa)$ is a kernel operator, there exists a function $\Delta(\kappa) = \det\{I - A(\kappa)\}$ [28] holomorphic in every finite part of the surface $\mathcal{H}$. The characteristic numbers $\kappa_N \in \Omega_N$ of the truncated (finite-dimensional) operators tend to the points $\kappa \in \Omega$ as $N \to \infty$. The rough estimate of the rate of convergence is given by the inequality $|\kappa_N - \kappa| \leq cN^{-1/2}$.

In the algorithms considered above we accounted for specific features of the grating under study. The method of generalized scattering matrices [175] is a universal approach, which allows one to study free oscillations in gratings (Fig.53c) and in more complicated structures (Fig.53,d,e). To demonstrate the advantages of this method, let us consider the determination of the fundamental frequencies for the grating formed by two "elementary" (strip) periodical inhomogeneities with a regular communication domain of the height $4\pi\delta$ and the layer permittivity $\varepsilon = 1$ (see Fig.53d). The spectra of elementary structures are discrete and have a finite multiplicity on the Riemann surface $\mathcal{H}$ ($|\kappa| < \infty$). It means, in particular, that the transmission, T, and reflection, R, matrices may be found as a result of the solution to the problem of the excitation of the planar grating by the complete set of spatial harmonics

$$U = \sum_{n=-\infty}^{\infty} U_n e^{i(\Phi_n y - \Gamma_n z)}, \quad z \geq 0.$$

They define the incident ($a = \{a_n\}$) and scattered ($b = \{b_n\}$) fields through the wave amplitudes $U = \{U_n\}$ by the relations

$$a = RU, \quad b = TU \tag{3.21}$$

that are, in fact, finite-meromorphic operator-valued functions of the parameter $\kappa \in \mathcal{H}$ acting on the pair of spaces $\tilde{l}_2 \to \tilde{l}_2$. Taking into account (3.21), one can reduce the spectral problem for the grating by the method of generalized scattering matrices to the homogeneous operator equation

$$d = [R(\kappa)E^\delta(\kappa)]^2 d, \quad d \in \tilde{l}_2. \tag{3.22}$$

The diagonal matrix $E^\delta(\kappa) = \{\delta_m^n e^{i4\pi\delta\Gamma_n(\kappa)}\}_{m,n=-\infty}^{\infty}$ defines a compact holomorphic operator-valued function of the spectral parameter $\kappa \in \mathcal{H}$. Hence, the spectrum forms not more than a countable set $\Omega \subset \mathcal{H}$ of finite-multiplicity spectral points. Similarly to the procedure described above, the convergent numerical algorithm is constructed on the basis of the finite-dimensional analogue of (3.22) The analysis is quite the same for the structures depicted in Fig.53,d,e. The dispersion relations and computational formulas become more complicated, because, in order to calculate matrices R and T, we introduce a virtual layer with $\varepsilon = 1$ separating the grating and the dielectric layer. Analytical properties of the final homogeneous operator equations of the second kind remain the same.

Here we consider semi-transparent structures having the plane of symmetry $z = 0$. Oscillations in such structures may be either symmetrical ($\frac{\partial v}{\partial z}\big|_{z=0} = 0$) or anti-symmetrical ($v|_{z=0} = 0$). Thus, the results obtained may be applied for the reflecting structures, in which the plane of symmetry $z = 0$ is replaced by a perfectly conducting plane (see Fig.53). We note that in the E-polarization case, the free oscillation spectrum in such semi-transparent structures consists of anti-symmetrical oscillations.

§ 13. Formation of fundamental frequencies and free oscillations in waveguide-type gratings

The completeness of the orthogonal system of eigenfunctions allows us to study basic qualitative regularities of the fundamental frequencies of closed structures with respect to varying geometrical and material parameters of the spectral problem [57, 58, 41]. A similar approach may be used in some particular situations for open structures, but very often analytical techniques should be essentially supplemented with the results of computer simulation that provide a qualitatively correct description of the phenomena considered.

Let us analyze the data of numerical experiments performed for model spectral problems of calculating free field oscillations in the classical waveguide-type grating (Fig.53b). The presence of the regular communication domain and the transmission and reflection zones enable us to apply all the basic results obtained for this grating to other types of gratings containing regular communication domains (see Fig.53,a,d,e) and to create a necessary background for the study of arbitrary periodical structures. The main attention is paid to the study of the variation of fundamental frequencies on the first sheet of the Riemann surface. The points on this sheet are the closest (with respect to an appropriate physical "metric") to the real values of the frequency parameter, for which the diffraction of waves excited by quasi-periodical monochromatic sources are considered. Taking into account the symmetry of the structure with respect to the plane $z = 0$, we will analyze symmetrical and anti-symmetrical H-type oscillations (defining two classes of the free field oscillation symmetry in this grating). They are analyzed separately because the corresponding spectral problems are completely independent, as well as the dynamics of the fundamental frequency variations.

Spectral lines 1 and 2 in Fig.54 correspond to the function describing the position of the fundamental frequencies with respect to the relative height δ for the four lowest types of free oscillations on the plane $\mathrm{Re}\,\kappa$, $\mathrm{Im}\,\kappa$ ($\mathrm{Im}\,\varepsilon = 0$). The curve segments lying on the first sheet of the surface $\mathcal{H}$ are depicted by bold lines. The dash-dotted lines with one and two points correspond to the second and the third sheets, respectively. The order of the sheets (except for the first one) is conditional and depends on the number of cuts intersecting the curves $\kappa(\delta)$ that go over from one sheet to another. The cut projections are also shown by the bold lines starting in the branch points κ_0, $\kappa_{\pm 1}$ situated on the real axes of the surface $\mathcal{H}$.

The curves $\varepsilon[(\mathrm{Re}\,\kappa)^2 - (\mathrm{Im}\,\kappa)^2] - m^2/(4\theta^2) = 0$, $m = 1, 2, \ldots$ ($\mathrm{Im}\,\kappa < 0$) are depicted by the dotted lines and may be considered as the continuations to the complex domain of the cut-off points $\mathrm{Re}\,\omega_m = \mathrm{Im}\,\omega_m = 0$ of the waveguide channels connecting the reflection and transmission zones of the grating. Here $\omega_m(\kappa) = \left(\kappa^2\varepsilon - m^2/(4\theta^2)\right)^{1/2}$. The method of continuation of this function to the complex domain κ (note that the function is initially defined on the real axis, where $\mathrm{Re}\,\omega_m \geq 0$ and $\mathrm{Im}\,\omega_m \geq 0$) is considered in the previous section. The parameters of the problem are chosen in the following way: the line that starts from the first cut-off point $\omega_1 = 0$ ($\kappa \approx 0.58$) goes to

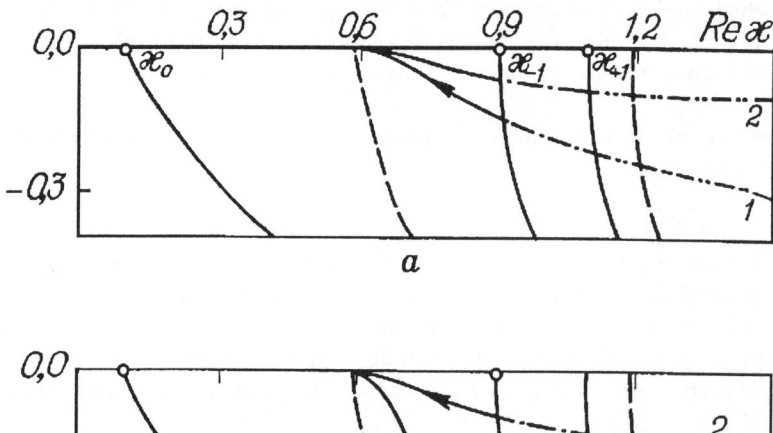

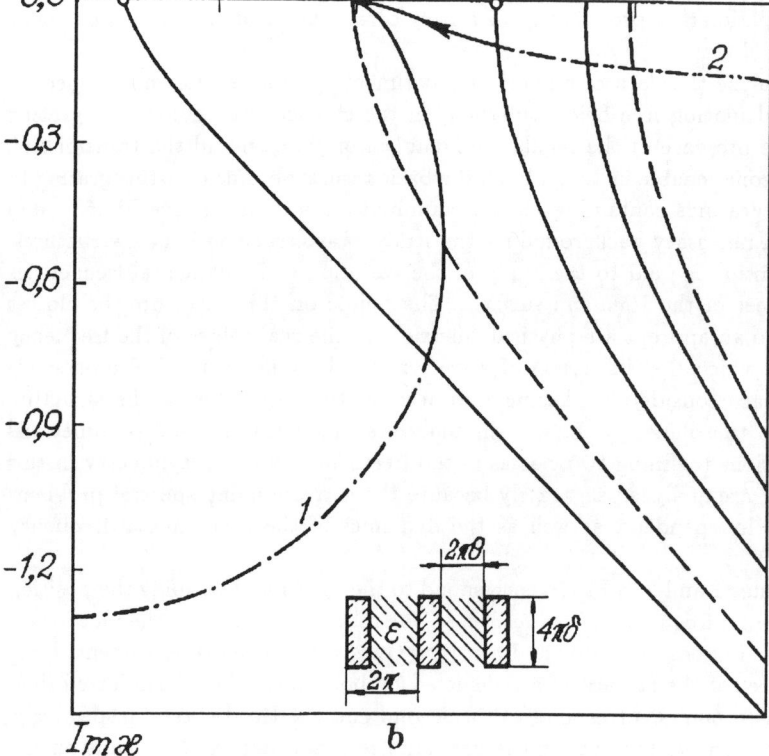

Figure 54.

the domain of the first sheet between the cuts corresponding to the branch points κ_0, $\kappa_{\pm1}$. When δ increases, all the spectral curves in this domain tend to the corresponding cut-off point and form one family marked by the index one (that is, by the number of the threshold point). The propagation of H_{01} waves is possible at the frequencies situated on the real axis to the right of the point $\omega_1 = 0$. These waves run to meet each other in the waveguide channels connecting the radiation zones. Their superposition may form oscillations with symmetrical and anti-symmetrical (with respect to the plane $z = 0$) field distributions. At complex frequencies κ the free field oscillations are also formed due to the field interference. This statement is illustrated by the behaviour of the spectral curves in Fig.54 and the corresponding field distributions in Figs.55 and 56, where the lines $|E_x(y,z)| = $ const are presented. The coordinate y covers one period of the structure and the coordinate z is counted under the assumption that the origin is situated on the plane of symmetry $z = 0$. In all figures the upper part of the grating and the exit to the reflection zone are shown (the characteristic dimensions δ and θ are used as the scale). The complete picture can be obtained if to construct the mirror reflection with respect to the plane $z = 0$. The lines along the curves $|E_x(y,z)| = $ const indicate the direction of the decrease of the field amplitude, and the the maximal values of $|E_x(y,z)|$ are marked by crosses. It is typical for every oscillation that the number of field variation in the regular communication channels is uniquely determined by the structure of the H_{01} field. The eigenfrequency curves presented in Fig.54 start from the points on the real axis that coincide with the H_{01} critical frequency. The spectral curves in Figs.55 and 56 are the same as in Fig.54. The only difference is that the positions of eigenfrequencies are governed by the relative height δ of the grating (the curves of the real and imaginary parts are shown separately). The arrows indicate the fragments corresponding to the calculated data. For large δ, when the diffractional q-factor is sufficiently high ($|\operatorname{Im}\kappa|$ is small), the structure of the free field oscillation is clearly expressed, and the maximal field concentration is observed close to the plane of symmetry of the communication domain. The diffractional q-factor decreases (and $|\operatorname{Im}\kappa|$ increases) together with δ. The points of the maximal field concentration in the communication and radiation domains approach each other, the field structure becomes less distinct and additional local maxima of $|E_x(y,z)|$ appear close to the grating. The latter is caused by the interference of the package of the higher order harmonics with the modes corresponding to one index $n = 0$ and propagating in opposite directions (this process is illustrated by the lower fragments in Fig.55).

In open structures the wave processes are observed in the conditionally bounded domains. Hence, only in specific situations (for example, when δ is sufficiently large and the fundamental frequency is close to the cut-off point) one can say that a resonance is observed in the communication domain of the transmission and reflection zones and that the formation of the free oscillation (that take place mainly in the part of the space directly adjoining the grating) is caused by the H_{01}-wave interference. In the latter case the degree of the H_{01}-wave transformation to spatial harmonics is low [14], and the wave processes in regular waveguide sections substantially influence resonant fields. For sufficiently large $|z|$ the level of $|E_x(y,z)|$ is completely determined by the value of $|\operatorname{Im}\Gamma_0(z)|$, because here only the dominant (zero) harmonic has the propagation constant (along the z axis) with the negative imaginary part.

According to the results of section 11, we have discovered the real fundamental frequencies in the interval $|\kappa| < \kappa_0$ situated on the real axis of the first sheet (see Fig.57). In this frequency range the energy is not radiated to the domain of large $|z|$ ($\operatorname{Re}\Gamma_n = 0$ and $\operatorname{Im}\Gamma_n > 0$ for all $n = 0, \pm 1, \pm 2 \ldots$). Therefore, free oscillations are formed in the communication domain $|z| < 2\pi\delta$. Let us choose the parameters in such a way that the first cut-off point of the waveguide channels (depicted by the straight line $\kappa \approx 0.278$) is lower than the first branch point of the surface $\mathcal{H}$ ($\kappa \approx 0.33$). Then the interference of running to meet each other H_{01} waves becomes possible without a pronounced interaction with the external space, and the superposition of the wave processes in the communication domain yields a formation of free oscillations. The

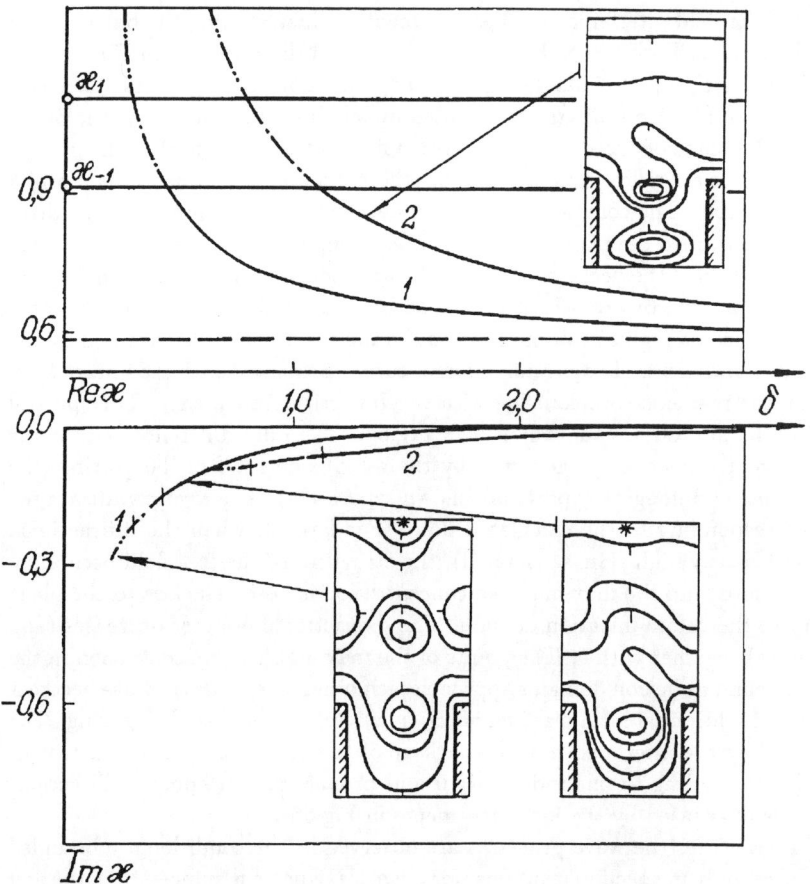

Figure 55.

spectral curves $\kappa(\delta)$ of these oscillations is presented in Fig.57. The field distributions (the lines $|E_x(y,z)| = \mathrm{const}$, $0 \leq z \leq 4\pi\delta$) correspond to the points marked by crosses. These resonances may be conditionally considered as the "pure" resonances of the communication domain. The reasons explaining such a classification are as follows: the field intensity exponentially falls in the direction off the grating, although the intensity is not equal to zero at the grating exit; the field structure is distinct and is uniquely determined by the fields of H_{01} waves; the spectral curves shown in Fig.57 approach the points on the real axis equal to the H_{01} critical frequencies when $\delta \to \infty$.

The oscillations considered above correspond to "pure" resonances of the communication domain, while the field distribution displayed in the left lower fragment of Fig.56 is close to field of the "pure" resonance of the external domain, The decrease of δ and the diffractional q-factor reduces the influence of the wave processes in the communication domain for H_{01} waves. In fact, the energy exchange between the reflection and transmission zones is supported by higher order waveguide waves that weakly attenuate at the distance of $4\pi\delta$ from the grating, and the field structure is determined by the wave interference in the domain $|z| > 2\pi\delta$. One can follow this dependence if to

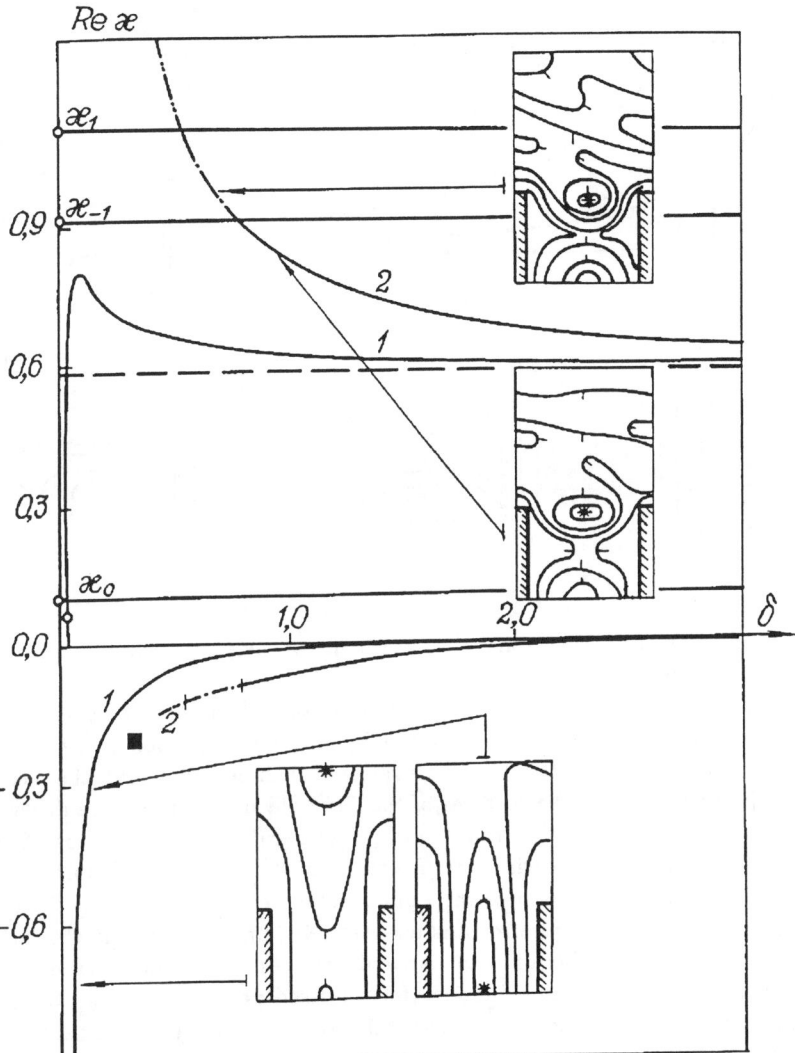

Figure 56.

move along spectral curve 1 (Fig.56) starting from the point $\delta = 0.12$ (that corresponds to the right lower fragment in Fig.56) in the direction of decreasing δ.

We have pointed out above that all the spectral curves from the first family tend to the cut-off point of the H_{01} wave as δ increases. However, the corresponding frequency is not the accumulation point of fundamental frequencies (for a fixed δ). Otherwise, it would contradict the statement concerning the absence of finite accumulation points in Ω. Really, for every finite δ one can always separate such a vicinity of the critical point, where there are no fundamental frequencies (see Figs. 55 and 57).

All the oscillations from the first family are non-degenerated and their spectral curves do not intersect. Variations of the field distributions are distinct and they are caused by the transformation of the resonances in the communication domain to that

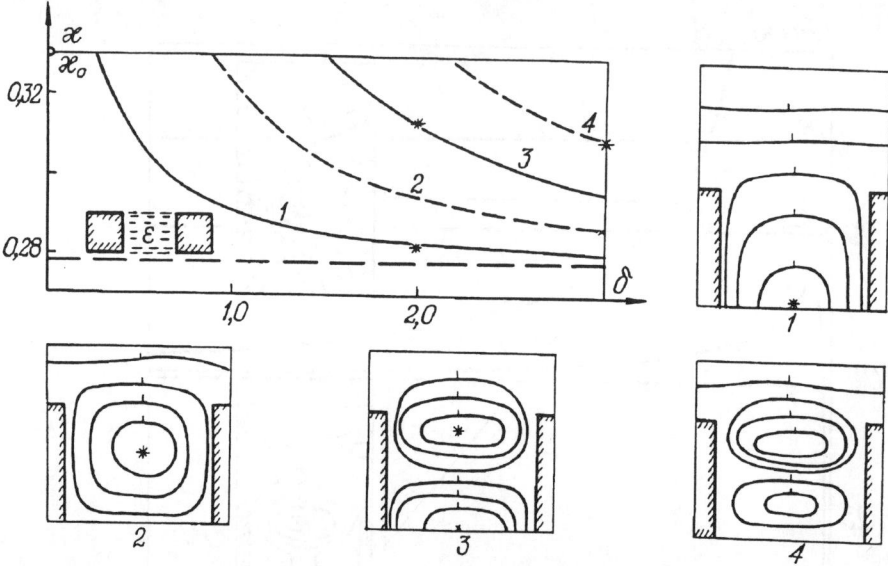

Figure 57.

of the reflection and transmission zones. The spectrum of the fundamental frequencies rarefies with decreasing δ.

One can increase or decrease the electric width of the communication channels in the grating (and to change the conditions of the H_{0m}-wave propagation in the regions $|z| < 2\pi\delta$) by varying the permittivity ε. As a result of such perturbations, the second spectral family may appear. The pictures of the free field distributions are evidence of the fact that the oscillations of this family are generated in the process of the interference of running to meet each other H_{02}-waves in the waveguide channels.

The dynamics of the variation of the fundamental frequencies forming the first family (and the free field configurations) is qualitatively similar to that in the latter case. However, there is a number of quantitative differences: the diffractional q-factor noticeably increases, the spectrum thickens (for the fixed parameter δ) and the real values of fundamental frequencies shift towards the low-frequency range. The diffractional q-factor of the free oscillations corresponding to the spectral points of the second family is two-three orders of magnitude greater than that of the oscillations from the previous family.

The free oscillation q-factors from the first and the second families (we will denote the corresponding fundamental frequencies by κ_1 and κ_2) are determined by the radiation loss. Qualitative properties of the q-factor established theoretically have been confirmed by numerical experiments. We calculate the amplitudes of waves forming the free field oscillations and estimate their contribution to the radiation field of the principal propagating harmonics of the spatial spectrum. The grating fundamental frequencies are also calculated with respect to the relative width θ of the communication channels.

It has been established that when the slots get narrow and the number of variations of the free field oscillations along the grating height decrease, the corresponding spectral curves $\mathrm{Re}\,\kappa$ approach the lines describing the positions of the critical points and the diffractional q-factor increases. Even the spectral curves $\operatorname{Re} \kappa$ of the oscillations with large numbers of the field variations along the grating height "feel" the directing influence of the critical points. Let us indicate the fundamental frequencies from the first family, for which the distributions (pictures) of the lines $|E_x(y,z)| = \text{const}$ are constructed at the critical point. The curves (corresponding to the real values of the fundamental frequencies) that belong to different families but one symmetry class intersect close to this point. It breaks the regular character of the spectral curve of the second family and the diffractional q-factor sharply increases. Note that a monotonic behaviour of the curve $\operatorname{Im} \kappa$ corresponding to the fundamental frequency of the oscillation from the first family is not violated.

We have demonstrated that there can be three lines "continuing" the critical points of H_{0m} waves with the indices $m = 1$ ($\kappa \approx 0.24$), $m = 2$ ($\kappa \approx 0.48$) and $m = 3$ ($\kappa \approx 0.71$). These lines are situated on the first sheet of the surface $\mathcal{H}$ between the first and the second cuts (see Fig.58). In the previous situations every such point was considered as an element of the family of free oscillations, and the field structure of each oscillation from this family was determined in the waveguide regions by the field of the corresponding H_{0m} wave (namely, by the number of the field variations along the waveguide width). This field structure (that defines the type of oscillations) is preserved along the spectral line.

The most important characteristics of the dynamics of fundamental frequencies and free oscillations are the points of a sharp (anomalous) increase of the oscillation q-factors. This effect is observed when the resonance of the communication domain is gradually transformed to the resonance of the external domain and the real parts of the fundamental frequencies belonging to different families and one symmetry class coincide.

Now we will apply the results of the above analysis to the discussion of the calculation data presented in Fig.58. The lines of the real values of the fundamental frequencies for symmetrical free oscillations are depicted in Fig.58 on the first sheet of the surface $\mathcal{H}$ ($\operatorname{Re} \kappa$ varies between the third critical point and the second branch point). The curves from the first family are shown by dotted lines. Note that the spectral lines $\operatorname{Re} \kappa$ of the free oscillations corresponding to one class of symmetry do not intersect. In a small vicinity of the possible intersection point the lines of one family are continuously transformed to the lines of another family and vice versa. Such an anomalous behaviour is caused by the intertype interaction of oscillations.

As has been shown in section 11, the spectral problems for waveguide-type gratings (presented in Fig.53,a,b) are equivalent to the problems on characteristic numbers for Fredholm matrix operator-valued functions. The dispersion relations for fundamental frequencies split in two independent equations corresponding to the symmetrical (antisymmetrical) field oscillations (with respect to the plane $z = 0$). Hence, the spectra of the E-polarized and H-polarized oscillations split in two classes of symmetry. Additional independent classes of symmetry [197] are caused by the symmetry of the grating sections with respect to the planes $y = \text{const}$. If to assume that varying parameters of the spectral problem do not violate the symmetry and the boundary conditions, then one can perform an independent analysis of the free oscillations belonging to different classes.

The results of calculations show that in the long and resonant wavelength ranges the spectrum of free oscillations from one symmetry class consist of the unit-multiplicity

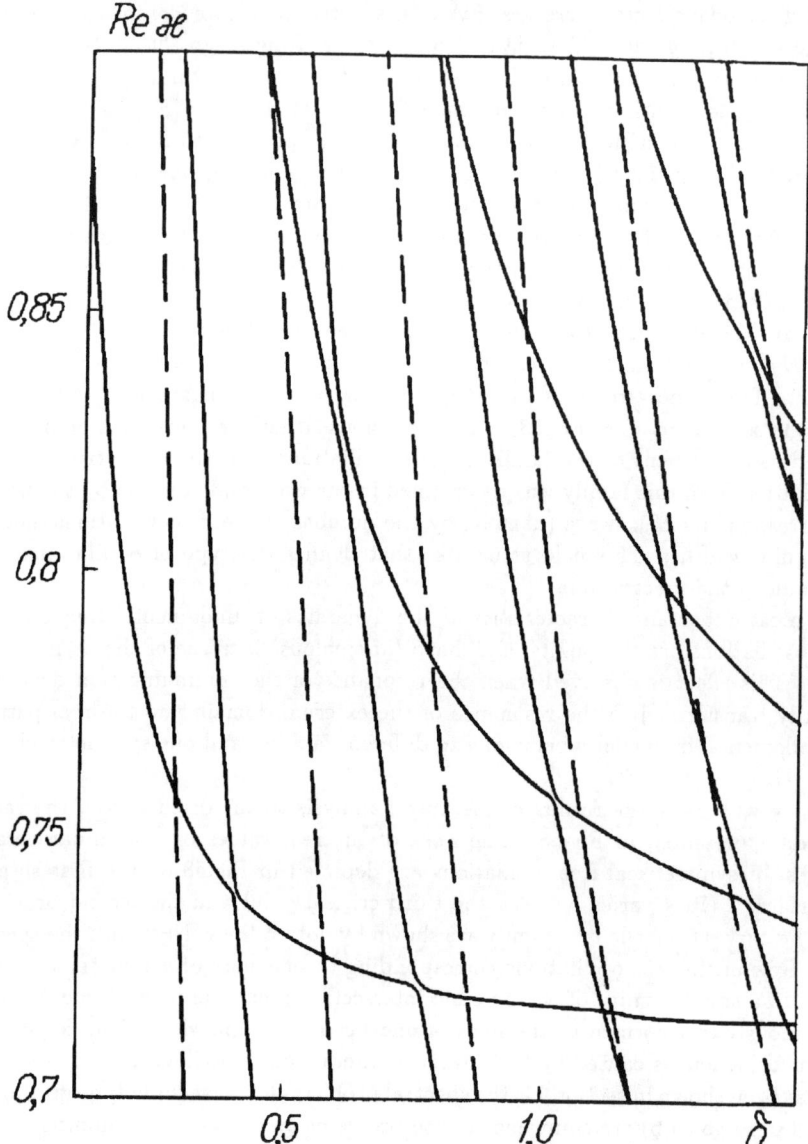

Figure 58.

points forming the spectral set Ω. Hence, we can specify the property of the resolvent $R(\kappa)$ of the corresponding boundary value problems established in sections 11 and 12, that is, to show that in these wavelength ranges the resolvent is a finite-meromorphic operator-valued function of the parameter $\kappa \in \mathcal{H}$ with simple poles. We have also proved the discreteness of the grating spectrum in every finite part of the surface $\mathcal{H}$. Therefore, the possibility appears to number the points from Ω, for example, with respect to the increasing absolute values of fundamental frequencies.

In order to correctly classify free field oscillations in open periodical resonators, it is necessary to know the position of the complex fundamental frequency on the sheet of the surface $\mathcal{H}$ and the number of this sheet. If to following the traditional way of the classification of free oscillations, which is based on the number of the field variations along one of the structure characteristic directions, one has to examine the whole two-dimensional region. Oscillations may be easily identified by the type of the wave (or its partial components) prevailing in the field (by the number of the family) and by the number of the field variations along the grating height (corresponding to the number of the oscillation in the given family). For example, the field configurations presented in Fig.57 correspond to the oscillations that may be classified in the following way: 1, H_{011}, 2, H_{012}, 3, H_{013}, and 4, H_{014}. Here H denotes the type of the waves forming the free oscillation fields (and the field polarization: H in the E-case and E in the H-case). The first, the second and the third indices denote, respectively, the number of the field variations along the axes x, y (along the grating period) and z. It is clear that the second and the third indices define the family corresponding to the given oscillation and the symmetry class (for arbitrary values of Φ). The Other types of the field oscillations characterized by other types of the field distributions may be classified in a similar manner. Variation of the grating parameters leads to a displacement of the spectrum points on the surface $\mathcal{H}$ and, hence, to the change of the conditions that govern the radiation to the free space. The symmetry class is a relatively stable characteristic of free oscillations, because usually it is not influenced by the varying parameters.

Table 6.

$|\kappa|$	$\operatorname{Re}\kappa$	$\operatorname{Im}\kappa$	$-\operatorname{Re}\kappa/(2\operatorname{Im}\kappa)$	H_{0nm}
0.3268	0.3268	$-2.66 \cdot 10^{-3}$	61.3	H_{011}
0.3570	0.3569	$-8.45 \cdot 10^{-3}$	21.0	H_{012}
0.4065	0.4062	$-1.38 \cdot 10^{-2}$	14.7	H_{013}
0.4717	0.4714	$-1.75 \cdot 10^{-2}$	13.4	H_{014}
0.5483	0.5480	$-1.97 \cdot 10^{-2}$	13.8	H_{015}
0.6327	0.6323	$-2.1 \cdot 10^{-2}$	15.0	H_{016}
0.6405	0.6405	$-1.22 \cdot 10^{-5}$	$2.61 \cdot 10^{4}$	H_{021}
0.6603	0.66029	$-3.87 \cdot 10^{-5}$	$7.39 \cdot 10^{3}$	H_{023}
0.7357	0.7357	$-1.33 \cdot 10^{-4}$	$2.75 \cdot 10^{3}$	H_{024}
0.8147	0.8144	$-2.19 \cdot 10^{-2}$	18.5	H_{018}

The characteristics of free oscillations are sensitive to the position of the complex fundamental frequency, when the spectral line is continuously transformed from the direction of one family to the direction of another family. The type of a free oscillation is preserved in certain regions of the parameter variations, and these regions are sometimes very small. We note that the complete classification of the quasi-stationary field states in periodical open resonators is possible if all the parameters of the spectral problem are precisely defined. As an example, we refer to the first twelve eigenoscillations of the grating presented on Fig.53b (with the parameters $\Phi = 0.1$, $\theta = 0.8$, $\delta = 1.2$ and $\varepsilon = 3.89$). The corresponding values of the fundamental frequencies given in Table 6 are positioned in the fourth quarter on the first sheet of the surface $\mathcal{H}$ with respect to their increasing absolute values.

The longitudinal resonance corresponding to one of the modes propagating in the communication channel Q is generated in the regular communication domain of the radiation zone (shown in Fig.53).

Numerical experiments were carried out in the fourth quarter of the first sheet of the surface $\mathcal{H}$, since in this quarter the spectral points κ lay close to the real axis. We note that it is important to consider such points because the physically correct radiation condition is fulfilled at real frequencies, for which the scattering of waves excited by quasi-periodical monochromatic sources is considered.

§ 14. Interaction of free oscillations and the Morse critical points of gratings

Variations of spectral characteristics with respect to geometrical and material parameters of open periodical resonators may be violated in the domain of interaction of free oscillations (corresponding to different fundamental frequencies κ_1 and κ_2) where they approach each other. We will assume here that the interaction means qualitative and quantitative changes of the eigenoscillation under the conditions close to the existence of other oscillations. The present section is devoted to the analysis of such interaction phenomena.

Let us go back to Fig.58. One can separate a part of its enlarged fragment, where the spectral lines are continuously transformed in the domain of possible intersection and finally change their directions.

The imaginary parts of fundamental frequencies approach each other in such a way that, first, the orders and, second, the values of $\operatorname{Im}\kappa(\delta)$ become equal. Oscillations with practically equal q-factors begin to interact. As a result, the q-factor of one oscillation substantially increases, while the q-factor of another oscillation sharply falls. An anomalous variation of the imaginary parts is observed in the vicinity of the point of the maximal rapprochement of the curves $\operatorname{Re}\kappa(\delta)$. Then the spectral curves $\operatorname{Im}\kappa(\delta)$ of the interacting oscillations exchange their directions (that are characteristic for the fundamental frequencies of each particular family, for example, H_{0m}).

As a result of the interaction of two different types of free oscillations in the narrow δ-interval, the type of the free oscillation connected with a definite continuous spectral curve changes. Outside the interaction zone the field of one oscillation is determined, for example, by the H_{03} wave and the spectral curve $\operatorname{Re}\kappa(\delta)$ (the H_{031} oscillation) comes to the critical (cut-off) point of the H_{03} wave. After the intersection the type of the oscillation is basically preserved. Other changes are more intensive and one can observe them by analyzing the equal contour lines of H_{023} oscillations. These changes are more distinct for greater values of δ, and the resulting free oscillation may be classified as the H_{023} oscillation. Note that the oscillation corresponds to the other spectral curve before entering the interaction zone.

Hence, the interaction yields the exchange of characteristics connected with the dynamics of fundamental frequencies and the types of oscillations described by different spectral lines. The spectral line having the definite direction as well as the corresponding oscillation may be classified by the cut-off points of the H_{0m}-curves to which they approach (with respect to the curve number in the family, which is equal to the number of the field variations in the communication channel).

Results of the analysis allow us to call these phenomena the intertype interaction of oscillations. In open structures the spectral problems are essentially non-self-adjoint, the fundamental frequencies are the complex numbers and eigenfunctions do not form the basis. Hence, one cannot apply the models of Refs. [57, 58] to consider the intertype interaction in open structures from the viewpoint of the classical theory of partial differential equations. Results of Refs. [57, 58] confirm the universal character of this phenomenon, which occurs in all known types of open resonators. The spectral curves corresponding to different types of free oscillations (from the same symmetry class) and having comparable diffractional q-factors may diverge in complex space without intersections. Intertype interactions eliminate the degeneration, i.e., the existence of different oscillations with equal fundamental frequencies. The deviation of resonator boundaries and material parameters from the ideal values described in Ref. [29] occurs due to the intertype interaction of oscillations. The degenerate oscillation instability is proved in Ref. [103]. These two results enable us to consider the intertype interaction as a behaviour of non-ideal systems (open periodical resonators) when they approach non-stable states.

The interaction of two oscillations from the same class of symmetry takes place when their diffractional q-factors become equal. Therefore, for complex fundamental frequencies only those oscillations may interact, which are formed by the running to meet each other higher order H_{0m} waves ($m \gg 1$). The spectral lines corresponding to free oscillations from different symmetry classes intersect, and the picture may appear which is similar to the one observed in the intertype interaction domain (see Fig. 59) where the H_{021} oscillation is transformed to the H_{013} oscillation. As δ increases, the H_{021}-type on this curve is again replaced in the vicinity of $\delta \approx 0.95$ by the H_{015} oscillation of the first family having five field variations.

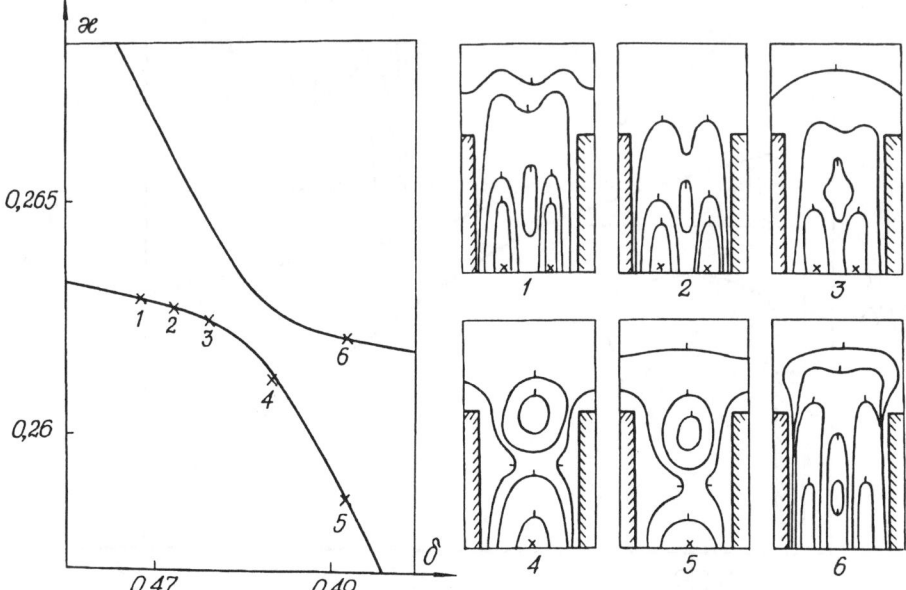

Figure 59.

Hence, the oscillations from the first family interact at real fundamental frequencies. These oscillations are formed by the lower order waves of the channel Q, and

for the real values of κ and Φ their fields and the "correct" (surface) eigenwaves propagating in the grating (considered as open periodical waveguides) are identical. The
study of the dynamics of eigenwaves and the fields of free oscillation opens additional
possibilities to apply their interaction in various microwave devices. The intertype
interaction reveals the possibility of the degeneration in the "ideal" situation, i.e., the
existence of several independent quasi-stationary free field states corresponding to one
fundamental frequency and occurring at definite (ideal) values of parameters. A search
for degenerate field states in open systems is always connected with the necessity of
analytical continuation of solutions to spectral problems into the complex domain of
non-spectral parameters. In the simplest situations one can model the elimination of
degeneration and transition to intertype interaction in the "physical" domain of the
parameter variation. A deviation from the ideal values that violates the symmetry
and leads to the removal of degeneration occurs when two oscillations belonging to the
same class of symmetry begin to interact. One can observe this interaction by varying
the quasi-periodicity parameter Φ in the structures having the planes of symmetry
$y = \text{const}$ (see Fig.53).

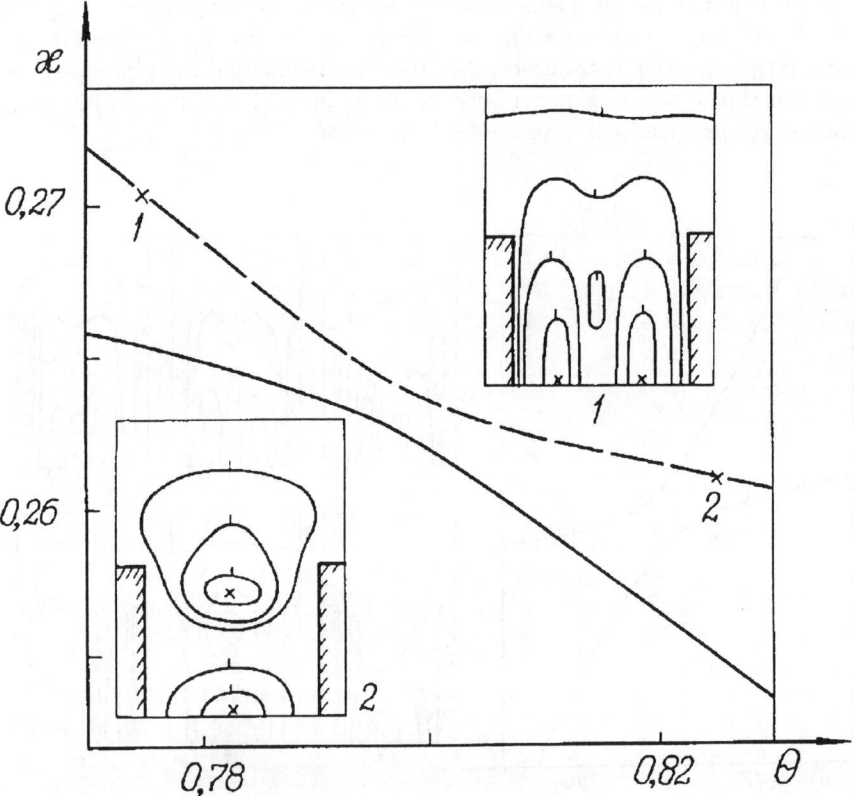

Figure 60.

The intertype interaction is sufficiently stable with respect to the substitution of
independent parameters (that serve as arguments in the functions describing spectral

curves). This stability is illustrated by Fig.60, where the lines $|E_x(y,z)| = $ const are calculated at the points marked by crosses. The presence of losses with $\mathrm{Im}\,\varepsilon \leq 0.05$ does not influence $\mathrm{Re}\,\kappa(\delta)$. The q-factors $(\mathrm{Im}\,\kappa(\delta))$ decrease proportionally to $\mathrm{Im}\,\varepsilon$. Variations of the basic spectral characteristics in the interaction zone for $\mathrm{Im}\,\varepsilon > 0$ are similar to the behaviour in the case of the ideal dielectric: oscillations interact, i. e., they exchange their types $(H_{031} \leftrightarrow H_{023})$.

In the domain of the frequency spectral parameter κ given by the conditions $\mathrm{Re}\,\kappa > 0$, $\mathrm{Im}\,\kappa \leq 0$, $\mathrm{Re}\,\Gamma_0 > 0$, $\mathrm{Re}\,\Gamma_n \leq 0$ and $n \neq 0$ (the first sheet of the surface $\mathcal{H}$ between the first and the second cuts), only one channel is open for the radiation into the free space, i.e., only one component of the free oscillation radiation field (the principal harmonic) carries the energy to the region $|z| > 2\pi\delta$. The interaction of eigenoscillations in this domain may lead to different results depending on their diffractional q-factors and on the difference between the excitation levels of the free fields for the dominant and background waves. Let us consider the interaction occurring as a result of the parameter δ variation. In the first case oscillations from the third (H_{035}) and the second (H_{029}) families interact (these oscillations are always symmetrical). Their diffractional q-factors have a difference of about one order before the spectral curves enter the interaction zone, and the oscillations are independent (they do not exchange types). The interaction leads to small local variations of the spectral characteristics of the oscillations from the third family. Relative variations in the second family are more essential: in the center of the interaction zone the field structure is not very distinct (a hybrid oscillation), the field is strongly influenced by the H_{03} wave (which is a background wave for this oscillation) and sharp (anomalous) changes of the diffractional q-factor take place. As a result of interaction in the physical domain of the parameter variation, free oscillations with real fundamental frequencies emerge (the super-high q-factor oscillations). According to the results of section 11, such a possibility is not eliminated when $\mathrm{Im}\,\varepsilon(g) \equiv 0$ and the open channels $\mathrm{Re}\,\Gamma_{nj} > 0$, $j = 1, 2, \ldots, k$ are closed $(a_{nj} = b_{nj} = 0)$. When oscillations interact, the radiation channel is closed due to the mutual radiation of the prevailing (dominant) and background waves which determine the structure of the fields of interacting oscillations.

One could expect the occurrence of various interaction phenomena in the domain where one of the radiation channels is open to the free space. The super-high q-factor free oscillations $(\mathrm{Im}\,\kappa = 0$, $\mathrm{Im}\,\Phi = 0)$ coincide with the "correct" (surface) waves propagating along a grating without attenuation. The domain with one or several open radiation channels is described by the inequality $\kappa > |\Phi|$ for $|\Phi| \leq 0.5$ (the basic interval of the variation of the propagation constant), and this domain is usually classified as a "prohibited" region for the "correct" (real) slow waves. If the radiation channel is open only for the principal spatial harmonic $(\mathrm{Re}\,\Gamma_0 > 0)$, then the amplitudes a_0 and b_0 of the plane waves (the partial field components of the super-high q-factor free oscillation) vanish. In this case the domain $\kappa > |\Phi|$ does not differ from $\kappa < |\Phi|$ (the latter interval corresponds to the domain of slow eigenwaves) if to compare the rates of the energy exchange between the grating and free space in both domains. In this way the existence of slow waves does not seem unusual.

Thus, the above results enable one to broaden the boundaries of the domain where the slow surface waves of open periodical waveguides may exist. The basic qualitative feature of the effects discovered is their "point" character. Namely, these interaction phenomena are observed when at least two conditions are combined which determine the spectral lines of open periodical waveguides with closed radiation channels. These qualitative differences play an important role in various applied problems, in particular, when gratings are used as decelerating systems.

The study of the grating spectral characteristics may be reduced to the problem on (isolated complex) characteristic numbers for kernel matrix finite-meromorphic operator-valued functions $\{I + A(\kappa)\} = 0$, $\kappa \in \mathcal{H}$. In the process of numerical experiments, when the problems in question were reduced to finite-dimensional equations, the roots of the dispersion relations with the multiplicities greater than one were not found. If to assume that the resolvent $\{I + A(\kappa)\}^{-1}$ has simple poles, the increase of multiplicities means the degeneration of fundamental frequencies κ. Usually this (ideal) situation is not realized when the parameters vary in the physical domain. The interaction of free oscillations may lead to local or global changes of their spectral characteristics. Violation of the regular behaviour of corresponding curves means that a trajectory $\kappa(\tau)$ approaches the critical point of the mapping $f(\kappa, \tau)$.

The interaction of free oscillations is characterized by changes in the dispersion law, i.e., by a specific behaviour of two fundamental frequencies as functions $\kappa(\tau)$ for $\operatorname{Im} \tau = 0$ (see Figs.58–60). The critical point (κ_0, τ_0) of the mapping $f(\kappa, \tau)$ lies in the interaction zone considered for complex values of τ, where

$$\frac{\partial f}{\partial \kappa} = \frac{\partial f}{\partial \tau} = 0, \quad \{\kappa, \tau\} = \{\kappa_0, \tau_0\}.$$

If (κ_0, τ_0) is a non-degenerated and isolated point (the Morse critical point), then

$$\frac{\partial^2 f}{\partial \kappa^2} \frac{\partial^2 f}{\partial \tau^2} - \left(\frac{\partial^2 f}{\partial \tau \partial \kappa}\right)^2 \neq 0, \quad \{\kappa, \tau\} = \{\kappa_0, \tau_0\},$$

and the equation $f(\kappa, \tau) = 0$ with the accuracy up to cubic terms $(\kappa - \kappa_0)^j (\tau - \tau_0)^k$, $j + k = 3$ may be written in the vicinity of (κ_0, τ_0) as a quadratic form (see Chapter 1). Four-parameter families of curves obtained with the help of this quadratic form contain spectral curves typical for interacting oscillations (see Figs.58–60).

The existence of the Morse critical point of the mapping $f(\kappa, \tau)$ yields two solutions of the initial spectral problem in the vicinity of (κ_0, τ_0). Their behaviour with respect to the non-spectral parameter τ is completely determined by the quadratic equation.

If $f(\kappa_0, \tau_0) = 0$, the trajectory for $\kappa(\tau)$ goes through the Morse point and the root $\kappa_0(\tau_0)$ has a double degeneration. If $f(\kappa_0, \tau_0) = 0$ and $\operatorname{Re} \tau \neq 0$, the trajectory does not go through the Morse point and a small displacement of $\operatorname{Re} \tau$ may cause the degeneration of the fundamental frequency. If $f(\kappa_0, \tau_0) \neq 0$, there is no degeneration whether or not the κ-trajectory goes through the Morse point.

Interaction of free oscillations is one of the most interesting phenomena considered in the electrodynamical theory of gratings. At the same time, the corresponding mathematical problems are weakly studied. The analysis performed in this section covers a limited domain of the parameter variation (when one radiation channel is open to the free space). The cases when three or more oscillations interact, were not considered. The results obtained are sufficient to discover and explain more complicated effects and phenomena, in particular, the formation of super-high q-factor oscillations in the gratings with several open radiation channels. Application of the results obtained in the theory of singularities of smooth mappings may considerably facilitate the solution to these problems. This approach is a universal tool that enables one to describe all kinds of anomalies occurring in various physical systems, in particular, electrodynamical.

§ 15. Scattering of surface waves of dielectric waveguides by strip gratings

Mathematical (and physical) problems of the resonant plane wave scattering by gratings with a simple geometry are studied in Refs. [5, 8, 24, 156, 158]. The results obtained qualitatively and quantitatively characterize practically all periodical structures that are widely applied in the millimeter and submillimeter technology, vacuum electronics and antenna engineering. One can simulate the behaviour of scattered fields on the basis of these results and perform the corresponding optimization.

But a number of important problems are not solved concerning the nature of anomalous scattering regimes in the resonant frequency domain, regularities of diffractional characteristics, excitation of a definite type of oscillation, etc. We apply the spectral theory to analyze the resonant wave scattering on gratings, in particular, the Wood anomalies, the total reflection and transmission in semi-transparent structures and the energy canalization at one of the higher order spatial harmonics. The spectral methods, which differ from the approaches that are traditionally applied in the diffraction theory, allow us to consider these phenomena from the qualitatively new positions and to obtain new data for creating a complete physical picture.

In this section we will consider in details only one problem: the wave scattering by a strip grating in a dielectric waveguide. Other scattering problems are studied, for example, in Refs. [24, 11].

We assume that it is given an inhomogeneous wave propagating along the (two-dimensional) plane dielectric waveguide and one has to determine the field scattered by an inhomogeneity. The distance between the grating and the waveguide (the aimed distance) controls the "degree of communication" in this source–scatterer system. We will apply the methods of the rigorous diffraction theory (for example, the method of the Riemann-Hilbert problem [5]) for the solution. Our basic task here will be to model the transformation of surface waves to volume waves by diffractional gratings.

Let the strip metallic grating be placed at the distance b from the dielectric waveguide in the domain $x < 0$ (the strips are perpendicular to the axis Oz, the stripwidth is d and the grating period is l). The origin of the coordinate system (x, y, z) lies on the waveguide axis and the grating axis goes through the strip center. The slow waves

$$H_y^0 = ikg\varepsilon_1 \left\{ {\cos \atop \sin}(ga) \right\} e^{p(x+a)i\gamma z}, \quad E_y^0 = -ikg \left\{ {\cos \atop \sin}(ga) \right\} e^{p(x+a)i\gamma z} \qquad (3.23)$$

propagate along the waveguide with the width $2a$ and permittivity ε_1.

Consider the diffraction of waves (3.23) by the strip grating. The scattered field exists at both sides of the grating. Let us separate the domains I: $-a - b < x < -a$ and II: $x > -a - b$. We will look for the field in the form $\mathbf{E} = \mathbf{E}^0 + \mathbf{E}^p$, $\mathbf{H} = \mathbf{H}^0 + \mathbf{H}^p$ in the domain I and $\mathbf{E} = \mathbf{E}^p$, $\mathbf{H} = \mathbf{H}^p$ in the domain II, where $\mathbf{E}^0, \mathbf{H}^0$ is the given field and $\mathbf{E}^p, \mathbf{H}^p$ is the (unknown) scattered field. The influence of the scattered field on the source eigenwave is not taken into account.

In the case when the total field is determined by the E_y-component, we have

$$E_y = -ikg \left\{ {\cos \atop \sin}(ga) \right\} e^{p(x+a)i\gamma z} + E_y^p, \quad E_x = E_z = 0, \qquad (3.24)$$

where $p = \sqrt{\gamma^2 - k^2}$ and $g = \sqrt{k^2 \varepsilon_1 - \gamma^2}$ are the propagation constants along Ox and Oy. The sought-for function E_y^p satisfies the Helmholtz equation $\Delta v + k^2 v = 0$, the boundary conditions, the edge and the radiation conditions.

We will represent the scattered fields in the form of Fourier series (for the H waves)

$$E_y^{pI} = -ikg \left\{ \begin{matrix} \cos \\ \sin \end{matrix} (ga) \right\} e^{-pb} \sum_{n=-\infty}^{\infty} a_n e^{ip_n(x+a+b)i\gamma_n z},$$

$$E_y^{pII} = -ikg \left\{ \begin{matrix} \cos \\ \sin \end{matrix} (ga) \right\} e^{-pb} \sum_{n=-\infty}^{\infty} b_n e^{-ip_n(x+a+b)i\gamma_n z},$$

(3.25)

where a_n, b_n are unknown Fourier coefficients, $\gamma_n = \gamma + 2\pi n/l$ and $p_n = \sqrt{k^2 - \gamma_n^2}$ are the propagation constants of the field Fourier components along the axes Oz and Ox, respectively (here $\sqrt{A}$ denotes the branch with $\operatorname{Im}\sqrt{A} > 0$ and $\operatorname{Re}\sqrt{A} > 0$ for $\operatorname{Im}\sqrt{A} = 0$). Representations (3.25) satisfy the uniqueness conditions. Other components of the scattered field are obtained from the Maxwell equations.

Fields (3.25) consist of an infinite number of plane waves going off the grating to the domains $x < 0$ and $x > 0$. For the real p_n, when $\left(\gamma + 2\pi n/l \right)^2 < k^2$, i.e., when the radiation conditions hold, the field harmonics are transformed to the homogeneous plane waves. Since for the given inhomogeneous wave we have $\gamma < k$, this condition holds for the harmonics with $n < 0$. Their propagation angles are determined from the relation

$$|\cos \theta_n| = \frac{1}{k}\left(\gamma - \frac{2\pi n}{l}\right) \leq 1,$$

(3.26)

which may be also considered as the radiation condition. If $n \geq 0$, then $|\cos \theta_n| > 1$, and such waves should be treated as inhomogeneous plane waves propagating along Oz with the phase velocities $v_\phi < c$. Hence, the diffractional field is a superposition of plane waves. A part of these waves are the volume waves going off the grating at the angles (3.26), while the rest are localized close to the grating and are similar to slow surface waves. The choice of the grating parameters defines a required number of the required type of waves.

Let us find the unknowns a_n and b_n. Using the boundary conditions and introducing the parameters

$$\nu = \frac{\gamma l}{2\pi}, \quad \kappa = \frac{kl}{2\pi}, \quad u = \cos\frac{\pi d}{l}, \quad \xi = \frac{2\pi z}{l}, \quad \varepsilon_n = 1 + i\sqrt{\frac{\kappa^2}{(n+\nu)^2} - 1}, \quad x_n = \tilde{a}_n(n+\nu),$$

(3.27)

we get a system of coupled summation-type equations with trigonometric kernels

$$\begin{cases} \sum_{n=-\infty}^{\infty} x_n e^{in\xi} = 0, & \frac{\pi d}{l} < |\xi| < \pi, \\ \sum_{n=-\infty}^{\infty} x_n \frac{|n+\nu|}{n+\nu} e^{in\xi} = \frac{pl}{2\pi} + \sum_{n=-\infty}^{\infty} x_n \varepsilon_n \frac{|n+\nu|}{n+\nu} e^{in\xi}, & |\xi| < \frac{\pi d}{l}, \\ \sum_{n=-\infty}^{\infty} (-1) \frac{x_n}{n+\nu} = 0. \end{cases}$$

(3.28)

These functional equations are reduced to the infinite system of linear algebraic equations

$$x_n = \sum_{s=-\infty}^{\infty} A_{ns} x_s + B_n,$$

(3.29)

by the method of the Riemann-Hilbert problem [5]. Here

$$
A_{ns} = \begin{cases}
\frac{|s+\nu|}{s+\nu}\varepsilon_s W_\sigma^{-s}, & -\infty < s < -j-1, \\[2mm]
\frac{|s+\nu|}{s+\nu}\varepsilon_s W_\sigma^{s}, & 0 < s < \infty, \\[2mm]
\left(\frac{|s+\nu|}{s+\nu}\varepsilon_s - 2\right)W_\sigma^{s}, & -j < s < -1,
\end{cases}
$$

$$
B_n = QW_\sigma^0, \quad W_\sigma^s = V_n^s - V_\sigma^s R_n R_\sigma^{-1}, \quad Q = \frac{pl}{2\pi},
$$

and $V_n^s(u)$, $V_\sigma^s(u)$, $R_n(u)$ and $R_\sigma(u)$ are given in Ref. [24]. The series $\sum_{n,s}|A_{ns}|^2$ and $\sum_n |B_n|^2$ converge and, hence, the matrix of the system (3.29) defines a compact operator in the Hilbert space l_2 of the sequences $\{x_n\}$. (3.29) is an infinite Fredholm system of the second kind. One can prove the existence and uniqueness of the solution to (3.29) and the possibility to apply the truncation method for its numerical analysis. For $\kappa < 1$ and $|u| \sim 1$ the inequality $\|A_{ns}\| < 1$ holds. One can derive explicit expressions for a_n and b_n obtained with the help of the methods described in Ref. [5] by successive approximations. As an example, let us write down the single non-zero radiating H_y harmonic

$$
H_y^{\mathrm{II}} = \frac{ik\varepsilon\left\{{\cos \atop \sin}(ga)\right\}e^{-pb}\left[u + (1+u)\xi \ln \frac{1+u}{2} - 1\right]}{2\left(1 + i\sqrt{\kappa^2 - \xi^2}\ln \frac{1+u}{2}\right)}.
\tag{3.30}
$$

Here $\kappa < 0.5$, $\xi < 0.5$. The energy conservation law for this problem yields the complex power theorem:

$$
\frac{cl}{8\pi}g\varepsilon_1 \cos ga(ke^{-2pb})\left\{\sum_{\mathrm{Im}\ \sqrt[n]{A}=0} p_n|\tilde{a}_n|^2 + \sum_{\mathrm{Im}\ \sqrt[n]{A}=0} p_n|\tilde{b}_n|^2 + 2e^{pb}\,\mathrm{Im}\,\tilde{a}_0\right\} = 0.
\tag{3.31}
$$

We will define for the E wave the energy of the diffractional field radiating to the free space as

$$
\mathrm{Re}\,\tilde{S}_\perp = Q^2 e^{-2pb}\sum_{\mathrm{Im}\ \sqrt[n]{A}=0} p_n|\tilde{b}_n|^2.
\tag{3.32}
$$

The power flow density for the single radiating harmonic with the number $n = -1(S_{-1})$ as a function of the transversal waveguide dimensions is presented in Fig.61 (curves 1 and 2 correspond to the E and H waves, respectively). One can see that S_{-1} has a resonant behaviour and vanishes when $2a$ tends to zero. The maximal value of S_{-1} for the H waves is much less than that for the E waves because the magnetic waves are more "pressed" to the waveguide than the electric waves and the energy propagating along the waveguide is concentrated inside it (see Ref. [8]). The dependence of S_{-1} on $u = \cos(\pi d/l)$ shows that there exists the "optimal" grating providing the maximal radiation from the structure. max S_{-1} is reached for the E waves is in the domain $(-u)$, i.e., when the larger section of the period is covered by the metal. For the H waves max S_{-1} is observed in the region of narrow slots ($u > 0$). Physical reasons explaining such a behaviour of $S_{-1}(u)$ are connected [8] with the current distributions on metallic strips: for the H-polarization the currents on strips and slots have a break since $j_z \neq 0$, while for narrow slots $|E|$ reaches its maximum between the slots. The

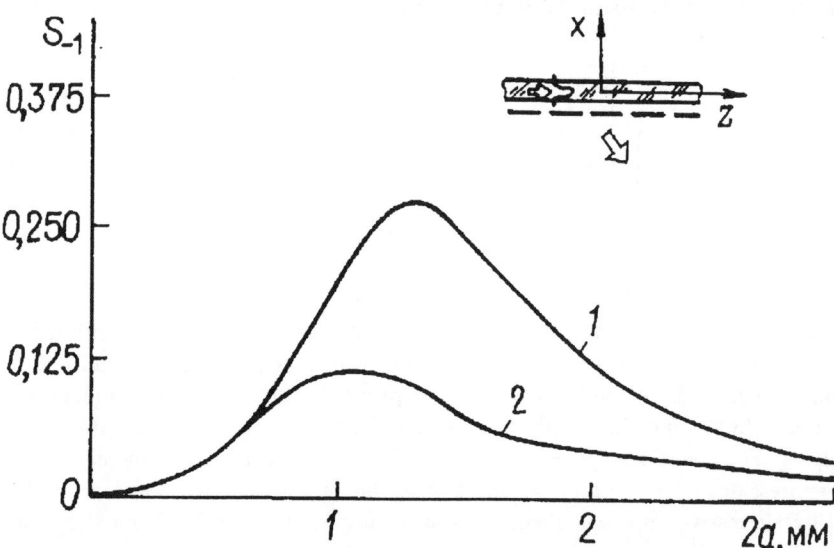

Figure 61.

symmetry of the Maxwell equations with respect to the electric and magnetic currents enables one to verify that the maximum of the radiated power is attained in the region of narrow strips for the opposite polarizations.

The multi-wave field character that affects the minus first harmonic, causes the presence of additional local maxima of S_{-1} and leads to the generation of higher order radiating harmonics (when the grating period l increases). The points on the graphs of $S_{-1}(l)$ corresponding to these harmonics are the singularities where the second derivative is discontinuous (analogous effects are observed in the case of the volume wave diffraction). One can determine the energy radiating to the free space, which decreases with respect to the increasing index. An anomalous growth (decrease) of S_{-1} corresponds to the even (odd) radiating harmonics, which coincides with the results of Ref. [24]. A resonant behavior of $S_{-1}(\kappa)$ reflects the dispersion in plane dielectric waveguides: for small wavelengths the phase velocity of the scattering wave is small and the energy is concentrated in the effective cross-section of this wave. The energy carried by the external field increases together with the wavelength. A further increase of the wavelength leads to a sharp growth of the effective waveguide dimensions and causes the radiation to the external space.

Let us consider the amplitude-phase distribution of the scattered field close to the grating. As follows from (3.25), the scattered field consists of an infinite number of spatial harmonics with complex amplitudes, and the radiating harmonics may appear in the scattered field under certain conditions. The near-field structure is determined by the interaction of spatial harmonics localized close to the grating with volume waves. The lines of equal phases are perpendicular to the lines of equal amplitudes at large distances from the grating $(2 \div 3\lambda_0)$. In the domains where lines of equal phases intersect and their absolute values cannot be determined, the energy vortexes emerge. The amplitude in these vortexes attain the minimum value, the energy flow circulates and does not go out to the free space. Hence, the field energy is localized at the distance of about λ_0 to the grating.

When the radiating field with $n = -1$ appears at the distance $3 \div 4\lambda_0$ from the grating where attenuating harmonics are small, one can write the representation

$$H_y = b_{-1} e^{-ix\sqrt{k^2 - (\gamma - \frac{2\pi}{l})^2}} e^{iz(\gamma - \frac{2\pi}{l})}.$$

Here the equal phase configuration sharply changes and becomes non-symmetrical, the circulation domains disappear and the flow of energy to the free space is observed. At large distances from the grating the lines of equal phases and amplitudes coincide and produce a plane wave. The non-symmetrical localization of the lines of equal phases in case of radiation to the free space occurs at the distance less than λ_0 from the grating (if there is no radiation, this distance may be arbitrarily large). The radiation angle measurements (in the domain where the plane wave exists) by the lines of equal phases give the results similar to (3.26). The field picture becomes considerably complicated when two harmonics radiate at different angles.

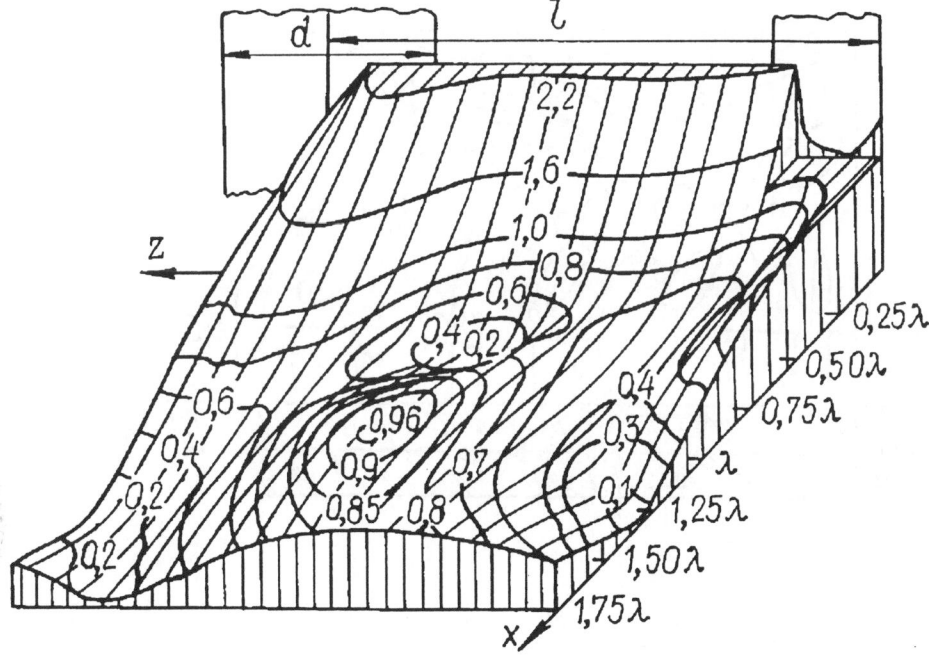

Figure 62.

Systematical computations and an analysis of amplitude-phase distributions show that the vortexes act as transformators that transform the surface wave energy to the energy of outgoing waves. When "approaching" the condition (3.26) (when the radiation angle is real), the vortexes disturb the field and force it to tear off the grating in the form of plane waves. One can assert that the closer the vortex to the grating, the larger is the volume wave amplitude. In fact, this increase can be explained by the fact that the vortex comes to the region of large amplitudes. The H_y-distributions for the fields of surface waves transforming to volume waves are presented in Fig.62. The transformator there is formed by the dielectric waveguide ($2a = 1$ mm, $\varepsilon_1 = 2.25$) and the grating ($\kappa = 1$, $u = 0.4$). The distance between them is $b = 0.5$ mm. One can clearly observe the field cavity containing the energy flow vortex and the "hill" that indicates the "tearing" field.

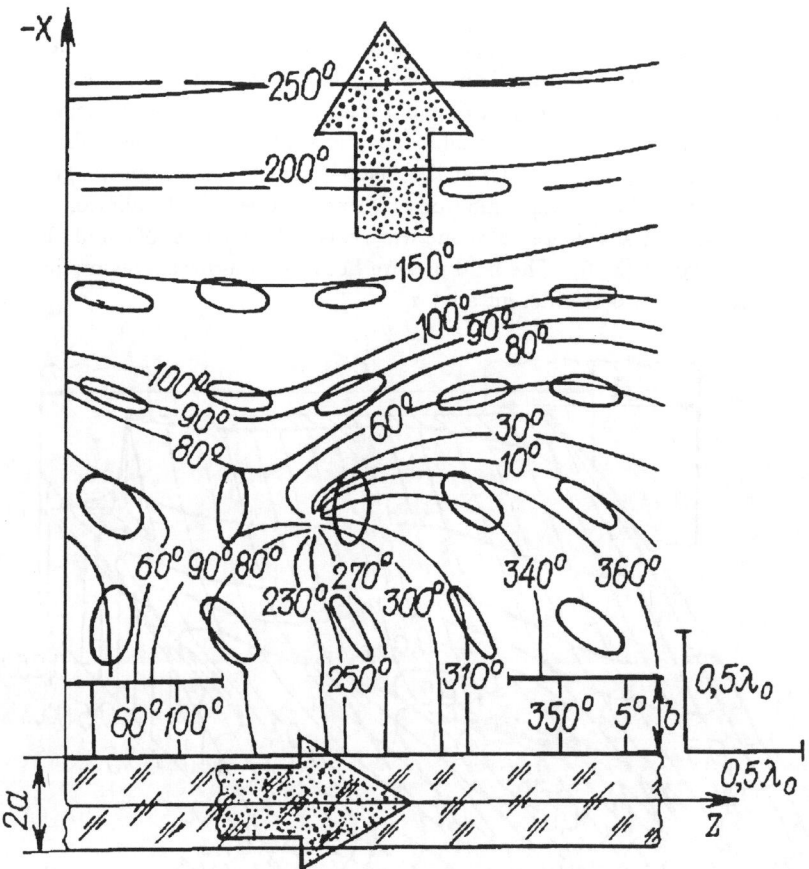

Figure 63.

The phase portrait of the surface-volume wave transformation in the dielectric waveguide with the grating is presented in Fig.63. The field has the distinct electric polarization close to the grating and is similar to a surface wave. At larger distances the polarization ellipses tend to rectilinear intervals and at the distance of about $3\lambda_0$ the linearly polarized wave appears. Its field has been transformed by this waveguide-grating transformator to a volume wave (or, more precisely, to the plane wave propagating in the normal direction).

§ 16. Experimental study of diffractional gratings

The program of experiments described in this section deals with different variants of periodical structures, excitation sources, devices for analyzing the scattered field, etc.

Experimental studies of the surface-volume and volume-surface wave transformation are of considerable interest in the millimeter and submillimeter wavelength ranges, because they stimulate the development of the vacuum electronics and the antenna and measurement technology.

The study of the spectral, amplitude-phase and polarization characteristics of the electromagnetic fields in the process of the surface-volume wave transformation is connected with the determination of their antenna properties. It causes the necessity to measure the near and the far fields. Experimental methods must take into account the features of the millimeter and submillimeter wavelength ranges, which is guaranteed by the use of the transformations considered in the scheme of the experimental plant.

The antenna system operating on the basis of the surface-volume wave transformation consists of two elements: the inhomogeneous plane wave source and a periodical scatterer in the form of a grating placed in the field of the surface wave (the dimensions of the domain where the communication between the surface wave and the grating elements occurs are much greater than the wavelength, and this domain may have the form of a radiating aperture). Various dielectric waveguides are used as the surface wave sources. If the thickness of a plane dielectric waveguide is less than the wavelength, a considerable part of the energy is radiated to the space close to the waveguide surface and the field intensity decreases exponentially (as a function of the distance to the waveguide boundaries). Dielectric waveguides with certain cross-sections (for example, circular, elliptical and rectangular) may have the characteristics which are close to that of the plane waveguide, if their transversal dimensions are of equal order. In particular, the the surface-volume wave transformation is observed when a circular (or an elliptic) dielectric waveguide is used as a source of surface waves.

The transmission and the reflection gratings are used for the surface-volume wave transformation. For the first type of gratings a directed radiation appears in the upper and in the lower half-spaces, while for the second type of gratings the radiation is registered only in the upper half-space. These gratings may consist of different elements: metallic strips, bars (rectangular or circular), jalousie-type or echelette-type elements, etc. Strip gratings allow us to study the volume waves excited by the surface waves of a dielectric waveguide applying a minimal number of parameters, which, in turn, opens the way to use more complicated gratings, in particular, the transmission and the reflection gratings made of bars. It is essential that, independently of the shape of the grating elements, its main characteristic, the radiation direction (3.26) of the far-field pattern maximum, are the same for all such gratings.

Amplitude-phase and polarization characteristics are determined by the shape of separate scatterers, the distance between the waveguide and the grating (the aimed distance) and the dielectric waveguide parameters. Experimental calculations of these values is first of all connected with the measurements of the field intensity close to the structure. We begin with the study of the radiation properties of the grating elements (strips, bars) and then to consider the grating itself. A complete picture of the waveguide surface field transformation to the grating volume waves may be obtained when the polarization and energy characteristics will be determined close to the structure and in a far zone (with the help of the antenna measurements). We will describe the experiments that are carried out under the assumption of the given field in the process of the surface-volume transformation. In addition to this, one must take into account that real structures have bounded dimensions, finite conductivity of

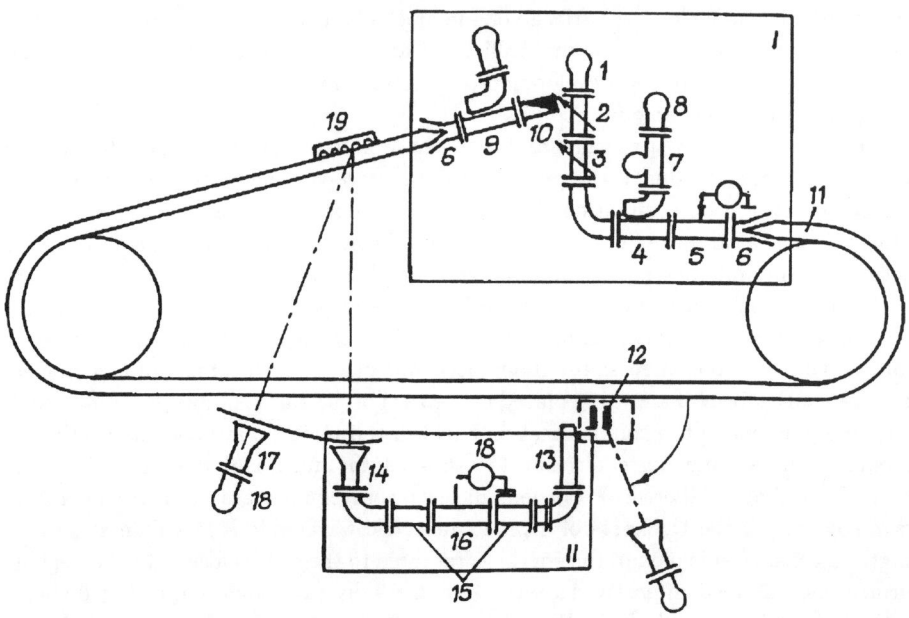

Figure 64.

metallic elements and lossy dielectric media.

The experimental stand for the near-field and the far-field measurements in the process of the surface wave transformation consists of two large blocks (Fig.64). Consider the one intended for far-field measurements, i.e., for registering the antenna characteristics. This stand provides the polarization measurements and the transformation of surface waves to volume waves. The spatial field distributions are registered in the far zones of the H and E planes, when the observation angle varies from 0° to 180° with an absolute error of ±0.5°. The generator operation wavelength is controlled, the deceleration in a dielectric waveguide is determined and the system waveguide–grating is coordinated with the waveguide section. The millimeter-wave generator 1 feeds waveguide 11 through unbinding 2 and measuring 3 attenuators, the regular part of coupler 4, measuring line 5 and the angle junction 6. The second waveguide channel is connected by the same junction with the regular part of coupler 9 and the coordinated load 10. A part of the energy is taken by coupler 4 to the wave-measuring device 7 and to the frequency control system 8. Measuring line 5 controls the coordination. The dielectric waveguide is fixed by two foam plastic discs with the radius $R \geq 40\lambda_0$. The grating lies on a table which is moved by the micro-meter screw and a special stop. The grating is pressed to the stop by a holder. These devices enable one to control the aimed distance. An adjustment unit allows one to incline the grating plane with respect to the waveguide axis by rotating the plane around the main axis. This part of the plant is used for the wave transmission and is placed on a moving platform rotating in two perpendicular planes. Potential-metric sensors and reverse engines allow one to use two-coordinate continuously moving ribbons as the recorder elements. The receiving part of the plant consists of an antenna with a detector section placed on a rotating bar at the required distance from the object. The receiving antenna is also used for the polarization measurements.

Far field patterns are measured by the mobile antenna method [143]. The relative sidelobe level, the width of the main radiation maximum and radiation directions are determined by the graphically constructed far-field patterns with the amplitude calibration over the angle coordinate. The width of the main lobe is taken to be equal to the length (in degrees) of the chord joining the points of the main lobe at the 3-dB level. The chord center defines the radiation direction for the main lobe. The levels of lateral maxima are determined directly by the amplitude scale. In order to decrease the angle coordinate error, the rotation axis in the E plane is chosen to intersect the radiating aperture along the grating axis. The same axis in the H plane coincides with the longitudinal waveguide axis. In our experiments the grating apertures are taken to be equal to $20 \div 30\lambda_0$. It allows us to carry out measurements in the four-millimeter and two-millimeter wavelength ranges directly on the laboratory table.

An analysis shows that effective measurements should be carried out by means of elliptical dielectric waveguides with $v_\phi/c \approx 0.65 \div 0.85$, when propagating modes have stable polarization characteristics and the E component is identical to the corresponding field component in a plane waveguide. The latter is achieved by applying specific excitation conditions for an elliptic waveguide: in the experiment the waveguide is excited by a smooth junction in the form of an angle exciting device fed by a standard metallic H_{10} rectangular waveguide (the big axis of the ellipse is parallel to the wide wall of the metal waveguide). The ratio of the big and small axes is 0.5 for $\varepsilon = 2.25$ and $v_\phi/c \approx 0.7 \div 0.8$. The effective diameter of such a waveguide is $5 \div 5.5$ mm, the big axis is 4.4 and $\lambda_0 = 4.0 \div 4.51$.

It is theoretically proved in [24] that when surface waves of a plane dielectric waveguide are scattered by a grating, the radiation energy flow monotonically decreases with respect to increasing b. The existence of the saturation domain, where the radiating power is constant, is established experimentally.

The detailed study of the surface wave transformation includes the radiation power measurements. The leaky-wave antennas energy properties are closely connected with their directional characteristics. Absolute values of the radiation field **E** with respect to the angle coordinates in the E and H planes are usually measured for far-field patterns with the successive integration over the considered domain.

This method may be used in the microwave frequency range if appropriate measuring set-ups are available. In the millimeter and submillimeter wave technology, there are no such devices that would allow one to measure the required energy parameters with sufficiently small errors. The special methods [6, 143] enable one to perform relative measurements (of the ratio of the powers of propagating waves registered with and without the grating). Let us describe the corresponding scheme of measurements. The power P_0 arrived to a coordinated load (connected with the waveguide) is measured without separation of the radiation power (i. e., without the grating). The grating placed in the waveguide surface field leads to the radiation of a part of energy, P_{rad}. It is important to determine the dependence of $P_0 - P_{rad}$ on the parameters of the open system waveguide–grating and to calibrate the data by the scaled attenuator 3 (Fig.65). The value of $P_0 - P_{rad}$ is measured by the directional coupler with the straight shoulder adjusted to the coordinated load 10.

We have also performed experimental studies of the near field in the radiation system formed by the dielectric waveguide and the grating. These investigations are

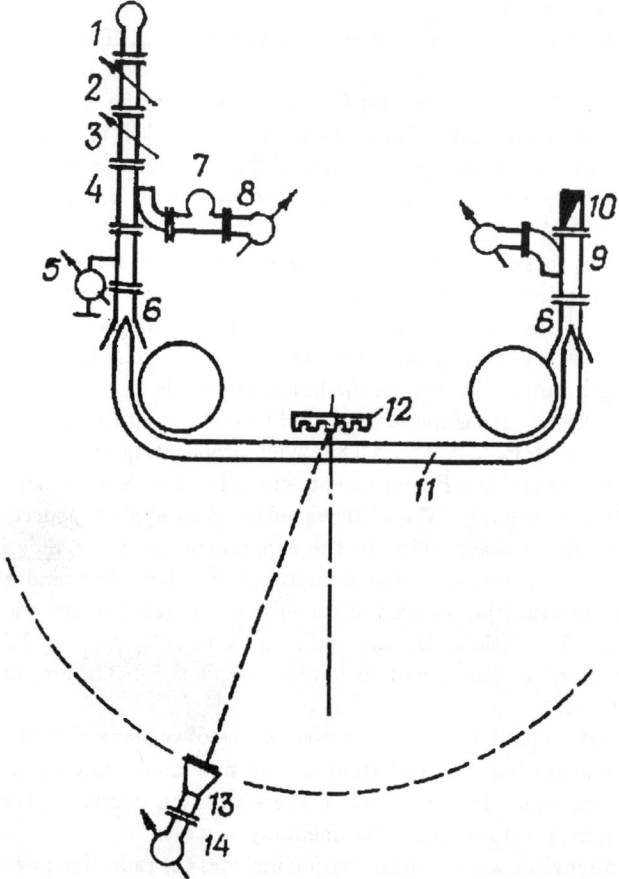

Figure 65.

important because surface-volume transformations occur in the domain occupied by this open system. Let us consider at first the fields of the surface wave diffraction by a single obstacle, a strip and a bar (of the rectangular and circular cross-sections).

In classical methods of the near-field measurements [200] small sensors are applied that have the form of electric dipoles and magnetic loops. The "layered" electric power is sent to a registering device by microwave cables. It is practically impossible to repeat such measurements in the millimeter and submillimeter scale modeling. At the same time, the known methods may be modified in such a way that the energy transmission from a sensor to an indicating device (in the microwave channel) is excluded. Here a small dipole $(0.1 - 0.2\lambda)$ antenna is applied. Electromagnetic fields radiated by the dipole are received by an additional high-directed antenna (for example, by an angle-type antenna). The position of this antenna is determined by the spatial structure of the radiated field (the butt-end radiation must be taken into account). The position of the receiving antenna is shown in Fig.66 (dipole 1 is fixed by the quartz pivot, the pivot and the receiver are adjusted with the plate of the scanning device and the signal is registered synchronously with respect to the plate movements in the plane yOz).

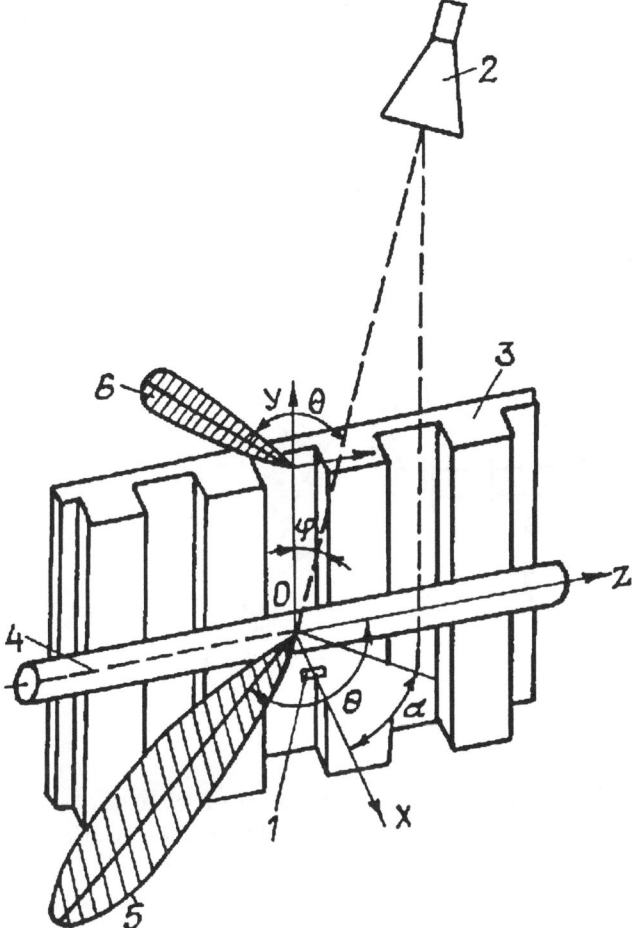

Figure 66.

Since the sensor is situated in the field of the system waveguide–grating, it is natural to expect its influence on the measured field. In order to register the corresponding noise factor, a sequence of far-field patterns is measured when the sensor is placed at different points along the aperture. It is established that the $0.5\lambda_0$ sensor practically does not disturb the near field.

In the millimeter and submillimeter wavelength ranges we use the interferometer methods to measure the scattered surface waves of a dielectric waveguide and the generated volume waves. The measurement devices are modified in the following way: The supporting signal channel consisting of the transmitting (19) and receiving (14) antennas (see Fig.67) is mounted on the second block of the measuring stand (Fig.64). The first channel is made in the form of a comb excited by the surface wave of the dielectric waveguide. The grating parameters (the number of periods, the channel depth and the aimed distance) provide the formation of the far-field patterns in the E plane with the width $6° - 8°$). It allows us to register signals with constant amplitudes and phases in the sector $2°$ containing the radiation direction of the main maximum

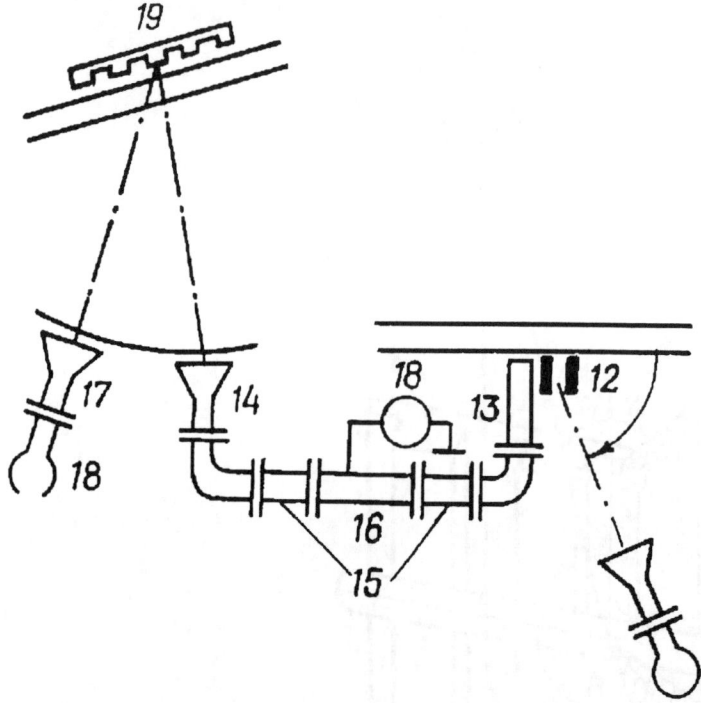

Figure 67.

(corresponding to the $\pm 4\lambda_0$ linear displacement at the distance $R \approx 100\lambda_0$ from the radiator). We note that here it is sufficient to measure amplitude-phase distributions over the grating period $l \approx \lambda_0$. Such a design of the supporting channel is convenient to adjust the set-up in order to compare the measurement data, since small variations of the aimed distance yield the changes of the supporting signal. The measuring waveguide line 16 is taken as the element for comparisons. The ferrite valves 15 exclude multiple reflections of the supporting and registered signals.

The waveguide measuring unit of the experimental stand (Fig.67) is constructed on a separate plate with adjustment units. The sensor has the form of the open end of the rectangular empty metallic 3.6×1.8 mm² waveguide narrowed in the E plane up to the size 3.6×0.1 mm² (providing the visualization accuracy $0.025\lambda_0$ in the four-millimeter range). The initial (zero) sensor position is fixed on the aperture plane by an electric contact (the ohm-meter) placed between the sensor and the object.

When the measuring line is applied for the phase determination, we find the standing wave minimum by fixing the coordinate of the $|\mathbf{E}|$ minimum. If to change the phase of one of the interfering waves, the resulting displacement of the minimum coordinate Δz [mm] will be proportional to the waveguide phase shift $\Delta\psi$: $\Delta\psi = 2\pi\Delta z/\lambda_g$ (λ_g is the wavelength). The phase measured by this method is negative when the minimum coordinate is in the supporting shoulder.

The experimental plant (Fig.64) for measuring the far and the near fields of independent scatterers and periodical systems of obstacles placed in the field of the surface wave of a dielectric waveguide must provide parallel displacements, coincidence of the rotation axes of receiving antennas and radiating apertures and coordination of the main waveguide section with power units. The plate with the measuring waveguide section must move parallel to the waveguide axis (the deviation does not exceed 0.1 mm and the linear displacement accuracy of the order of 0.01 mm is controlled by a watch-type indicating head).

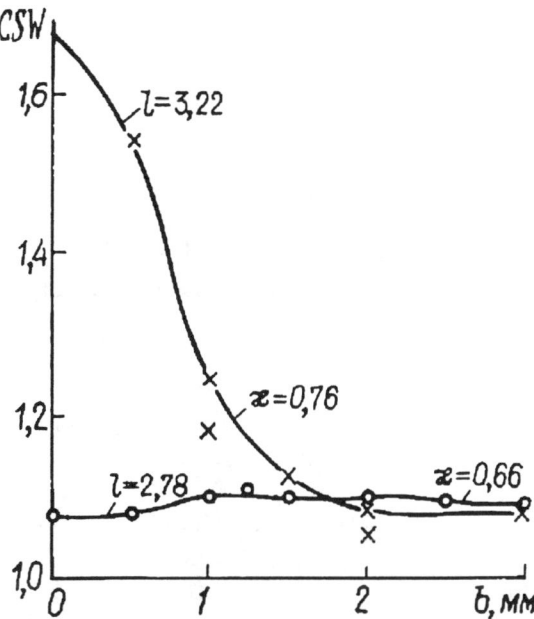

Figure 68.

A correct coordination of the transmitting antenna with the dielectric waveguide blocks the reflected waves that perturb the amplitude-phase field distribution. One of the main reasons of unsatisfactory coordination is the volume wave radiation in the direction perpendicular to the grating. This situation occurs in the experiment when $l/\lambda_0 = v_\phi/c$, i.e., when the Bragg's diffraction condition [83] holds ($\kappa = m\beta/2$, $m = 1, 2, \ldots$, is the Bragg's diffraction order). The SWR values against the aimed parameter for different values of the grating period when $\beta = 0.78$ (obtained for the comb with the parameters $l = 3.22$ mm, $\kappa = 0.76$ and $l = 2.78$ mm, $\kappa = 0.66$), are presented in Fig.68. One can see that in this case a proper coordination is violated.

Hence, the diffractional grating forming the wave in the supporting channel of the measuring line may have the form of the comb with the parameters $l = 2.78$ mm, $\kappa = 0.66$ and $\beta = 0.78$ (for the waveguide). This structure eliminates the Bragg's diffraction and its energy reflection coefficient (defining radiation into the channel) is very low and equal to 0.0475 (the SWR value is 1.1). This data is obtained by sensor 16 moving parallel to the waveguide surface. Here the violation of coordination is confirmed by the diffraction theory.

The position of the transmitting antenna in the supporting channel is chosen in such a way that the main lobe of the far-field pattern is perpendicular to the axis of the longitudinal displacement. Note that the second unit is placed along the same axis and the adjustment is provided by the additional antenna (17) placed on a rotating bar.

Experimental studies are performed for the dielectric waveguide with the known characteristics obtained by theoretical and experimental methods [143]. Experimental dependence of the phase shift registered as a function of the distance z shows that the phase front has the form of a plane in the central domain of the far-field pattern.

Let us consider the surface-volume waves transformation for separate grating elements.

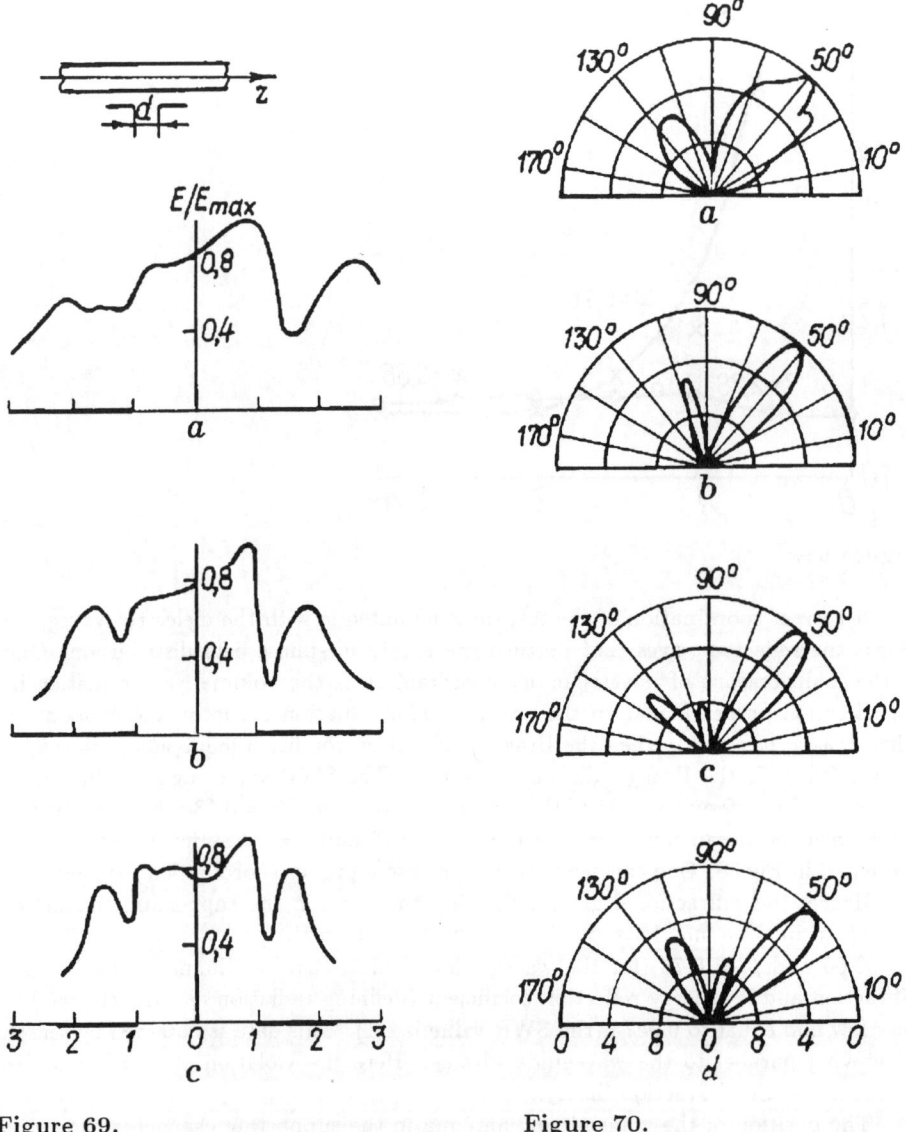

Figure 69.   Figure 70.

Consider first amplitude-phase characteristics and far-field patterns of separate metal strips. The near-field amplitude distributions for two strips having the width 2 mm for different distances between them are presented in Fig.69 when the aimed distance is fixed (letters a, b and c correspond to the distances 1.4 mm, 2.3 mm and 3.3 mm, respectively). The far-field patterns for the same scatterers with a dielectric waveguide are depicted in Fig.70 (the far-field pattern for the single strip is presented in Fig.70a). The analysis of far-field patterns shows that in the strip grating one strip acts as a single radiator. In fact, the main lobe of the separate strip has the angle coordinates of about 50° and the far-field patterns of two strips are stable with respect to the increasing distance between them, although sidelobes may appear (see Fig.70,c,d). Here the strips are very close to the waveguide surface $b \approx 0.1 \div 0.2\lambda_0$ and, therefore, the intensive radiation is observed. Note that it is necessary to eliminate the influence of the waveguide waves of the dielectric waveguide on the sensor field, when the sensor is close to the strips.

Two bars form one radiating element for the grating consisting of (rectangular) bars. The field distribution in the vicinity of two bars and their far-field patterns confirm the generation of an additional waveguide wave which contributes to the near field of the considered structure. The waveguide surface wave is scattered by the bar edges turned to the waveguide surface. The field phases on the edges are different, which yields excitation of the higher order waveguide waves in the cavity between the bars.

The far-field pattern with the field maximum oriented along the the slot axis corresponds to the amplitude distribution in the interval $d < \lambda_0/2$. Two aperture maxima divide the main part of the far-field pattern in two lobes and cause the angle variation ($\pm 45° - 50°$) in the direction of radiation.

Pictures of amplitude-phase distributions on the aperture between two bars are obtained for two important cases. For $d' = d/\lambda_0 = 1.85$ the phase front in the vicinity of the aperture strongly differs from the plane front and corresponds to the wave radiating along the slot axis. At the distance of about λ_0 the phase front becomes homogeneous and determines the single direction of radiation to the free space. When d' increases up to the value of 0.9, two types of oscillations appear in the waveguide region. These oscillations are shown on the phase portrait. The phase splash indicates the minimal field amplitude and divides the phase picture in two parts with smooth phase variations. Consequently, the energy flow consists of two tubes and each is transformed to separate lobes of the far-field pattern.

We have also investigated how surface waves propagating in the dielectric waveguide are transformed to volume waves by two circular bars.

It is known (see section 15) that the problem of the surface wave diffraction by a strip grating is solved under the assumption of the given surface field. We treated this problem experimentally and registered the amplitude and phase field distributions for the grating with the period $l = 4$ mm ($\kappa = 0.95$), the stripwidth $l - d = 0.8$ mm, $\lambda_0 = 4.2$ mm and $\beta = 0.79$. In line with condition (3.26), two volume waves with the indices $n = -1$ and $n = -2$ must radiate to the free space for the grating with such a set of parameters. Experimental pictures of the lines of equal phases and amplitudes show qualitative and quantitative coincidence of the experimental and calculated data (similar to the results of section 15), in particular, the spatial domains where the singularities of the scattered field appear. Note that such "accumulation points" of the field determined experimentally are very close to the calculated points. The lines of equal phases perpendicular to the waveguide plane correspond to the surface wave phase fronts. The presence of two radiating harmonics with $n = -1, -2$ leads to the field interference: the tubes of the phase lines condensation appear as well as the amplitude minima, which is also confirmed experimentally. At the distance of $1 - 1.5\lambda_0$ from the grating the radiation volume field is formed, which is separated from the surface field of the dielectric waveguide. For smaller distances the amplitude-phase distribution has a more complicated character and it is difficult to reconstruct far-field patterns of the grating if to use these data.

We note that the amplitude-phase distribution close to this open structure looks like the picture an the oblique plane wave diffraction by the strip diffractional grating [5].

The reflecting comb gratings made of rectangular bars find broad applications in the millimeter and submillimeter wave technology. The elements of such gratings have the form of waveguide cavities and may produce resonant fields and, consequently, far-field patterns radiating in the direction perpendicular to the grating. The simplest method to study the comb radiation is based on the leaky wave model of the open structure waveguide–grating. One of the results obtained within the frame of this model is the decrease of the incident field amplitude along the axis directed from one element to another, which causes the non-symmetrical amplitude field distribution. The latter may be approximately described by the expression $E = E_0 e^{-\delta 2z/a}$, where δ defines the field weakening in the interval $L = 2a$ (L is the longitudinal comb size). Here the gradient of the near field weakening along the radiating comb depends on the degree of electrodynamical communication between the grating and the waveguide. Therefore, let us first of all investigate experimentally the near field characteristics of the grating aperture with respect to the parameter describing the degree of communication.

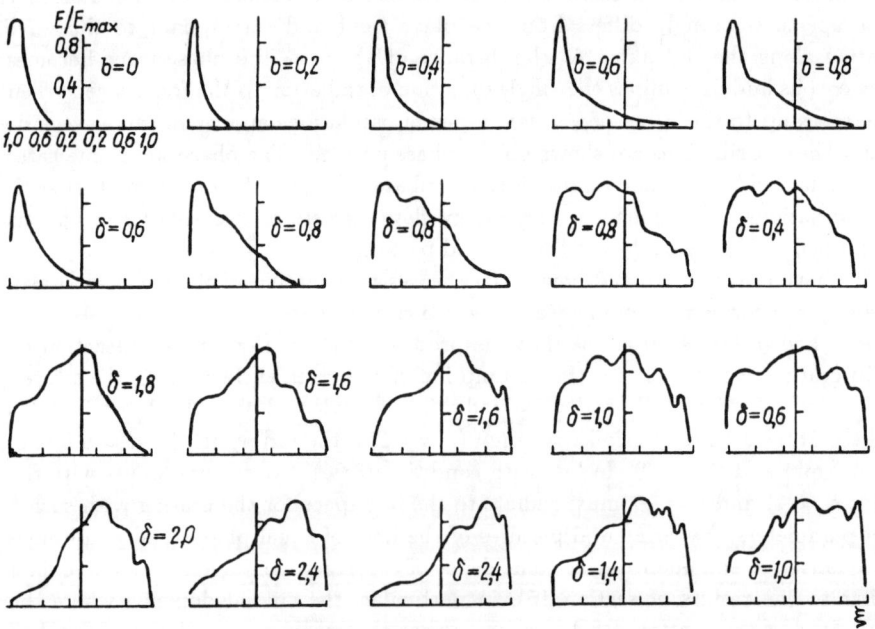

Figure 71.

It is known that the degree of this electrodynamical communication is determined by the distance between the objects and the comb element and is proportional to the amplitude of the waveguide surface wave. This amplitude decreases exponentially outside the comb surface. Experiments are carried out with the help of the plant presented in Fig.67 (where the comb aperture is $L = 120$ mm $\approx 30\lambda_0$, $l = 3$ mm, $\kappa = 0.75$, $h/\lambda_0 \approx 0.7$ and $\beta = 0.76$). The measurement results that are presented in Fig.71 (where $\xi = 2z/L$) correspond to the case when the surface wave propagates from $-\xi$ to ξ.

We divide the curves in Fig.71 in two families. In the first family corresponding to the homogeneous communication the amplitude field distribution is caused by the variation of the degree of communication between the waveguide and the grating due to the comb displacements parallel to waveguide surface. In the second family (the inhomogeneous communication) the amplitude distribution is determined by the simultaneous parallel and oblique (wedge-type) comb displacements.

In the first case (the first line in Fig.71) the energy is taken from every element under equal conditions and is determined by the radiator position with respect to the surface wave field. The first field distribution corresponds to the maximal communication ($b = 0$) when the comb lies on the dielectric waveguide and the sharp amplitude increase is observed in the beginning of the communication domain. The substantial field decrease along the aperture is evidence of the strong communication in the structure waveguide–grating, which leads to a practically complete radiation of the energy transferred to the grating by the waveguide. Note that the field covers approximately one-third of the comb, while the rest of the comb does not participate in the radiation process. When b increases, the field "spreads" on the aperture surface and its decrease becomes less pronounced. The number of the elements involved in the surface-volume wave transformation increases, as well as the grating operation coefficient. However, the main part of the amplitude distribution (for example, at the level 0.5 of the maximum) is concentrated in the small region close to the comb aperture. Hence, the homogeneous communication between the grating and the waveguide does not allow one to obtain optimal radiation characteristics, which might be compared with the radiation parameters of the mirror, lens and other types of antennas.

The inhomogeneous communication considerably broadens the limits of variation of the amplitude distributions on the grating aperture. This type of communication is observed when the comb (or any other grating) is placed at a certain angle to the dielectric waveguide. The influence of the wedge-type communication is illustrated by the data presented in the vertical column in Fig.71 (for $b = 0$ and $\delta = 0.6, 1.8, 2.0$): the aperture field maximum changes but the lateral distribution is always steep and non-symmetrical. One can decrease the slope steepness and create the symmetrical field distribution by varying the aimed distance b and the comb inclination δ. Other field pictures in Fig.71 confirm this assumption.

One can introduce additional quantitative criteria for the analysis of experimental data. In fact, the surface-volume wave transformation is suitable for practical applications only in small intervals of the variation of parameters b and δ (at $\xi = 0$).

The radiation direction for the main maxima of far-field patterns and spatial field distributions in the far zone with respect to the parameter defining the degree of communication are determined with the help of the plant depicted in Fig.67 (the comb with $L = 120$ mm forms far-field patterns with the width $2\Delta_{3dB} \approx 3° \div 4°$ in

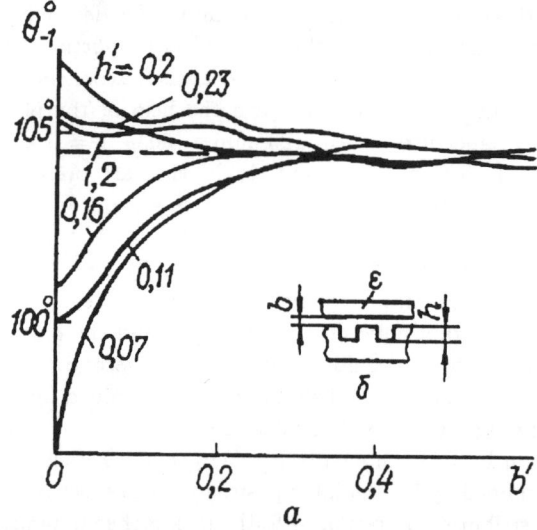

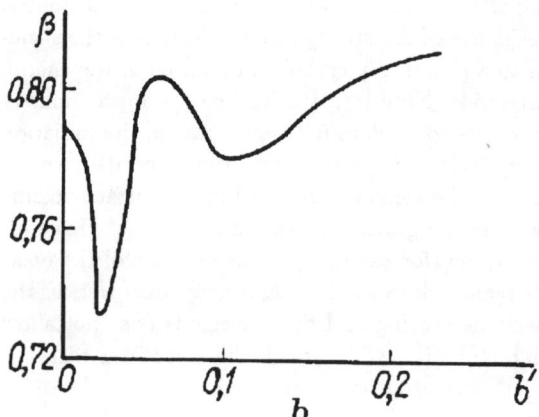

Figure 72.

the E plane). The main maximum is well distinguished on the background of lateral radiation. The radiation parameters have been measured for the system waveguide–grating formed by the elliptic waveguide with the big axis 4.4 mm and $\beta = 0.82$ and the comb grating. The results are presented in Fig.72a for a sequence of combs with the equal cavity depth and the slotwidth $d = 1$ mm (for different values of $h' = h/\lambda_0$ at $\kappa = 0.7$). The dashed lines correspond to the calculated values of the radiation angles and divide the family of curves in two groups situated above and below the calculated curve. The radiation direction (for the given κ) is determined by the phase velocity of the waveguide surface wave. Therefore, all the curves in Fig.72 situated close to each other (with the difference $\pm0.5°$ for $b' = b/\lambda_0 > 0.4$) are evidence of the weak interaction between surface waves and the grating.

The influence of the comb on the waveguide characteristics is checked by measuring $\beta = v_\Phi/c$ as a function of b'. The maximal value of β, which corresponds to the regular waveguide and the radiation angle θ_{-1}, is depicted by the dashed lines in Fig.72a. Here we perform an experimental simulation of the influence of metallic walls on the phase velocity v_Φ in a dielectric waveguide.

If to place the comb in the field of a surface wave, we obtain more complicated dependence of all characteristics that determines the generation of volume waves in the open structure waveguide–grating (due to the presence of resonant cavities). Resonances result in the maximal radiated power. Calculations for the comb with the slotwidth $d = 1$ mm and $\lambda_0 = 4.235$ mm show that $h_{res} = 0.96$ mm. One can see in Fig.72a that the upper region of the curves contain the value h which is a little less than h_{res} ($h > h_{res}$). It may be connected with the assumption concerning the absence of the inverse radiation communication. When analyzing the results presented in Fig.72, we assume that the value of h_{res} lies in the interval 0.7 mm $< h_{res}^{real} < 0.85$ mm. This assumption may be justified by rather simple considerations based on the analysis of concentrated constants: when the depth varies in the vicinity of the resonant value from $h \sim \lambda_0/4$ to $h \sim \lambda_0/2$, the influence of the grating on the waveguide is less pronounced (if to compare with the case $0 < h < h_{res}$).

Non-symmetrical amplitude distributions on the aperture affect the lateral radiation. The "diffused" minimum appears in the far-field pattern (in place of zero) corresponding to the non-symmetrical (exponential) aperture field distribution. The waveguide field non-symmetry is accompanied by the decrease of the minimum depth and stronger lateral radiation [142]. These results stress the necessity of experimental investigations of the secondary radiation characteristics in the structure waveguide–grating, since the secondary radiation is always present and may be caused by non-controlled reflections from surrounding objects.

In various applications one has to provide low lateral radiation and a narrow main lobe (of about 2°) of the radiating structure waveguide–grating. We have established theoretically and experimentally that for definite values of b and δ such far-field patterns may be obtained if to control the degree of communication by varying the slot depth and other parameters of the structure. Similar possibilities appear when this structure operates as a receiver, i.e., when the energy is brought into the waveguide by the comb.

Let us describe the energy characteristics of the structure waveguide–grating. We determine experimentally the dependence of the radiation power (taken from the grating by the comb) on the structure parameters, in particular, on the parameters defining the type and degree of the communication between the dielectric waveguide and the grating.

Let us establish first the influence on the radiated power of the parameter describing the homogeneous communication of the comb placed in the field of the dielectric waveguide. The curves in Fig.73a are constructed for different values of h, the dashed line is calculated by the formula $P \approx e^{-2\alpha\beta}$, where $\alpha = 2\pi\sqrt{\beta^{-2} - 1}/\lambda_0$ is the transversal wavenumber of the waveguide wave (this relation determines the field decay in the direction off the waveguide and does not take into account the comb influence on the total field).

It is convenient to analyze the experimental data by dividing the curves in Fig.73a in two groups, the initial, where $P_{rad}/P_0 \approx 1$ and the final, where the radiated power decreases. The first group remains constant in a certain interval of the values of $b' = b/\lambda_0$, which is caused by the character of the electromagnetic communication

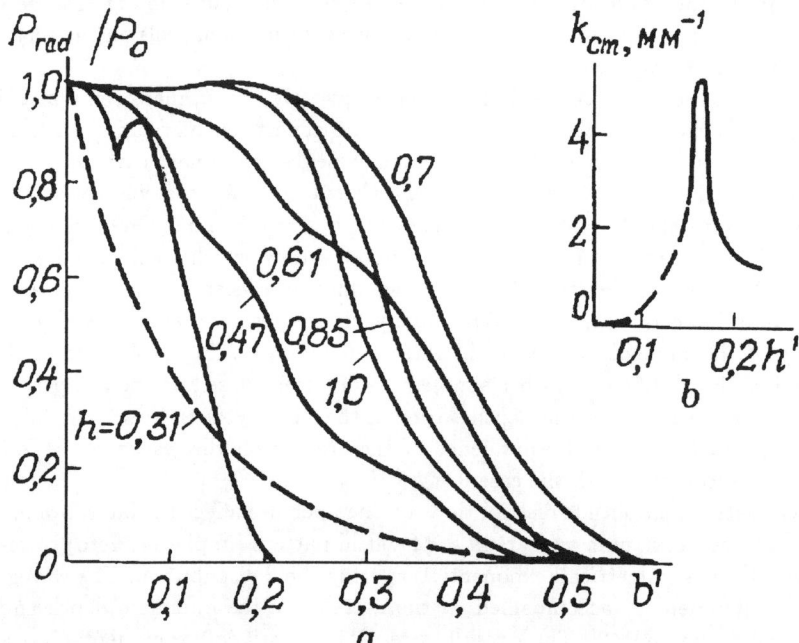

Figure 73.

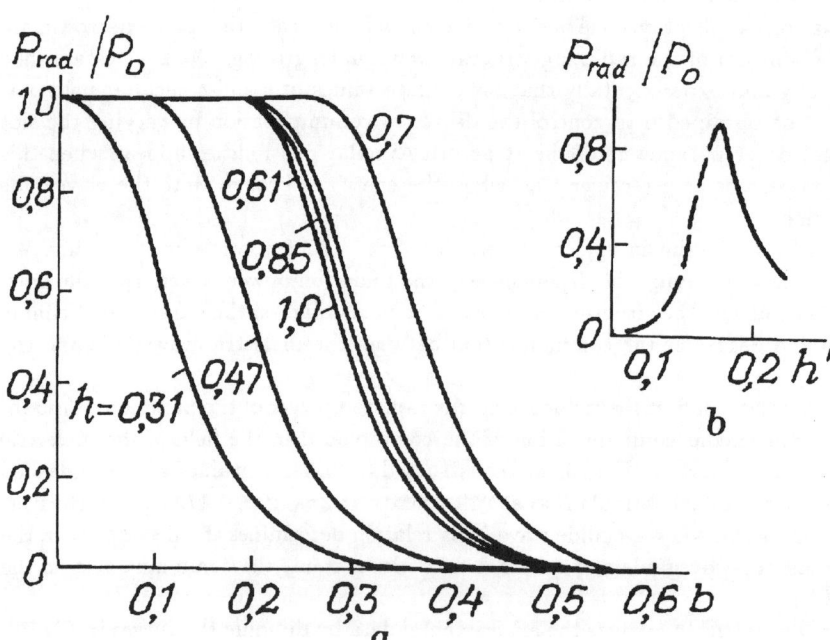

Figure 74.

between the grating and the waveguide. When the degree of communication increases, all the waveguide power may be radiated from the first region of the grating, and the near field in this region will have the form of a spot. Further weakening of the communication (as a result of the variation of b' or h') yields a greater efficiency of the communication domain. The fall of the radiated power is observed when all the grating elements take part in the radiation process. The experimental curves in Fig.74 show the power radiated by the Nth comb elements ($h' = 0.16$). If the communication is strong (note that its level is controlled only by b'), it is sufficient to take the comb with ten or twelve elements when $b' = 0 \div 0.07$ and the whole power will be radiated out of the waveguide. In the case of weak communication the number of elements participating in the radiation process increases.

The curves in Fig.73a (calculated for $P_{\mathrm{rad}}/P_0 = f(b')$) show that the size of the saturation domain depends on the depth of the slot. One can approximately analyze this process by applying the coupling wave method and eliminating the inverse communication between the waveguide and the grating. Simple computations give the formula $P_{\mathrm{rad}} = P_0(1 - e^{k_{\mathrm{cm}} L e^{-2\alpha x}})$, which determines the power radiated by the grating of the length L situated at the distance $x = b$ from the waveguide surface for a certain value of the communication coefficient k_{cm}. The experimentally obtained values of k_{cm} for different combs are listed below.

$h' = h/\lambda_0$	0.07	0.11	0.14	0.16	0.20	0.23
b'	0.52	0.85	1.32	1.58	1.28	1.24
k_{cm}	0.054	0.22	1.64	5.0	1.38	1.17

The oscillating curve of the function $k_{\mathrm{cm}} = f(h')$ shown in Fig.73b corresponds to the resonances of the separate grating cavity.

The radiation power of the structure waveguide–grating is calculated with the help of the values of k_{cm}. A comparison of experimental (Fig.73) and theoretical curves reveals their coincidence. Hence, for the given comb with the length L and the known value β of the waveguide we find the coefficient k_{cm} that defines the degree of communication of the dielectric waveguide with the grating. Then we calculate by the formula for P_{rad} the effective energy absorption of the comb having an arbitrary length (and the same values l and h). Note that the degree of the surface-volume wave transformation is 90% in the wide range of b and δ. In practice one can reach the 100% transformation by using the specially chosen grating placed close to the waveguide.

It is known [201] that when the waveguide surface wave is scattered by the grating and the condition $2\kappa = n\beta$ holds, the Bragg's diffraction takes place and the propagating surface wave is completely reflected from the grating. This phenomenon has a resonant character and is observed experimentally, in particular, when the SWR is measured. These measurements are carried out in the regular waveguide section by the plant depicted in Fig.67 with some modifications. The use of the modified plant enables one to take into account the junction between the generator and the regular part of dielectric waveguide, which is introduced in order to eliminate the waveguide excitation by the "additional grating" (here we have an example of the solution to the "inverse" problem, when the grating volume wave is transformed to the waveguide surface wave). The degree of the waveguide–grating coordination is determined by the aimed distance b and the slot depth h. The SWR is measured for different values of b and the generator parameter λ_0.

The measurement data obtained for the comb with $l = 1.7$, $d = 0.5$ and $\lambda_{0\,\text{res}} = 4.14$ mm reveal the influence of the slot depth on the SWR values. For small depths ($h = 0.2$ mm) the SWR is also small, although this ratio has a distinct resonant character and attains the maximum at $\lambda_0 = 4.35$ mm ($b = 0.2$ mm) shifting in the direction of greater wavelengths. The increase of the slot depth is accompanied by increasing interaction between the waveguide and the grating, which is confirmed by the SWR behaviour in the same wavelength interval. According to the Bragg condition, the maximal interaction is possible when the slot depth takes the resonant value $h = 0.8$ mm (corresponding to the resonance with respect to the spectral parameter h/λ_0. The SWR level of about 20 dB is registered in the comparatively small wavelength range ($\lambda_0 \approx 4.16$ mm). Here the resonances of the Bragg's diffraction are mixed with the resonances of the comb cavities. Our experiments show that the correct choice of the resonant parameters allows us to obtain high reflection coefficients for a fixed wavelength. The resonant behaviour of the reflection coefficient $\Gamma = (\text{SWR} - 1)/(\text{SWR} + 1)$ in the considered open structure finds specific applications in the millimeter and submillimeter wave technology.

We use the Bragg's diffraction to form resonant far-field patterns with the main maximum directed perpendicular to the comb. Here the resonant scattering occurs as a result of the transformation of surface waves to volume wave when $k_{\text{cm}} = \beta$ and, according to formula (3.26), the minus first radiation harmonic radiates at the angle $90°$. The Bragg radiation yields the generation of the strong wave in the communication domain, which is reflected back to the dielectric waveguide and radiates to the free space. The radiation directions of the propagating and reflected waves coincide, which yields the superposition of their far-field patterns.

The degree of the waveguide–grating coordination and the far-field parameters (in particular, the width of the far-field pattern at $\lambda_0 = 3.7 \div 5.1$ mm) are registered by different types of sensors. The results reveal the resonance character of the reflection coefficient as a function of λ_0. Deviations of $2\Delta\theta$ from the average value is determined by the superposition of the far-field patterns of the direct and reflected waves at the moment when the reflected wave intensity is high. A further increase of λ_0 destroys the far-field pattern of the reflected wave, and only one far-field pattern is left. If to consider the surface-volume transformation from the antenna viewpoint, the radiation of volume waves perpendicular to the surface is undesirable. The operation domain $\theta_{-1} \neq 90°$ is preferable and especially when the antenna must be coordinated with a feeder.

The radiation direction of the main maximum of the grating far-field pattern is determined by known relations [6]. Let us use these relations to find the radiation angle as a function of the wavelength applying the results of measurements in the eight-millimeter wavelength range for the comb with $L \approx 30\lambda_0$. Since the dielectric waveguide is characterized by the dispersion, we consider the structure waveguide-grating in terms of the angle-frequency sensitivity. The corresponding theoretical curve is depicted in Fig.75 by the bold line and the experimental data is marked by points. The angle sensitivity is determined in Ref. [200] as

$$A = \frac{\partial\theta}{\left(\frac{\partial\lambda}{\lambda}\right)} - \frac{0.537}{\cos\theta(\sin\theta - \gamma_b)},$$

where $\gamma_b = c/v_b$ is the group velocity deceleration of the surface wave (the coefficient 0.537 is used to transfer the angle-frequency sensitivity to the relative percent units)

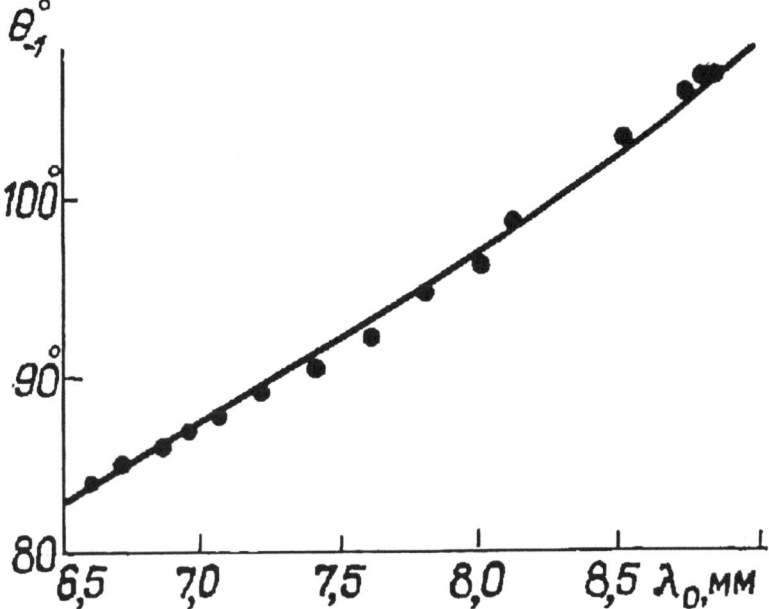

Figure 75.

and γ_b is determined from the dispersion relation of the dielectric waveguide. In our experiments λ_0 was taken to be equal to 8.07 mm. The angle deviation sector for the far-field patterns is $\theta_{max} - \theta_{min} = 18°$ (when λ_0 has a 10% variation). Here the angle-frequency sensitivity A is 0.8 relative percent units of the frequency interval. Note that the angle-frequency sensitivity of the waveguide-slot antenna made on the basis of the rectangular H_{10} waveguide oscillates from one tenth to several units of degrees in the interval corresponding to one percent of the frequency variation. We may conclude that the open structure formed by the dielectric waveguide and the grating may be successfully applied in the millimeter and submillimeter wavelength ranges as an antenna that provides the possibility of the frequency scanning.

References

1. R.G. Mirimanov. *Millimeter and submillimeter waves.* Izd. Inostr. Literatury, Moscow, (1959) (in Russian).
2. N.D. Devyatkov and M.B. Golant. The ways of development of electronic devices in the millimeter and submillimeter wavelength ranges. *Radiotekhnika i Elektronika.* **12**, 1973-1978 (1967).
3. L.A. Weinshten and V.A.Solncev. *Lectures on high-frequency electronics.* Sov. Radio, Moscow, (1973) (in Russian).
4. A.Ya. Usikov. Investigations and research projects in microwave electronics carried out at the Institute of Radiophysics and Electronics of the Ukrainian Academy of Sciences. *Elektron. Tekhn. Elektronika SVCh.* **1**, 39-49.
5. V.P. Shestopalov. *Method of Riemann-Hilbert problem in the theory of diffraction and propagation of electromagnetic waves.* Kharkov University Press, Kharkov, (1971) (in Russian).
6. V.P. Shestopalov. *Diffractional electronics.* Kharkov University Press, Kharkov, (1976) (in Russian).
7. V.P. Shestopalov. *Summation-type equations in the modern theory of diffraction.* Naukova Dumka, Kiev, (1983) (in Russian).
8. V.P. Shestopalov. *Physical foundations of millimeter and submillimeter technology. Vol. 1. Open structures.* Naukova Dumka, Kiev, (1985) (in Russian).
9. V.P. Shestopalov. *Physical foundations of millimeter and submillimeter technology. Vol. 2. Sources. Element base. Radio-systems.* Naukova Dumka, Kiev, (1985) (in Russian).
10. V.P. Shestopalov. *Spectral theory and excitation of open structures.* Naukova Dumka, Kiev, (1987) (in Russian).
11. V.P. Shestopalov. *Morse critical points of dispersion equations.* Naukova Dumka, Kiev, (1992) (in Russian).
12. V.P. Shestopalov, A.A. Vertij, G.P. Ermak, B.K. Skrynnik, G.I. Khlopov and A.I. Tsvik. *Generators of diffractional radiation.* Naukova Dumka, Kiev, (1985) (in Russian).
13. V.P. Shestopalov. On non-linear processes in generators of diffractional radiation-lasers on free electrons. *Dokl AN SSSR.* **261**, 1115-1118 (1981).
14. L.A. Weinschtein. *Open resonators and open waveguides.* Sov. Radio, Moscow, (1966) (in Russian).
15. Yu.V. Shestopalov. Justification of the spectral method for calculating normal waves of transmission lines. *Diff. Uravnenija.* **16**, 1504-1512 (1980).

16. Yu.V. Shestopalov. Properties of the spectrum of a class of non-self-adjoint boundary value problems for the systems of Helmholtz equations. *Dokl. AN SSSR.* **252**, 1108-1111 (1980).

17. Yu.V. Shestopalov. The spectrum of normal waves of slotted transmission lines. *Radio Eng. and Electron. Phys.* **26**, 2064-2073 (1981).

18. A.S. Il'inskij and Yu.G. Smirnov. Mathematical modelling of the propagation of electromagnetic waves in a slotted transmission line. *USSR Comput. Maths. Math. Phys.* **27**, 163-170 (1987).

19. Yu.V. Shestopalov. On the theory of cylindrical resonators. *Math. Methods in the Appl. Sciences.* **14**, 335-375 (1991).

20. A.S. Il'inskij and Yu.V. Shestopalov. *Applications of the methods of spectral theory in the problems of wave propagation.* Moscow University Press, Moscow, (1989) (in Russian).

21. A.E. Poyedinchuk. Point spectrum of one class of open cylindrical structures. *Dokl. AN USSR. Ser. A.* no. 8, 48-52 (1983).

22. A.E. Poyedinchuk. On a spectral theory of open two-dimensional resonators with dielectric inclusions. *Dokl. AN USSR. Ser. A.* no. 1, 46-49 (1984).

23. A.V. Brovenko. Numerical algorithm for calculating spectral characteristics of two-dimensional open resonator with inhomogeneous dielectric inclusions. *Dokl. AN USSR. Ser. A.* no. 1, 51-54 (1988).

24. V.P. Shestopalov and Yu.K. Sirenko. *Dynamical theory of gratings.* Naukova Dumka, Kiev, (1989) (in Russian).

25. V.N. Koshparenok, P.N. Melezhik, A.E. Poyedinchuk and V.P. Shestopalov. Interaction of waves in open resonators. *Dokl. AN SSSR.* **279**, 1114-1118 (1984).

26. V.N. Koshparenok, P.N. Melezhik, A.E. Poyedinchuk and V.P. Shestopalov. Spectral theory of open two-dimensional resonators with dielectric inclusions. *Zhurn. Vych. Matem. i Matem. Fiziki.* **25**, 562-573 (1985).

27. Yu.V. Svishchev, Yu.A. Tuchkin and V.P. Shestopalov. On the resonance readjustment of oscillations in the open resonator with spherical mirrors. *Dokl. AN SSSR.* **312**, 1111-1114 (1990).

28. I.Ts. Gohberg and M.G. Krein. *Introduction to the theory of linear non-self-adjoint operators.* Nauka, Moscow, (1965) (in Russian).

29. M.G. Krein and G.Ya. Lubarskij. On the theory of transmission strips of periodical waveguides. *Prikladnaya Matem. i Mekhanika.* **25**, 24-37 (1961).

30. R.A. Valitov. *Submillimeter wave technology.* Sov. Radio, Moscow, (1969) (in Russian).

31. R.A. Valitov, S.F. Dubko, V.V. Kamyshan and V.P. Sheiko. On a method of the measurement of the field distributions in an open resonator. *Zhurn. Experim. i Teoretich. Fiziki.* **47**, 1173-1177 (1964).

32. A.A. Petrushin, I.M. Balaklitskij and V.P. Shestopalov. The plant for investigation of the electromagnetic field in the open resonator of the millimeter wavelength range. *Pribory i Tekhnika Experimenta.* no. 2, 147-149 (1970).

33. D. Auston, R. Primich and R. Khayami. The use of Fabry–Perot resonators for plasma microwave diagnostics. In: *Quasioptica,* 387-423. Mir, Moscow, (1966).

34. A.P. Bazhulin, E.A. Vinogradov, N.A. Irisova and S.A. Fridman. Obtaining of visual image of radio-radiation in the millimeter wavelength range. *Pis'ma v Zhurn. Experim. i Teoretich. Fiziki.* **8**, 261-263 (1968).

35. E.A. Vinogradov, N.A. Irisova, A.A Lazarev et al. Application of thermo-luminophor sensor for investigation of the fields in open resonators. *Radiotekhnika i Elektronika.* **23**, 936-939 (1978).

36. N.A. Irisova. Metric of submillimeter waves. *Vestnik AN SSSR.* **63**, 63-71 (1968).

37. R.A. Valitov, S.F. Dubko, B.I. Makarenko et al. *Measurements in millimeter and submillimeter waves.* Radio i Svyaz', Moscow, (1984) (in Russian).

38. A.A. Vertij. *Resonance method for the study of fields in the millimeter and submillimeter wavelength ranges. Candidate Thesis,* (1973) (in Russian).

39. G.A. Andreev, O.V. Bazarnij, A.S. Glauberman et al. Visualization and transmission of fields in the millimeter wavelength range. *Zarubezhn. Radioelektronika.* no. 11, 3-27 (1984).

40. H. Reichardt. Ausstralungsbedingungen für die Wellengleichung. *Ab. Math. Sem. University Hamburg.* **24**, 41-53 (1960).

41. H. Hönl, A. Maye and K. Westphal. *Theorie der Beugung.* Springer-Verlag, Berlin, (1961).

42. H. Schwartz. Über die 1, 2 und 3 aubere Randwertahchfgabe der Schwingungsgleichung. *Mathematische Nachrichten.* **28**, 337-363 (1965).

43. V.P. Koshparenok, P.N. Melezhik, A.E. Poyedinchuk and V.P. Shestopalov. Rigorous theory of open two-dimensional resonators with dielectric inclusions. *Izv. Vuzov. Radiofizika.* **28**, 1311-1321 (1985).

44. V.P. Koshparenok, P.N. Melezhik, A.E. Poyedinchuk and V.P. Shestopalov. Method of Riemann–Hilbert problem in the spectral theory of open two-dimensional resonators. *Radiotekhnika i Elektronika.* **31**, 271-278 (1986).

45. I.T. Selezov and V.V.Yakovlev. *Diffraction of waves by symmetrical inhomogeneities.* Naukova Dumka, Kiev, (1971) (in Russian).

46. N. Moskalev and A.M. Stefanovskij. *Plasma diagnostics by means of open cylindrical resonators.* Energoatomizdat, Moscow, (1985) (in Russian).

47. M. Reed and B. Simon. *Methods of modern mathematical physics. IV: Analysis of operators.* Academic Press, New York, (1978).

48. B.V. Shabat. *Introduction to complex analysis. Vol. 1.* Nauka, Moscow, (1976) (in Russian).

49. G.M. Vainikko and O.O.Karma. On the rate of convergence of approximate methods in eigenvalue problem non-linear with respect to the spectral parameter. *Zhurn. Vych. Matem. i Matem. Fiziki.* **14**, 1393-1408 (1974).

50. I.Ts. Gohberg and E.I. Sigal. Operator generalization of logarithmic residue theorem and Roushet's theorem. *Matem. Sb.* **84**, 607-629 (1971).

51. R. Courant and D. Hilbert. *Methods of mathematical physics. Vol. II.* Interscience Publ, USA, (1962).

52. L.V. Kontorovich and G.P. Akilov. *Functional analysis.* Nauka, Moscow, (1984) (in Russian).

53. V.M. Babich and V.S. Buldyrev. *Asymptotical methods in the problems of short wave diffraction.* Nauka, Moscow, (1972) (in Russian).

54. B.F. Lazutkin and B.Ya. Terman. On a number of "bouncing ball" modes. *Zap. nauchn. seminara LOMI. Matem. voprosy teorii rasprost. voln.* **117**, 172-181 (1981).

55. A.G. Fox and T. Li. Resonant modes in master interferometer. *Bell. Syst. Techn. J.* **40**, 453-488 (1961).

56. Yu.A. Anan'ev, V.V. Lubimov and I.B. Orlov. The mode deformation in open resonators with plane mirrors. *Zhurn. Tekhn. Fiziki.* **39**, 1872-1880 (1969).

57. V.B. Shteinshleiger. Phenomena in electromagnetic resonators close to the points of coincidence of fundamental frequencies. *Dokl. AN SSSR.* **59**, 669-672 (1949).

58. V.B. Shteinshleiger. *Phenomena of the wave interaction in electromagnetic resonators.* Sov. Radio, Moscow, (1966) (in Russian).

59. D.M. Eidus. Principle of limiting amplitude. *Uspekhi Matem. Nauk.* **24**, 91-156 (1969).

60. A.G. Sveshnikov. On the radiation principle. *Dokl. AN SSSR.* **75**, 137-141 (1950).

61. B.R. Weinberg. On eigenfunctions corresponding to the poles of resolvent's analytical continuation through the continuous spectrum. *Matem. Sb.* **87**, 293-308 (1972).

62. V.N. Koshparenok, P.N. Melezhik and V.P. Shestopalov. On intertype communication of oscillations in the cylinder with longitudinal slots. *Radiotekhnika i Elektronika.* **24**, 2350-2353 (1979).

63. P.N. Melezhik, A.E. Poyedinchuk, Yu.A. Tuchkin and V.P.Shestopalov. On analytical nature of the phenomenon of intertype communication of eigenoscillations. *Dokl. AN SSSR.* **300**, 1336-1340 (1988).

64. B.V. Shabat. *Introduction to complex analysis. P. II. Functions of several variables.* Nauka, Moscow, (1976) (in Russian).

65. V.I. Arnold, A.N. Varchenko and S.M. Gussein-Zade. *Singularities of differentiable mappings.* Nauka, Moscow, (1982) (in Russian).

66. J. Milnor. *Singular points of complex hypersurfaces.* Princeton University Press, Princeton, N.Y., (1968).

67. I.M. Balaklitskij, V.G. Kurin and B.K. Skrynnik. The plant for the study of the millimeter-wave field distributions in open resonators. *Pribory i Teknika Experimenta.* no. 5, 107-109 (1974).

68. A.P. Koretskij and V.P. Shestopalov. Resonant properties of elliptical open resonator. *Dokl AN SSSR.* **307**, 857-861 (1989).

69. G.D. Bayd and J.P. Gordon. Confocal multimode resonator for millimeter waves through optical wavelength measure. *Bell Syst. Techn. J.* **40**, 489-508 (1971).

70. A.A. Vertij and V.P. Shestopalov. Visualization of the amplitude-phase structure of electomagnetic fields in the millimeter and submillimeter wavelength ranges. *Pribory i Teknika Experimenta.* no. 2, 145-147 (1973).

71. V.M. Ginzburg and V.M.Meschankin. Measurement of amplitude-phase radio-field distributions by holographical methods. *Radiotekhnika i Elektronika.* **18**, 221-225 (1973).

72. A.A. Vertij, V.N. Derkach, I.V. Ivanchenko et al. New method for the measurement of the phase field structure in open resonators. *Vestnik Kharkov. Universiteta. Radiofizika i Elektronika.* **7**, 37-40 (1978).

73. A.A. Vertij, I.V. Ivanchenko V.P. Shestopalov et al. Method for the measurement of the phase field structure in quasi-optical open resonators. *Dokl AN USSR. Ser. A.* no. 10, 992-925 (1977).

74. V.S. Letokhov. Diffractional losses of open resonator and mirrors with absorbing strip. *Zhurn. Tekhn. Fiziki.* **35**, 809-812 (1965).

75. A.A. Vertij, V.N. Derkach, N.A. Popenko et al. Generator of diffractional radiation with diffractional output of energy. *Dokl. AN USSR. Ser. A.* no. 4, 354-356 (1976).

76. E.L. Kosarev and Yu.M. Tsipenyuk. Forsed oscillations of an open resonator connected with the waveguide by a small hole. *Elektronika Bol'shikh Moshchnostej.* no. 5, 105-116 (1968).

77. I.V. Ivanchenko. Automatic measuring device of the amplitude-phase field structure in open resonators. *Pribory i Tekhnika Experimenta.* no. 3, 135-136 (1979).

78. A.A. Vertij, V.N. Derkach and V.P. Shestopalov. Method for obtaining spectra of the spatial structure of the millileter and submillimeter electromagnetic fields. *Dokl. AN USSR. Ser. A.* no. 3, 247-250 (1978).

79. A.A. Vertij, V.N. Derkach, V.B. Krasuyk et al. The study of degenerated oscillations of quasi-optical open resonator formed by rectangular mirrors with the aperture. *Izv. Vuzov. Radiofizika.* **24**, 76-83 (1981).

80. A.I. Goroshko and E.M. Kuleshov. The use of empty dielectric beam waveguides of the millimeter and submillimeter wavelength ranges. *Radiotekhnika.* **21**, 215-219 (1972).

81. N.P. Gavelya, A.D. Istrashkin, Yu.K. Muravjev et al. *Antennas.* Voen. Krasnoznam. Acad. Svyazi, Leningrad, (1963) (in Russian).

82. D.B. Kanareikin, N.F. Pavlov and V.A. Potekhin. *Polarization of radar signals.* Sov. Radio, Moscow, (1966) (in Russian).

83. M. Born and A. Wolf. *Foundations of optics.* Nauka, Moscow, (1970) (in Russian).

84. A.A. Vertij, I.V. Ivanchenko, Yu.P. Popkov et al. *Polarimetry of resonance quasi-optical pencils. Preprint № 134.* Inst. of Radiophysics and Electronics, Ukr. Acad. of Sciences, Kharkov, (1979) (in Russian).

85. A.A. Vertij, A.O. Petrushin and V.P. Shestopalov. Novel experimental technique for millimeter and submillimeter waves. *Vestnik AN SSSR.* no. 9, 13-20 (1974).

86. V.G. Andreev. Open volume resonator for electomagnetic oscillations in the radio wavelength range. *Zhurn. Tekhn. Fiziki.* **36**, 1851-1859 (1966).

87. Yu.V. Troitskij. Open resonator with absorbing metallic film. *Radiotekhnika i Elektronika.* **14**, 1641-1647 (1969).

88. V.S. Averbakh, S.N. Vlasov and N.N. Sherokova. Experimental investigation of an open resonator with non-spherical mirrors. *Izv. Vuzov. Radiofizika.* **11**, 1393-1397 (1968).

89. O. Flamer. *Tables of wave spheroidal functions.* Svyaz', Moscow, (1962) (in Russian).

90. A.A. Vertij, I.M. Karnaukhov and V.P. Shestopalov. *Polarization of atomic nuclei by millimeter waves.* Naukova Dumka, Kiev, (1990) (in Russian).

91. P.E. Krasnushkin. Normal waves. In: *Physical Aencyclopedical Dictionary.* Vol. 3, 435-439. Entsiklopediya, Moscow, (1963).

92. B.Z. Katsenelenbaum. *High-frequency electrodynamics.* Nauka, Moscow, (1966) (in Russian).

93. A.I. Nosich, A.E. Poyedinchuk and V.P. Shestopalov. Discrete spectruum of eigenwaves of the open partially shielded dielectric pivot. *Dokl. AN SSSR.* **283**, 1165-1168 (1985).

94. G.C. Sherman and C.H. Hennessy. Complex TM modes in a generalized Goubau line with arbitrary conductivities. *IEEE Trans. Anten. and Propag.* **37**, 553-562 (1983).

95. I.Ts. Gohberg. On linear operators analytical with respect to the parameter. *Dokl. AN SSSR.* **178**, 629-633 (1951).

96. A.I. Nosich. Calculation of spectra of electromagnetic eigenoscillations and eigenwaves in open waveguides. *Dokl. AN USSR. Ser. A.* no. 5, 51-55 (1986).

97. V.V.Shevchenko. Object classification of waves guided by regular open waveguides. *Radiotekhnika i Elektronika.* **14**, 1768-1772 (1969).

98. B.Z. Katsenelenbaum. Symmetrical excitation of infinite dielectric cylinder. *Zhurn. Tekhn. Fiziki.* **19**, 1168-1181 (1949).

99. A.A. Kirilenko, S.L. Senkevich, Yu.K. Sirenko and B.G. Tysik. On reconstruction of scattering matrices of waveguides and periodical structures by the spectrum of complex frequencies. *Radiotekhnika i Elektronika.* **33**, 468-473 (1989).

100. A.I. Nosich, A.E.Svezhentsev and V.P. Shestopalov. Spectrum of quasi-T-waves in partially shielded dielectric pivot. *Dokl. AN SSSR.* **289**, 88-92 (1986).

101. A.I. Nosich, A.E.Svezhentsev and V.P. Shestopalov. Electrodynamic analysis of discrete spectrum of eigenwaves of partially shielded dielectric pivot. *Radiotekhnika i Elektronika.* **32**, 2262-2272 (1987).

102. A.I. Nosich and A.E.Svezhentsev. Physical features of the wave propagation in cylindrical slot and strip lines. *Radiotekhnika i Elektronika.* **34**, 1633-1641 (1989).

103. L.A. Weinshtein. *Electromagnetic waves.* Sov. Radio, Moscow, (1953) (in Russian).

104. G.I. Veselov, S.G. Semenov and A.A. Blagoveshchentskij. Propagation of hybrid waves in the circular waveguide with a dielectric pivot. *Radiotekhnika i Elektronika.* **28**, 2186-2194 (1983).

105. G.I. Veselov, S.G. Semenov and A.A. Blagoveshchentskij. Dispersion properties of the circular waveguide strongly perturbed by a dielectric pivot. *Radiotekhnika i Elektronika.* **31**, 815-817 (1986).

106. A.E. Svezhentsev. *Investigation of a spectrum of eigenwaves of multi-layered cylindrical slot and strip transmission lines. Preprint № 95.* Inst. of Radiophysics and Electronics, Ukr. Acad. of Sciences, Kharkov, (1986).

107. A.E. Svezhentsev. Propagation of electromagnetic waves in multi-layered metallic-dielectric transmission lines. *Vestnik Kharkov. Univers.. Radiofizika i Elektronika.* **307**, 33-37 (1987).

108. A.I. Nosich, A.E. Poyedinchuk and V.P. Shestopalov. Existence of discrete spectrum and singular types of waves in open waveguides. *Dokl. AN SSSR.* **293**, 587-591 (1986).

109. A.E. Svezhentsev. Phenomenon of intertype interaction of surface waves in partially shielded dielectric pivot. *Dokl. AN USSR. Ser. A.* no. 7, 58-62 (1986).

110. V.V. Kryzhanovskij, A.I. Nosich, A.E. Svezhentsev and V.P. Shestopalov. Theoretical and experimental study of eigenwaves in cylindrical strip line. *Dokl. AN USSR. Ser. A.* no. 10, 41-45 (1986).

111. L. Felsen and N. Marcuvitz. *Radiation and scattering of waves. Vol. 2.* Mir, Moscow, (1978) (in Russian).

112. E.I. Veliev, A.I. Nosich and V.P. Shestopalov. Propagation of electrodynamic waves in circular waveguide with longitudinal slot. *Radiotekhnika i Elektronika.* **22**, 466-473 (1977).

113. V.V. Nikol'skij. *Electrodynamics and propagation of radio waves.* Nauka, Moscow, (1973) (in Russian).

114. G.I. Komar' and V.P. Shestopalov. *Methods of investigating the characteristics of slot transmission lines. Preprint № 229.* Inst. of Radiophysics and Electronics, Ukr. Acad. of Sciences, Kharkov, (1983).

115. T. Tamir and A.A. Oliner. Guided complex waves. Part 2. Fields at an interface. *Proc. IEE.* **110**, 30-324 (1963).

116. V.V. Sedykh. *Strip lines and high-frequency devices.* Vishcha Shkola, Kharkov, (1974) (in Russian).

117. G.I. Komar'. Radiation of leaky waves from circular cylinder with longitudinal slot. In: *Physics and technology of the millimeter and submillimeter waves,* 138-146. Naukova Dumka, Kiev, (1983).

118. G.I. Komar', A.I. Nosich and V.P. Shestopalov. Image slot line. *Dokl. AN USSR. Ser. A.* no. 5, 57-60 (1983).

119. I.M. Frank and I.E. Tamm. Coherent radiation of fast electrons. *Dokl. AN SSSR.* **14**, 109-112 (1937).

120. G.I. Komar' and V.P. Shestopalov. Leaky waves of open waveguide structures. *Dokl. AN SSSR.* **254**, 85-91 (1984).

121. A.I. Nosich. Influence of resonant regimes on scattering characteristics of unclosed screens. *Radiotekhnika i Elektronika.* **23**, 1733-1737 (1973).

122. B.M. Bolotovskij and V.L. Ginzburg. The Vavilov-Cherenkov and Doppler effects for the sources moving with the speed greater than the speed of light in vacuum. *Uspekhi Fiz. Nauk.* **106**, 577-592 (1972).

123. A. Sommerfeld. *Vorlesungen über theoretische Physik. Bd. 3.* Wiesbaden, Leipzig, (1948).

124. A. Einstein. *Collection of scientific works.* Nauka, Moscow, (1965) (in Russian).

125. M. Kadaton. Film technology and microwave integrated schemes. In: *Technology of thick and thin films,* 45-58. Mir, Moscow, (1972).

126. V.I. Vol'man. *Reference book for calculation and construction of microwave strip devices.* Radio i Svyaz', Moscow, (1982) (in Russian).

127. S.B. Cohn. Slot line on a dielectric substrate. *IEEE Trans. Microwave Theory and Techn.* **17**, 768-778 (1972).

128. E.A. Mariani, G.P. Heinzman, J.P. Agrios and S.B. Cohn. Slot-line characteristics. *IEEE Trans. Microwave Theory and Techn.* **17**, 1091-1112 (1969).

129. J.B. Knorr and P.M. Shayda. Millimeter-wave fin-line characteristics. *IEEE Trans. Microwave Theory and Techn.* **28**, 737-749 (1977).

130. K.O. Lagerlöf. Redged waveguide of planar microwave circuits. *IEEE Trans. Microwave Theory and Techn.* **21**, 499-500 (1977).

131. V.F. Vyaz'mitinov. *Dielectric waveguides.* Sov. Radio, Moscow, (1970) (in Russian).

132. C. Somekh. Optical dielectric couplers. In: *Introduction to integrated optics,* 194-226. Mir, Moscow, (1977).

133. V.F. Vzyatyshev. *Dielectric waveguides.* Sov. Radio, Moscow, (1970) (in Russian).

134. P.N. Melezhik. Spectral characteristics of coupled open resonant volumes. *Dokl. AN USSR. Ser. A.* no. 2, 65-67 (1980).

135. V.P. Koshparenok, P.N. Melezhik and V.P. Shestopalov. Quasi-dipole radiation of two circular cylinders with logarithmic slots. *Pis'ma v Zhurn. Tekhn. Fiziki.* **4**, 1145-1147 (1978).

136. V.I. Kalinin and G.M. Gershtein. *Introduction to radiophysics.* Gostekhizdat, Moscow, (1957) (in Russian).

137. A.A. Fel'dshtein, L.R. Yavich and V.P. Smirnov. *Reference book on elements of waveguide technology* . Sov. Radio, Moscow, (1967) (in Russian).

138. A.M. Goncharenko. *Gaussian light pencils.* Nauka i Tekhnika, Minsk, (1977) (in Russian).

139. V.V. Kryzhanovskij. On the Sommerfeld wave propagation in metallic strips. In: *Physics and technology of millimeter and submillimeter waves*, 146-149. Naukova Dumka, Kiev, (1983).

140. S.D. Andrenko, V.V. Kryzhanovskij, A.I. Nosich and V.P. Shestopalov. Experimental investigation of the electromagnetic wave propagation in cylindrical slot lines. *Radiofizika i Elektronika.* **28**, 888-893 (1983).

141. S.D. Andrenko and V.P. Shestopalov. *Experimental study of the surface-volume wave transformation in the millimeter wavelength ranges. Preprint № 43.* Inst. of Radiophysics and Electronics, Ukr. Acad. of Sciences, Kharkov, (1975).

142. S.D. Andrenko, Yu.B. Sidorenko and V.P. Shestopalov. On the problem of transformation of surface waves to volume waves. *Dokl. AN USSR. Ser. A.* no. 2, 156-159 (1976).

143. S.D. Andrenko, V.G. Belyaev, S.A. Provalov et al. Transformation of millimeter and submillimeter electromagnetic waves and application of this phenomenon in physics and technology. *Vestnik AN USSR.* no. 1, 8-22 (1977).

144. R. Norton, B. Easter and A. Gopinath. Variation of microstrip losses with thickness of strip. *Electron. Lett.* **7**, 490-491 (1971).

145. T. Citazawa, Y. Hayashi and M. Suzuki. Analysis of the dispersion characteristic of a slot line with thick metal coating. *IEEE Trans. Microwave Theory and Techn.* **28**, 387-392 (1980).

146. V.V. Vorobjov. Slot transmission lines and co-planar waveguides for integrated microwave schemes. *Zarubezhn. Radioelektronika.* no. 5, 93-116 (1972).

147. R. Gard and I.J. Bahl. Characteristics of coupled microstrip line. *IEEE Trans. Microwave Theory and Techn.* **27**, 700-705 (1979).

148. B.E. Spielman. Dissipation loss effects in isolated and coupled transmission lines. *IEEE Trans. Microwave Theory and Techn.* **25**, 648-655 (1977).

149. S.D. Andrenko, V.G. Belyaev, N.D. Devyatkov and V.P.Shestopalov. On diffractional input of energy into dielectric waveguide. *Dokl. AN SSSR.* **247**, 73-79 (1979).

150. S.D. Andrenko, V.G. Belyaev, A.P. Evdokimov and V.V. Kryzhanovskij. On an influence of grating's length on the efficiency of diffractional input. In: *III All-Union Symposium on millimeter and submillimeter waves.* Vol. 1, 85-94. Gor'kij, (1980).

151. Yu.I. Glotov, A.G. Kovalenko, G.I. Khlopov et al. *Propagation of electromagnetic waves in cylindrical slot line. Preprint № 241.* Inst. of Radiophysics and Electronics, Ukr. Acad. of Sciences, Kharkov, (1984).

152. A.F. Chaikovskij, V.M. Puzikov and A.V. Semenov. Precipitation of diamond films from the carbon ionic pencils. *Kristallographiya.* **26**, 219-227 (1981).

153. L.Z. Rumshinskij. *Mathematical processing of the measurement data.* Nauka, Moscow, (1971) (in Russian).

154. N.G. Bova and I.B. Laikhtman. *Measurement of parameters of waveguide elements.* Tekhnika, Kiev, (1968) (in Russian).
155. G.I. Komar' and V.P. Shestopalov. Cylindrical and image slot lines. *Radiotekhnika i Elektronika.* **30**, 13-47 (1985).
156. V.P. Shestopalov, L.N. Litvinenko, S.A. Masalov and V.G. Sologub. *Diffraction of waves by gratings.* Kharkov University Press, Kharkov, (1973) (in Russian).
157. E.I. Nefedov and A.N. Sivov. *Elecprodynamics of periodical structures.* Nauka, Moscow, (1977) (in Russian).
158. V.P. Shestopalov, A.A. Kirilenko, S.A. Masalov and Yu.K. Sirenko. *Diffractional gratings.* Naukova Dumka, Kiev, (1986) (in Russian).
159. A.S. Il'inskij and A.G. Sveshnikov. Numerical methods in the problems of diffraction by inhomogeneous periodical structures. *Applied Electrodynamics.* no. 1, 51-93 (1977).
160. R. Petit. *Electromagnetic theory of gratings.* Springer-Verlag, New York, (1980).
161. T.N. Galishnikova and A.S. Il'inskij. *Numerical methods in diffraction problems.* Moscow University Press, Moscow, (1987) (in Russian).
162. V.P. Shestopalov, A.A. Kirilenko and S.A. Masalov. *Convolution-type matrix equations in diffraction theory.* Naukova Dumka, Kiev, (1984) (in Russian).
163. S.A. Masalov, Yu.K. Sirenko and V.P. Shestopalov. On applicability of the truncation method for some infinite systems of equations. *Dokl. AN USSR. Ser. A.* no. 6, 539-543 (1977).
164. S.A. Masalov and V.G. Sologub. Rigorous solution to the problem of electromagnetic wave diffraction by certain strip structure. *Zhurn. Vych. Matem. i Matem. Fiziki.* **10**, 693-715 (1970).
165. Yu.K. Sirenko. Justification of semi-inversion method of matrix operators in the problems of wave diffraction. *Zhurn. Vych. Matem. i Matem. Fiziki.* **23**, 1381-1391 (1983).
166. R. Mittra and S.W. Lee. *Analytical techniques in the theory of guided waves.* The McMillan Co., New York, (1971).
167. R. Mittra. *Computational electrodynamics.* Mir, Moscow, (1977) (in Russian).
168. Z.S. Agranovich, V.A. Marchenko and V.P. Shestopalov. Diffraction of electromagnetic waves by plane metallic gratings. *Zhurn. Techn. Fiziki.* **32**, 381-294 (1962).
169. Yu.K. Sirenko and V.P. Shestopalov. Rigorous theory of wave scattering by diffractional echelette-type grating with absorbing sides. *Dokl. AN SSSR.* **263**, 851-854 (1982).
170. V.G. Sologub. On a method of the study of the diffraction probelm on a finite number of strips situated in one plane. *Dokl. AN USSR. Ser. A.* no. 6, 550-554 (1975).
171. T.V. Gavrilova. Scattering of modulated signals by periodical obstacles. *Izv. Vuzov. Radiofizika.* **23**, 1067-1074 (1980).
172. S.M. Zhurav. Diffraction by the periodical structure formed by conducting plane of finite thickness. *Izv. Vuzov. Radiofizika.* **19**, 1848-1853 (1976).
173. J. Carson and A.E. Heins. The reflection of an electromagnetic plane wave by an infinite set of plates. *Quart. Appl. Math.* **4**, 313-329 (1974).
174. F. Berz. Reflection and refraction of microwaves on a set of parallel metallic plates. *Proc. IEEE.* **98**, 47-55 (1951).

175. V.P. Shestopalov and V.V. Shcherbak. Matrix operators in diffraction problems. *Izv. Vuzov. Radiofizika.* **9**, 285-306 (1968).

176. Yu.K. Sirenko. On a possibility of analytical solution to some classical problems of diffraction theory. *Zhurn. Vych. Matem. i Matem. Fiziki.* **23**, 202-208 (1983).

177. L.A. Weinschtein and V.I. Vol'man. On rigorous solution of one diffraction problem. *Dokl. AN SSSR.* **286**, 1360-1364 (1986).

178. V.M. Astapenko and G.D. Maluzhinets. Diffraction of the plane sound wave by fine-tooth periodical grating. *Akust. Zhurn.* **16**, 354-362 (1970).

179. L.M. Brekhovskikh. Diffraction of waves by rough surfaces. *Zhurn. Experim. i Teoretich. Fiziki.* **23**, 275-304 (1952).

180. E.N. Podol'skij. Justification of short-wave asymptotics in the problem of plane electromagnetic wave diffraction on the periodical grating formed by metallic strips. *Radiotekhnika.* no. 1, 58-68 (1965).

181. V.G. Sologub. Diffraction of the plane wave by a strip grating in the case of short wavelengths. *Zhurn. Vych. Matem. i Matem. Fiziki.* **12**, 974-989 (1972).

182. B.R. Weinberg. *Asymptotical methods in the equations of mathematical physics.* Moscow University Press, Moscow, (1982) (in Russian).

183. S.A. Masalov. On a possibility of application of echelette in generators of diffractional radiation. *Ukr. Fiz. Zhurn.* **25**, 570-574 (1980).

184. R.A. Silin and V.P. Sazonov. *Deceleration systems.* Sov. Radio, Moscow, (1966) (in Russian).

185. L.A. Weinshtein and A.B. Manenkov. Excitation of open waveguides. In: *Lectures on microwave electronics and radiophysics*, 141-197. Saratov University Press, Saratov, (1986).

186. A. Hessel and A.A. Oliner. A new theory of Wood's anomalies in optical gratings. *Appl. Optics.* **4**, 1275-1297 (1965).

187. V.G. Sologub and V.P. Shestopalov. Resonance phenomena in the plane *H*-polarized wave diffraction on the grating formed by metallic bars. *Zhurn. Vych. Matem. i Matem. Fiziki.* **38**, 1505-1520 (1968).

188. E.I. Veliev and V.P. Shestopalov. Effect of total long-wave reflection of plane waves by a diffractional strip grating. *Pis'ma v Zhurn. Tekhn. Fiziki.* **6**, 1327-1330 (1980).

189. J.R. Andrewarsha, J.R. Fox and I.J. Wilson. Resonance anomalies in the lamelian grating. *Opt. Acta.* **26**, 69-89 (1979).

190. S.V. Sukhinin. Qualitative problems of the scattering theory for periodical cylindrical obstacles. *Dynamics of solid medium.* no. 67, 118-134 (1984).

191. M.V. Keldysh. On a completeness of eigenfunctions of some classes of non-self-adjoint linear operators. *Uspekhi Matem. Nauk.* **26**, 15-41 (1971).

192. E. Sanchez-Palencia. *Non-homogeneous media and vibration theory.* Springer-Verlag, New York, (1980).

193. A. Hurvitz and R. Courant. *Theory of functions.* Springer-Verlag, New York, (1963).

194. E.C. Titchmarsch. *Eigenfunction expansions associated with second-order differential equations. 1, 2.* Clarendon Press, Oxford, (1946).

195. N.M. Günter. *Théorie du potentiel.* Gauthier-Villars, Paris, (1934).

196. A.B. Akimov and B.A. Iskenderov. Principles of limiting absorption, limiting amplitude and partial radiation conditions for boundary value problem in n-dimensional layer for Helmholtz equation. *Diff. Uravnenija.* **13**, 1503-1505 (1977).

197. Yu.K. Sirenko. *Some mathematical questions in problems of the wave diffraction by waveguide-type gratings. Preprint № 103.* Inst. of Radiophysics and Electronics, Ukr. Acad. of Sciences, Kharkov, (1978) (in Russian).

198. S.A. Masalov, Yu.K. Sirenko and V.P. Shestopalov. Solution to the problem of the plane wave diffraction by knife-type gratings with a complicated structure of the period. *Radiotekhnika i Elektronika.* **23**, 481-487 (1978).

199. I. Poston and I. Stuart. *Catastrophe theory and its applications.* Mir, Moscow, (1980) (in Russian).

200. R. King and Tai Tsun Wu. *The scattering and diffraction of waves.* Harvard University Press, Cambridge, Massachusetts, (1959).

201. M. Barnoski. *Introduction to integrated optics.* Nauka, Moscow, (1974) (in Russian).